NEW PRINCIPLES IN PLANNI

T0250914

New Principles in Planning Evaluation

Edited by

ABDUL KHAKEE
Royal Institute of Technology, Sweden

ANGELA HULL
Heriot-Watt University, UK

DONALD MILLER
University of Washington, USA

JOHAN WOLTJER
University of Groningen, Netherlands

Routledge
Taylor & Francis Group

LONDON AND NEW YORK

First published 2008 by Ashgate Publishing

2 Park Square, Milton Park, Abingdon, Oxon OX14 4RN
711 Third Avenue, New York, NY 10017, USA

Routledge is an imprint of the Taylor & Francis Group, an informa business

First issued in paperback 2016

British Library Cataloguing in Publication Data
New principles in planning evaluation. - (Urban planning
  and environment)
  1. City planning - Evaluation
  I. Khakee, Abdul
  307.1'216

Library of Congress Cataloging-in-Publication Data
New principles in planning evaluation / by Abdul Khakee ... [et al.].
    p. cm. -- (Urban planning and environment)
  Includes index.
  ISBN 978-0-7546-7507-5
  1. City planning--Evaluation. 2. Land use, Urban. 3. Business
logistics. 4. Planning. I. Khakee, Abdul. II. Series.

  HT166.N443 2008
  307.1'216--dc22

                                                                    2008024719

  ISBN 978-0-7546-7507-5 (hbk)
  ISBN 978-1-138-27215-6 (pbk)

Contents

PART I: SOCIO-ENVIRONMENTAL PRINCIPLES

PART 2: SOCIO-INSTITUTIONAL PRINCIPLES

PART 3: INTERACTIVENESS/COMMUNICATION PRINCIPLES

List of Figures

List of Tables

List of Contributors

Maurizio d'Amato is Associate Professor in the Faculty of Engineering, of the Polytechnic University of Bari. His research interests are property valuation and investment.

Angela Barbanente is Professor in Urban and Regional Planning in the Taranto School of Engineering of the Polytechnic University of Bari. She is on leave of absence and is currently Deputy Director for the Commission on land use planning and social housing in the Apulian Regional Government.

Tom Bauler is Research Fellow in the Institute for Environmental Management and Land Planning of the Free University of Brussels (Université Libre de Bruxelles), in Belgium.

Alessandro Bonifazi is a PhD candidate in Urban and Regional Planning in the Department of Architecture and Urban Planning of the Polytechnic University of Bari. His research activity focuses on strategic environmental assessment and sustainability evaluation.

Dino Borri is Professor of Urban Planning Techniques and Director of the Department of Architecture and Urban Planning, of the Polytechnic University of Bari. His research interests are artificial intelligence, spatial planning and planning evaluation.

Roberto Camagni is Professor of Urban Economics at the Polytechnic of Milan and Past-President of European Regional Science Association (ERSA).

Roberta Capello is Professor of Regional Economics at the Polytechnic of Milan, and Incumbent President of Regional Science Association International (RSAI).

Sylvia Dovlén is Research Associate in the Department of Planning and Environment, Royal Institute of Technology, Stockholm. Her research focuses on policy implementation and planning for sustainable development.

Roger Ellis OBE is Professor and Director of Research Policy in the Faculty of Health and Social Care of Chester University. He was the founder of the Social and Health Evaluation Unit at Chester University.

Luigi Fusco Girard is Professor in the Department of Conservation of Architectural and Environmental Heritage, University of Naples 'Frederico II'. He has extensively researched in the theory of planning and evaluation.

Anders Hanberger is Associate Professor in Political Science and Director of Umeå Centre for Evaluation Research, Umeå University. His research interests include policy analysis, policy and program evaluation methodology, legitimacy and democracy issues in evaluation, the functions and societal implications of evaluation.

Tuija Hilding-Rydevik is Associate Professor in the Royal Institute of Technology, Stockholm, and Research Leader in the Department of Urban and Rural Development, Swedish University of Agricultural Sciences, Uppsala. Her research focus is on the practice of sustainable development planning and impact assessment.

Elaine Hogard is the head of the Social and Health Evaluation Unit of the Faculty of Health and Social Care, Chester Univeristy. She has extensive experience of social research in Canada, the USA and Northern Ireland.

Angela Hull is Professor of Spatial Planning in the School of the Built Environment, Heriot-Watt University. She researches the theory and practice of urban planning systems investigating the interests that are involved in policy formulation and implementation.

Tom Kauko is Associate Professor in the Department of Geography, Norwegian University of Science and Technology (NTNU), and formerly researcher in the OTB Research Institute for Housing, Urban and Mobility Studies, at the Delft University of Technology. His research interests are real estate economics and evaluation.

Abdul Khakee is an honorary member of the Centre of Urban and Regional Studies (CUReS), Örebro University, and of the Centre of Evaluation Research (UCER), Umeå University. He is currently Professor Emeritus in the Department of Planning and Environment of the Royal Institute of Technology, Stockholm.

Willem K. Korthals Altes is Professor in Land Development in the OTB Research Institute for Housing, Urban and Mobility Studies at the Delft University of Technology. His research centres on the relationship between planning and land markets.

Donald Miller is Professor in the Department of Urban Design and Planning at the University of Washington, and has chaired the Department and the Seattle City Planning Commission.

Valeria Monno is Assistant Professor in Urban and Regional Planning in the Taranto School of Engineering of the Polytechnic University of Bari. Her research interests are social exclusion, public participation and strategic decision-making.

Domenico Patassini is Professor in planning evaluation techniques at the Faculty of Planning at the University of Venice. His recent research activities include Brownfield remediation and urban upgrading, capacity building in African countries and evaluation practices within urban management processes and governance.

Jenny Stenberg is a Researcher in the Department of Built Environment and Sustainable Development of the Faculty of Architecture at Chalmers Technical University, Göteborg. Her research interests are local partnership, social exclusion and urban renewal.

Carmelo M. Torre is Senior Researcher in the Department of Architecture and Urban Planning of the Polytechnic University of Bari. His research activity focuses on urban economics and planning and project appraisal methods.

Shinji Tsubohara is a PhD researcher working in the Faculty of Spatial Sciences at the University of Groningen. He has two masters degrees, from the University of Tokyo and Kobe University respectively. His key interests involve transportation planning and local democracy.

The late **Henk Voogd** was Professor of Urban and Regional Planning at the University of Groningen. He was one of the coordinators of the international network in planning evaluation and has made many valuable contributions in spatial planning and evaluation.

Johan Woltjer is an Associate Professor in the Faculty of Spatial Sciences, University of Groningen. He is involved in various research projects in regional planning, water management, evaluation, and public participation.

List of Abbreviations

ASSIA	Applied Social Sciences Indexes and Abstracts
B&W	(College Van Burgemeester En Wethouder)
BIDS	Bath Information and Data Services for the Social Sciences
CIE	Community Impact Evaluation
DG	Directorate General
DOT	Department of Transportation
EA	Environmental Authority
EA	evaluative argumentation
EC-IA	European Commission Impact Assessment
EIA	Environmental Impact Assessment
EPA	Environmental Protection Agency
EPI	Environmental Policy Integration
EU	European Union
ExIA	Extended Impact Assessment
FHWA	Federal Highway Administration
GAM	Goals Achievement Matrix
GDP	Gross Domestic Product
GMM	General Member Meeting
GOF	Groningen Entrepreneur Federation
HAT	Housing Action Trust
IA	Impact Assessment
ICA	International Communication Association
KvK	Chamber of Commerce
LDA	Local Development Agreements (Lokala Utvecklingsavtal)
LETS	Local Exchange Trading Systems
LIP	Local Investment Programme
MA	Managing Authority
MC	multicriteria
NDC	New Deal for Community
NEPA	National Environmental Policy Act
NUTEK	Swedish Business Development Agency
OECD	Organization for Economic Cooperation and Development
ORP	Operative Regional Programme
PBS	Planning Balance Sheet
PNT	Policy Network Theory
PvdA	Partij van de Arbeid
RDP	Regional Development Programmes
RDS	Regional Development Strategies
RGA	Regional Growth Agreements
RGP	Regional Growth Programmes

RIA	Regulatory Impact Assessments
SD	sustainable development
SEA	Strategic Environmental Assessment
SEL	Systeme d'Exchange Local
SF	Structural Funds
TIPs	Territorial Integrated Programmes
VCP	Verkeerscirculatieplan

Preface

Abdul Khakee, Angela Hull, Don Miller and Johan Woltjer

The title of this book *New Principles in Planning Evaluation* may give an impression that the book outlines some brand new or previously unheard of standards in planning evaluation. This is of course not the case. Planning evaluation has been an exciting field of research obtaining new impulses from planning as well as policy science research. Both of these research areas are being affected by far-reaching changes that are taking place in individual and societal values as a result of technological, socio-economic and environmental changes. It is against this background and with the results of the last twenty years of planning evaluation research at hand, that the book explores the latest ideas that have been developed in this field.

The forerunners of the chapters in this book were first presented at the sixth international workshop in planning evaluation held at the Division of Urban and Regional Studies, Royal Institute of Technology, Stockholm. Göran Cars, the head of the Division provided marvellous support in organising the workshop. With his keen interest in planning evaluation, Dino Borri has been a key actor in the organisation of the Stockholm workshop. The Department of Architecture and Urban Studies (DAU), University Polytechnic of Bari and the Swedish Research Council for Research in Environment, Agricultural Sciences and Spatial Planning (FORMAS) financially supported the workshop.

The book is dedicated to the memory of Henk Voogd, Professor of Urban and Regional Planning at the University of Groningen. Henk was one of the coordinators of the international network in planning evaluation and has made many valuable contributions in spatial planning and evaluation. His death in March 2007 has deprived us of his intellectual and professional contribution to the field of planning evaluation, as well as his cherished friendship and encouragement.

We wish to thank all of the contributing authors for their helpfulness through the editing of this book and the staff at Ashgate for their supportiveness during its production.

Chapter 1

Introduction:
New Principles in Planning Evaluation

Abdul Khakee, Angela Hull, Donald Miller and Johan Woltjer

Planning and evaluation have been described as inseparable concepts both from a theoretical and a practical point of view (Lichfield 2001; Khakee 1998). In fact there has been a growing appreciation of this interrelationship owing to the increased integration of democratic concerns in planning and a convergence of ecology with urban planning. In the following pages we trace this development with the help of a brief retrospect of the recent evolution of planning evaluation research and with reference to the institutional realities of planning evaluation practice.

Planning Evaluation in Retrospect

Planning evaluation has been an established field of research for a considerable number of years. Its development had been closely associated with changes in planning theory and practice as well as in policy analysis and programme evaluation. Planning evaluation most obviously refers to the making of normative judgements about the success (or otherwise) of the intervention outcomes of planning or assessing the success of the process of planning. As such, planning evaluation acquires knowledge from a vast number of disciplinary sources. It is difficult to track down all the changes that have taken place in the field of planning evaluation during these years. Very roughly, however, we can discern the development of planning evaluation research along three lines: the fundamentals and purpose of planning evaluation, the scope of planning evaluation, and the methodological innovations and improvements.

Fundamentals and Purpose of Planning Evaluation

Two distinct paradigms, consisting of clear and well-defined theoretical and empirical propositions, have determined planning theoretical research. These paradigms are rational planning and communicative (also referred to as 'deliberative') planning respectively. The two planning theories are both descriptive and normative. They not only explain the nature of planning and the process this involves but also guide various phases of the process including evaluation (Lichfield et al. 1975).

Rational planning is based on instrumental rationality and has dominated planning research for more than half a century. Instrumental rationality implies that the most favourable relationship between goal achievement and resource use is obtained. This requires that goals are carefully specified and that goal achievement implies the minimization of expenditure or the most effective use of resources. In policy programme evaluation the rational paradigm has found its expression in various measurement methods and goal achievement models. These have been characterized as the 'first' and the 'second' generation of evaluation methods (Guba and Lincoln 1989).

Several theoretical phenomena e.g. incrementalism, advocacy, implementation and strategic planning were developed to redress shortcomings in the rational planning model. They showed that in reality problems are poorly defined, many goals can only be formulated qualitatively, relationships between goals and means are poor on account of value uncertainty and scarcity of knowledge, only privileged groups can participate in policy formulation and goal formulation is not exclusively an analytical process – it contains a great deal of politics. The proponents of these alternative approaches were nevertheless in agreement with the advocates of the rational planning approach that despite limitations, evaluation should try to emulate an optimization procedure as far as possible (Faludi 1987). However as a result of these alternative approaches, optimization has given way to such concepts as 'satisficing' 'second-best solutions', 'political accord', etc. (Faludi and Voogd 1985; Khakee 1998).

The communicative or deliberative (sometimes even called collaborative planning) model not only brought to the fore the already existing recognition of the political nature of planning but also alternative objectives concerning the democratic principles for preparing and implementing planned interventions, mediating conflicts and organizational (or community) learning (Healey, 1997). Since communicative planning emphasizes both interaction and iteration, which takes place in an extensive institutional context, and where the aim is to obtain commitment and consensus among all the stakeholders, the principal aspects of evaluation centre around how best to organize an inclusionary discourse, to promote a learning process which is emancipatory and expedites progress, and to emulate political, social and intellectual capital (Khakee 2002; Davoudi 2005). A central aspect of evaluation is to focus on both the quality of the planning process and the programme of actions. Evaluation thus becomes a question not only of effectiveness and legitimacy but also of integrity and mutual understanding. Evaluation itself becomes a form of interactive discourse where all participants get involved in:

1) the opportunity presented for deciding on goals;
2) deciding on what the community or the planning organization's primary objectives should be;
3) the realization of the existence of important conflicts;
4) forging consensus or exposing existing conflicts;
5) providing information for market participants or government organizations;
6) developing bids for scarce resources;

7) helping to identify competing aims for the process of planning (Healey, 1993).

Communicative planning evaluation corresponds to what Guba and Lincoln (1989) describe as the 'fourth generation evaluation' that includes several approaches including Guba and Lincoln's own 'naturalistic responsive approach', the multiplist model (Cook 1985) and the design approach (Bobrow and Dryzek 1987). These approaches take up the post-positivist challenge of interactive participatory evaluation.

Whether communicative planning replaces the rational model or not is a controversial issue. Some researchers feel that it does so (Healey 1993; Innes 1995; Khakee 1998). Others maintain that the rational model at least in planning and evaluation practice is a robust and flexible instrument (Alexander 1998; Lichfield 2001). According to the latter, what is needed is an open and value related discourse in order to make the consequences of a plan or a policy measure as clear as possible. In practice there is however, a strong adherence to the rational approach and to the quantification of consequences, which has led to an emerging gap between research and practice (Henkel 1991; Khakee 2003).

The communicative logic has nevertheless compelled the advocates of the rational approach to recognize the need for making the evaluation process and evaluation results more transparent and to improve the communication between the evaluators and those who make use of or have a concern with evaluation studies. Since planning is a systematic attempt to construct frames for the justification of decisions, it may be that planning seems to perform to sufficient levels of plausibility despite poor planning methodology. This is a question of conformance rather than performance. The evaluation of performance therefore needs a new design that focuses on the arguments advanced during the justification of decisions (Faludi and Korthals Altes 1997).

The communicative model has also implied an extension of the purpose of planning evaluation to problem generation and definition. The evaluation of complex decision-making processes needs a broader policy analysis framework in order to have a description of the problems, their causes, the use of current policy, the preconditions for achieving desired scenarios and goal structures (Yewlett 1993; Rosenhead 2005). Planning can be viewed from several angles: as consensus building it can be analysed as a persuasive process in dealing with sensitive issues, as a learning process it can deal with 'wicked' problems and as a negotiation process for administrating distributional problems (Woltjer 2001).

Extending the Scope of Planning Evaluation

Some of the major factors that have led to the extension of the scope of planning evaluation include the idea of incorporating various ecological factors into the evaluation process through the integration of risk analysis and non-market values (Miller and Patassini 2005). The integration of ecological aspects in socio-economic planning and evaluation poses conceptual challenges with regard to interpreting concepts like 'sustainable development', 'biodiversity', and 'ecology' that are interpreted in a wide variety of ways. Evaluation research has contributed to the exposure of contradictions in these different interpretations and in the provision of guidelines when using 'environmental quality' as a yardstick to measure development (Barbanente et al. 1998).

The ecological dimension also reveals problems involved in challenging the conventional growth perspectives, especially the neo-classical economic growth paradigm. In assessing individual as well as community values for environmental quality and natural goods, it is necessary to go beyond the neo-classical social choice theory with its linear addition of individual utility. It is necessary to examine conflicts and complementarities between growth and nature conservation with particular attention to such issues as 'limits to growth', 'inter-generational rights and preferences' and 'time-horizon for more balanced development' (Macchi and Scandura 1997). In an empirical context, evaluation models need to consider biospheric quality and capacities and thresholds that limit the exploitation of natural resources (Davoudi 1997). Moreover it is necessary to pay attention to the role of externalities hiding the real value of resources, the lack of consideration given to the irreversibility of damage done to the environment, conflicting environmental resource values held by different social interests and the use of 'thematic maps' and other similar methods for assessing risk associated with natural disasters (Gentile et al. 1997).

Relevant in this context are the techniques for measuring and for estimating non-market values including land use and development performance indicators used by national planning departments, the incidence and degree of environmental impacts on various social groups, and survey opinion data concerning the nuisance and desirability effects of alternative designs for public facilities (Miller 2005). Throughout, the problems of complexity and resulting uncertainty are important. There is a strong tendency for decision makers to be more attentive to those aspects of options that are measured; making it important that valid metrics can be found or designed for all the objectives the decisions should address (Barbanente et al. 1998).

Incorporating questions of value into an account of justice has been a major issue in evaluation research dealing with the distributional impact of plans and policy measures. Social justice has always had to navigate between the individual and the collective. The inclusion of environment in evaluation implies not only intra-generational environmental equity but also between current and future generations as well as obligations human beings have to nature per se (Campbell 2006). So far evaluation research and practice have focussed on environmental equity within a community or a nation (Miller 2005). Hardly any environmental impact assessments have extended to the entire planet or across current and future generations; even less so when it comes to the human-nature relationship. The extension of the scope of environmental justice poses a tremendous challenge (O'Neill 1993).

Methodological Innovations and Improvements

The decreasing importance of conventional evaluation methods in spatial planning has resulted in methodological innovation according to several perspectives such as:

1) the use of an evaluation matrix in dialogue with stakeholders to account for values (Fusco Girard 1998);
2) the replacement of a narrow welfare perspective by an interactive 'positional' analysis integrating organizational and individual values, politics and ideology (Söderbaum 1998);
3) the replacement of instrumental rationality, with its quantitative and utilitarian grounds, by a radical generic approach for acknowledging differences in values (Barbanenete et al. 1998); and
4) the application of social constructivist ontology for bringing in new perspectives and values (Barbanente et al. 2001).

A major aim of these innovative methods is to overcome the mechanistic and reductionist approaches to evaluation, to combine the issues of effectiveness, efficiency, and equity, to surmount the disciplinary barriers and integrate the different forms of knowledge within the evaluation process. Specific models that have been developed in this context include:

- a multi-model system of sustainability indicators classification (Lombardi and Curwell 2005);
- a meta-analytic approach for analysing and comparing sustainable development policies, in terms of their common and divergent components, success factors as well as impediments (Bizzaro and Nijkamp 1998);
- Community Impact Analysis for identifying 'actual use' values and 'passive use' or 'altruistic' values for people who may not actually use natural goods but nevertheless gain satisfaction from the possibility of potential use (Lichfield 1998);
- the extension of impact assessment to evaluate social distributional effects, especially the environmental justice implications of development projects by integrating the technical analysis of the environmental spillovers of a project, and their impact on specific population groups, with the analysis of information from affected parties (Miller 2001);
- improvements to environmental impact analysis (EIA) through making ex post evaluation an intrinsic part of EIA practice, in order not only to improve individual EIA activities but also improve EIA practice more generally (Arts 2001);
- the introduction of a creative, conscious and interactive discussion of goals in community impact assessment in order not only to prepare the ground for the planning analysis but also for intelligent stakeholder participation (D. Lichfield 2001; N. Lichfield 2001);

- the integration of guidelines for substantive, procedural and policy integration in impact assessment and planning, specifically emphasizing the complementary and subsidiary nature of impact assessment and planning (Fusco Girard et al. 2005);
- the use of a retrospective sociological analysis of community-relevant environmental conflicts to complement the traditional community impact assessment (Selicato et al. 2001).

The shift from a rational to a communicative style of planning has raised the fear that such a development would mean throwing out 'the baby with the bathwater'. The fear is that rational choice and instrumental rationality would be subsumed to all manner of legitimizing decisions with fuzzy notions about interactive practices (Alexander 1998). Notwithstanding these fears, several approaches have been put forward to combine rational choice with interactive practice:

- Planning and Management Learning System as a continuous evaluation approach integrating methods and resource organization, taking into account value pluralism and enhancing participation (Lichfield and Prat 1998);
- Multi Criteria Analysis, making explicit evaluation assumptions, integrating empirical and experiential knowledge but at the same time exposing rhetorical expedients about classical and communicative rationalities respectively (Alexander 1998);
- Integrated Multi Criteria and Benefit Cost Analysis in order to take into account disaggregated and weighted analysis of the preferences of the people as well as equity considerations (Levent and Nijkamp 2005);
- 'will-shaping' in planning evaluation whereby attitudes and preferences are synchronized towards certain goals using either a 'strategic model' whereby a strategic plan is made the subject of public and political debate provided that it appeals to the public or an 'elaboration model' that illustrates operational alternatives to start public discussion (Voogd 1997);
- modifying benefit-cost analysis with the help of multi-objective decision-making (MODM). The model replaces a priori goal and criteria setting with sensitivity testing of a systematic set of goal-criteria weights reflecting alternative value orientations (Alexander 2001).

This brief exposé shows the tremendous importance of two factors on the planning evaluation research namely the shift in the theoretical perspective from the rational to the communicative or deliberative logic, and the environmental concern. However there are other changes that have also had an impact on this research. Notable among these are the emergence of the network society, market liberalism and its subsequent impact on public management (often described as a shift from 'government' to 'governance') and increasing ethical concerns owing to rapid changes in genetic engineering, diagnostics, information technology, etc. All these have had an impact on planning and evaluation research as well as practice. However, the shift in the theoretical approach and the increasing concern about the deterioration of the natural environment, have meant significant changes in the purpose, scope and methodology of planning evaluation theory and practice.

It is against this background that we now turn to the chapters that are included in this volume. We see as a crucial challenge here the need to understand how planning proposals are linked to their social and environmental context. Planning evaluation is concentrated increasingly on the influence of institutional contexts, on processes of plan making and implementation. This book tries to find out how new principles make reference to this awareness, and what principles have been developed in recent research in planning evaluation.

Evaluation and the Institutional Realities of Planning Practice

The chapters in this book illustrate, in a variety of ways, the importance and possibilities of taking institutional principles into consideration. The institutional context of planning initiatives consists of both formal and informal social characteristics, including legal frames of reference, but also values, norms, interests, perceptions and beliefs. These institutions influence the success and failure of planning decisions. The book tries to highlight the nature and role of evaluation in the context of the institutional realities of planning practice. It raises the issue of socio-environmental and socio-institutional principles for effective evaluation. It also shows how these issues shed a new light on the importance of interaction between key actors in specific planning situations. The main difference with the rational approach is that effectiveness is a matter of the extent to which planning proposals match the situated social and political processes of which these proposals are a part, not the extent to which they match given objectives.

The emphasis given here has a series of implications. First the integration of more socially orientated environmental considerations in planning. Planning proposals related to issues such as infrastructure projects, housing areas, and impact assessment studies have been increasingly embedded in broader policy settings including environmental, economic, and social sustainability at the same time. Part I of the book discusses related socio-environmental principles such as environmental justice, equity, and hedonic pricing. There is an emphasis on new linkages between social, economic and environmental issues. Planning evaluation then becomes an activity aimed at a multi-dimensional understanding of planning. The chapters in this part of the book show that the evaluation methods required for such an understanding imply triangulation, and carefully balanced assessment.

A second implication is that planning efforts can only be evaluated effectively if they make reference to their specific institutional context. Part II of the book deals with the importance of socio-institutional principles such as plausibility, capacity building, institutional anticipation, performance, and environmental integration. A key point for these chapters is that planning evaluation should help recognize and appreciate the influence of strategic contextual factors such as market development, regional change, and culture. Evaluation must handle the 'embeddedness' of planning within its wider range of social and economic processes. Accordingly, planning evaluation becomes an activity focused on methods to monitor these kinds of processes. At the same time, evaluation makes reference to the contextually specific positions and assumptions of those actors upon which the development and implementation of policies are based.

A third implication is that the increased reliance on evaluation approaches enables the capture of interactive processes between actors. Part III of the book shows, in a multiplicity of ways, how interaction principles related to citizenship, provider-recipient communication, policy networks, and dialogue, accentuate the necessity of including all relevant stakeholders within the evaluation. Planning evaluation, thus, recognizes the pertinence of stakeholder and citizen interests in plans, policies and projects, and seeks to interpret their perspectives, arguments and actions. Qualitative approaches like participatory evaluation, and communication and performance audits are crucial for making clear their influence on certain plans, projects or policies. Planning evaluation then produces forms of socially relevant knowledge, which informs action to improve plan effectiveness.

Socio-environmental Principles

One distinctive issue emphasizing the integration of socio-environmental considerations in planning is environmental justice. Environmental justice explicitly focuses on the adverse environmental impacts of a development project on low-income and minority population. The chapter by Don Miller proposes an innovative approach to the assessment of social justice in environmental planning. A key problem has been the separation between quantitative evaluation and impacts perceived, qualitatively, by citizens themselves. Miller combines quantitative estimations of population size with a qualitative assessment of perceived negative impacts derived from a community-based dialogue. His model is used to evaluate environmental justice in the case of the replacement of the Alaskan Way Viaduct and the seawall in the city of Seattle. His approach has some major advantages, including the fact that dialogue can help a more socially embedded assessment of likely impacts. Moreover, the dialogue provides a learning process for all the concerned parties, aimed at avoiding, minimizing, and mitigating negative impacts of a development project.

Jenny Stenberg highlights in her chapter the importance of social and institutional dimensions to sustainable development in Swedish housing. Like many other countries, Sweden has struggled with segregation and an unequal distribution of poverty concentrated in vulnerable housing areas. Also, environmental policy measures such as household waste separation have become more clearly dependent on the social attitudes, and ethnic and economic integration of the individuals involved. Stenberg proposes, therefore, that planning efforts based on sustainable development need multidimensional evaluation. The chapter follows an integrated socio-environmental conception in planning evaluation, using a specific model entitled Main^{TETRA}, illustrated by Swedish housing projects. A specific focus is on understanding impeding institutional conditions to implementation.

Tom Bauler, Alessandro Bonifazi, and Carmelo Torre analyse how the European Commission addresses equity issues in their impact assessments. Planning evaluation here refers to 'ex ante' judgements to inform European decision-makers on the positive and negative impacts of selected alternatives. The chapter considers the incorporation of inequality issues considered imperative for European policy making. It also calls attention to the increased coherence between policy fields such as transport, economy, and environment. The evidence presented suggests that the impact-assessment reports conform poorly to accepted guidelines on equity. The authors seek an explanation for this finding in the complexity of the supra-national scale (covering the EU as a whole), the integrated focus (on social, economic and environmental impacts) and the wide diversity of policy proposals. A more 'limited mandate' for establishing equity in European planning evaluation would help.

The chapter by Sylvia Dovlén and Tuija Hilding-Rydevik is also concerned with socially-oriented environmental planning, and new linkages between social, economic, and environmental issues. The main focus in their chapter, however, is on the integration of national directives on sustainable development in regional planning. Ecological sustainability, inter-generational responsibility, and North-South equality are presented as some of the major elements for sustainable development. Dovlén and Hilding-Rydevik identify a set of four specific discourses as a yardstick to their evaluation of the implementation of sustainable development policies, including 'an intellectual challenge', 'the goals are accepted', 'a negative stance towards the national directive', and 'the national task is uninteresting or impossible'. An evaluation based on these kinds of discourses can help make clear the organizational, cultural and social problems that arise as a result from implementing national policy goals.

A major issue in the evaluation of a large-scale development project is how to estimate prospective net benefits of such a project for the entire local community. A multi-dimensional understanding of planning is also a key here. Roberto Camagni and Roberta Capello in their chapter present such an understanding via an improved usage of the principle of hedonic pricing. A proposal for under-grounding a railway in the central business district of the City of Trento shows their attempt to specify as many variables of the possible parameters covering hedonic pricing as possible. The authors opt for flexibility in the choice of the hedonic function, avoid distortion arising from spatial inter-dependence and extend the geographical boundary of their survey in order to cover as many cases and conditions as possible. Planning evaluation studies like this one show how the linkages between overall communal benefits of urban projects and environmental, engineering and other costs, can be made clear.

Socio-institutional Principles

The key conclusion of the chapters in this section is the notion that an understanding of the cultural environment in which planning initiatives are operating is essential to their effectiveness. Increasingly, planning evaluators are seeking to understand the socio-institutional context including both the formal and informal social characteristics which structure the context within which the agency of actors takes place (Hull, 2006; Vigar et al. 2000). The 'structuring rules' include the legal frames of reference, the authority and position of different stakeholders, the distribution of responsibilities and resources, and the negotiation of possible outcomes. The values, norms, interests, perceptions and beliefs of actors also influence the success and failure of decision making through influencing capacity building, institutional anticipation, performance, and environmental integration in planning decisions.

Maurizio d'Amato and Tom Kauko use their evaluation of the real estate market in Bari, Italy to highlight the interaction between land use regulation and the real estate rental market. The liberalization and deregulation of the land market in Italy has set up a sequence of events in different districts of Bari to which both market and planning actors are responding. They use hedonic prices to understand and evaluate this behaviour and the increase and decline in rental values. They contend that that there is yet little mutual understanding between urban planners and real estate agents or realtors in the use of performance indicators of the real estate market to predict likely urban development in different districts of a city.

Domenico Patassini offers plausibility as a promising principle for planning evaluation. Taking the experience of urban planning in Ethiopia between donors and recipients as a foundation for his argumentation, Patassini emphasizes that the way that contact is established between key actors will directly determine the chance that effective partnerships and policy success will emerge. If processes of co-operation and programming disregard culture, then rejection, unwillingness, or mere passive acceptance could easily result. It is important, therefore, for evaluation to determine the plausibility of plans or programmes. A programme is plausible if it tries to understand and anticipate cultural factors such as differences in attitudes to state intervention, property, democracy or quality of life, and base the assessment of alternative options for action on some shared cultural attitudes or common language.

Planning evaluation, then, is about understanding and anticipating the cultural roots of partnership. Planning evaluation thus becomes a highly contingent, i.e. situation-specific, activity. Patassini uses a metaphor of the 'searcher' to explain the role of planners here. Searchers treat planning as a discovery process for plausible, that is socially or culturally valid, proposals. Using plausibility as a key principle, planning evaluation becomes a practical exercise of finding answers to questions such as why and to what extent actors agree on certain proposals, and how their cultural backgrounds have encouraged, or discouraged, mutual contact.

Angela Hull reflects on the challenge of undertaking an evaluation of the holistic regeneration of multiple deprived districts in five English cities. She draws attention to the quality of the evidence available, the steering role by the government client, and the ontological questions of understanding, and isolating, the mechanisms, which produce the 'additional' anticipated impacts from the interaction embedded within a complex web of socio-political structures existent in the area. Despite these problems the government client required measurement of the cost effectiveness or value-for-money of the public sector spend. The local regeneration agencies developed their own alternative quantitative and qualitative measures of performance, which were more meaningful and closely related to local objectives. These included long run measures to track the performance of residents through training and into employment and movement to better jobs. Hull's evaluation focuses on how the residents felt they had influenced the agenda, their own criteria to evaluate programme success and how the programme evolved in response to their views and the lessons learned.

Johan Woltjer, in his chapter, explores what he refers to as an entrepreneurial urban strategy for the Province of South-Holland, which attempts to anticipate societal change, opportunities and market trends. His interviews with key private and public actors highlight the importance of these criteria in evaluating strategic options, and the importance in their view of entrepreneurial approaches to stimulating actions and investments in partnership with parties other than the lead agency in preparing the plan. The strategic planning tasks involve measuring the strength of the regional development options to mobilize the capabilities of stakeholders, institutions, and their networks for decision making. The South-Holland case points out how evaluation should address the public-private sector partnership generation and delivery of strategic development issues at this spatial scale, and important interregional issues that need to be accounted for as well.

The interplay between national and local planning is the subject of Willem Korthals Altes's chapter. In his evaluation of Dutch national urbanization policies he addresses two questions: Did the national plan result in the desired changes? And did the plan result in better decision-making processes? He found that the answer to the *conformance* question was that developments over this period were largely in the designated concentration areas. However, it was less clear, on the *performance* question that the plan resulted in better decision making. Principally, the national plan was not effective in dealing with uncertainty. Contracts between the national government and local agencies did not respond to changes in demand in location for housing and housing sizes, and in fact he found that housing production stagnated in most urban areas in the Netherlands. Both the design of this evaluation and its application can provide useful guidance for undertaking other efforts to assess national plans that are necessarily implemented at the local level.

Angela Barbanente, Dino Borri and Valeria Monno evaluate how planning argumentation could be improved to integrate environmental issues more fully in policy development. They examine the micro-narratives of actors involved in the allocation of EU Structural Funds in Southern Italy and find that proposals are evaluated on the basis of rational management and efficient and effective procedures. Barbanente et al. conclude that, in circumstances where local practice and routines discourage new approaches, evaluation becomes a symbolic gesture in which authoritative actors close off dialogue and the development of learning processes. They suggest that with the move towards multi-agent interaction in policy development and delivery, we need new tools to evaluate the discursive argumentation of different actors and understand the different norms and values these contain.

Interactiveness/Communication Principles

The chapter by Luigi Fusco Girard poses some of the major issues concerning democratic interaction in decision-making. Specifically, Fusco Girard seeks to reconcile several uncertainties that emerge from trying some new forms of urban governance, and the pressures for effective action.

One of these uncertainties occurs when national interests have precedence over local interests, and as a consequence democratic participation may well lead to opposition to a project. Realistic planning evaluation then needs to account for both the larger-scale objectives, and localized impacts and concerns. 'Participatory evaluation' is explored as a means of developing broad understanding of the project, confronting the distribution of costs and benefits, and discovering ways to make the project locally acceptable.

A second uncertainty is how best to reflect institutional and cultural factors in planning evaluation. Frequently, short-term impacts of alternatives being assessed are seen as most important, because too little attention is paid to institutional history and past experience. To counteract this, Fusco Girard proposes 'deliberative forums' that build from citizenship principles and seek social cohesion in the decision-making process. Instruments with promise for accomplishing this include a 'participatory budget' that focuses on indicators for quality of life, an 'ecobudget' that makes explicit the environmental and social costs of choices regarding economy and land use, and the 'Local Agenda 21 for Culture' initiative to stimulate cultural awareness among citizens.

Too often, programme evaluation has failed to give adequate attention to communication between the providers and the recipients of public services. In the chapter by Roger Ellis and Elaine Hogard, the authors make the case for evaluating the communication processes between these actors, and whether the participants deem this communication effective. The technique of a communication audit is assessed as a means of accomplishing this, and it's application is illustrated using two cases. The first case involves introducing a clinical facilitator to improve communication in college-based and ward-based learning for nursing students. In the second case, a communication audit is used as part of the evaluation of two Sure Start educational programmes, especially the interaction among members of the multi-professional team working in this innovative environment.

The contribution by Shinji Tsubohara and Henk Voogd explores the application of Policy Network Theory (PNT) to evaluate decisions taken by the ruling party in Groningen, the Netherlands in introducing a traffic circulation plan. It focuses on the processes of collaboration and communication between officials of the Labour Party.

The authors conclude that PNT is difficult to apply successfully in ex-post planning evaluation, and needs to be adapted to show how personal linkages can affect policy outcomes as well as the transfer of policy ideas in society. Doing so could account for how planning is a continually changing deliberative act characterized by negotiations between stakeholders.

In the chapter by Abdul Khakee and Anders Hanberger, the authors point out the growing use of performance audits in local government as a means of holding public officials accountable to the electorate. Their critique of the 'public review' of environmental policies in Sweden highlights both the variability in how these are carried out in practice, and at the same time the emphasis on efficient management and monitoring processes. They conclude that these reviews need to be enhanced to provide stronger emphasis on environmental goals, and to include broader and more accessible dialogue with ordinary citizens. A principle issue that Khakee and Hanberger raise is the possibility of including accountability in planning evaluation.

Each of the chapters in this book illustrate in one way or another a stronger awareness for including institutional principles in planning evaluation. While a rational approach has tended to analytically separate planning actions from their institutional and social context, these chapters emphasize how some key social realities of planning practice need to be taken into account. These social realities include citizens' sensitivity on planning impacts, attitudes, perceived inequality, mutual understanding, trust, and accepted norms. Another set of principles that has come to the fore includes a multi-dimensional understanding of planning, emphasizing in evaluation new coherences between social, economic and environmental issues. A third new set of principles makes the point that planning evaluation should address the institutional 'embeddedness' of planning, emphasizing principles such as plausibility, accountability, capacity building, and the understanding of stakeholders' perspectives, arguments and actions. Clearly, these principles are still somewhat indeterminate, and the stronger institutional awareness mentioned earlier is wide-ranging. The chapters presented here do reveal, however, an increasing awareness and a shared conviction that institutional principles can help planning evaluation to establish more adequately informed decision making, help legitimize plans or projects politically, and make possible more effective planning intervention.

References

Alexander, E. (1998), 'Conclusions: Where do we go from here?', in N. Lichfield, A. Barbanenete, D. Borri, A. Khakee and A. Prat (eds), *Evaluation in Planning. Facing the Challenge of Complexity*, Dordrecht, Kluwer, pp. 355–74.

Alexander, E. (1998), 'Evaluation in Israeli Spatial Planning. Theory vs. practice', in N. Lichfield, A. Barbanenete, D. Borri, A. Khakee and A. Prat (eds), *Evaluation in Planning. Facing the Challenge of Complexity*, Dordrecht, Kluwer, pp. 299–312.

Alexander, E. (2001), 'Unvaluing Evaluation: Sensitivity analysis in MODM applications', in H. Voogd (ed.), *Recent Developments in Evaluation*, Groningen, Geo Press, pp. 319–40.

Arts, J. (2001), 'EIA Follow-up: Different approaches to ex-post evaluation in environmental impact assessment', in H. Voogd (ed.), *Recent Developments in Evaluation*, Groningen, Geo Press, pp. 37–66.

Barbanente, A., Borri, D., Concilio, S., Macchi, S. and Scandurra, E. (1998), 'Dealing with Environmental Conflicts in Evaluation. Cognitive Complexity and Scale Problems', in N. Lichfield, A. Barbanenete, D. Borri, A. Khakee and A. Prat (eds), *Evaluation in Planning. Facing the Challenge of Complexity*, Dordrecht, Kluwer, pp. 73–96.

Barbanente, A. Borri, D. and Concillo, G. (2001), 'Escapable Dilemmas in Planning: Decisions vs. transactions', in H. Voogd (ed.), *Recent Developments in Evaluation*, Groningen, Geo Press, pp. 355–76.

Bizzaro, F. and Nijkamp, P. (1998), 'Cultural Heritage and the Urban Revitalization: A meta-analytic approach to urban sustainability', in N. Lichfield, A. Barbanenete, D. Borri, A. Khakee and A. Prat (eds), *Evaluation in Planning. Facing the Challenge of Complexity*, Dordrecht, Kluwer, pp. 193–212.

Bobrow, D.B. and Dryzek, J.S. (1987), *Policy Analysis by Design*, Pittsburgh, PA, University of Pittsburgh Press.

Campbell, H. (2006), 'Just Planning: The art of situated ethical judgement', *Journal of Planning Education and Research* 26, pp. 92–106.

Cook, T.P. (1985), 'Postpositivist Critical Multiplism', in R.L. Shortland and M.M. Mark (eds), *Social Science and Social Policy*, Newbury Park, CA, Sage, pp. 129–46.

Davoudi, S. (1997), 'Environmental Considerations in Minerals Planning: Theory versus practice', in D. Borri, A. Khakee and C. Lacirignola (eds), *Evaluating Theory-Practice and Urban-Rural Interplay in Planning*, Dordrecht, Kluwer, pp. 157–66.

Davoudi, S. (2005), 'Towards a Conceptual Framework for Evaluating Governance Capacities in European Polycentric Urban Regions', in D. Miller and D. Patassini (eds), *Beyond Benefit Cost Analysis. Accounting for Non-market Values in Planning Evaluation*, Aldershot, Ashgate, pp. 277–94.

Faludi, A. (1987), *A Decision-centred View of Environmental Planning*, Oxford, Pergamon.

Faludi, A. and Voogd, H. (eds) (1985), *Evaluation of Complex Policy Problems*, Delft, Delftsche UM.

Faludi A. and Korthals Altes, W. (1997), 'Evaluating Communicative Planning', in D. Borri, A. Khakee and C. Lacirignola (eds), *Evaluating Theory-Practice and Urban-Rural Interplay in Planning*, Dordrecht, Kluwer, pp. 3–22.

Fusco, Girard, L. (1998), 'Conservation of Cultural and Natural Heritage: Evaluation for good governance and democratic control', in N. Lichfield, A. Barbanenete, D. Borri, A. Khakee and A. Prat (eds), *Evaluation in Planning. Facing the Challenge of Complexity*, Dordrecht, Kluwer, pp. 25–50.

Fusco Girard, L., Cerreta, M. and De Toro, P. (2005), 'Integrated Planning and Integrated Evaluation – Theoretical References and Methodological Approaches', in D. Miller and D. Patassini (eds), *Beyond Benefit Cost Analysis. Accounting for Non-market Values in Planning Evaluation*, Aldershot, Ashgate., pp. 173–204.

Gentile, F., Milillo, F. and Trisorio-Liuzzi, G. (1997), 'Planning in Urbanised Areas Under Natural Risk', in D. Borri, A. Khakee and C. Lacirignola (eds), *Evaluating Theory-Practice and Urban-Rural Interplay in Planning*, Dordrecht, Kluwer, pp. 231–45.

Guba, E.G. and Lincoln, Y.S. (1989), *Fourth Generation Evaluation*, Newbury Park, CA, Sage.

Healey, P. (1993), 'The Communicative Turn in Planning Theory and its Implication for Spatial Strategy Formation', in F. Fischer and J. Forester (eds), *The Argumentative Turn in Policy Analysis and Planning*, Durham, NC, Duke University Press, pp. 233–53.

Healey, P. (1997), *Collaborative Planning*, London, Macmillan.

Henkel, M. (1991), *Government, Evaluation and Change*, London, Jessica Kingsley.

Hull, A.D., (2006), 'Structures for Communication and Interpretation in Neighbourhood Regeneration: Mainstreaming the Housing Action Trust philosophy', *Urban Studies* 43(12), pp. 2317–50.

Innes, J. (1985), 'Planning Theory's Emerging Paradigm: Communicative action and interactive practice', *Journal of Planning Education and Research* 14, pp. 183–90.

Khakee, A. (1998), 'Evaluation and Planning. Inseparable Concepts', *Town Planning Review* 69(4), pp. 359–74.

Khakee, A. (2002), 'Assessing Institutional Capital Building in Agenda 21 Process in Göteborg', *Planning Theory and Practice* 3(1), pp. 53–68.

Khakee, A. (2003), 'The Emerging Gap between Evaluation Research and Practice', *Evaluation* 9(3), pp. 340–52.

Levent, T.B. and Nijkamp, P. (2005), 'Evaluation of Urban Green Spaces', in D. Miller and D. Patassini (eds), *Beyond Benefit Cost Analysis. Accounting for Non-market Values in Planning Evaluation*, Aldershot, Ashgate, pp. 63–88.

Lichfield, D. (1998), 'Integrated Planning and Environmental Impact Assessment', in N. Lichfield, A. Barbanenete, D. Borri, A. Khakee and A. Prat (eds), *Evaluation in Planning. Facing the Challenge of Complexity*, Dordrecht, Kluwer, pp. 151–76.

Lichfield, D. (2001), 'Community Impact Assessment and Planning. The Role of Objectives in Evaluation Design', in H. Voogd (ed.), *Recent Developments in Evaluation*, Groningen, Geo Press, pp. 75–84.

Lichfield, N. (2001), 'The Philosophy and Role of Community Impact Evaluation in the Planning System', in H. Voogd (ed.), *Recent Developments in Evaluation*, Groningen, Geo Press, pp. 153–74.

Lichfield, N. and Prat, A.(1998), 'Linking Ex Ante and Ex Post Evaluation in British Town Planning', in N. Lichfield, A. Barbanenete, D. Borri, A. Khakee and A. Prat (eds), *Evaluation in Planning. Facing the Challenge of Complexity*, Dordrecht, Kluwer, pp. 283–98.

Lichfield, N., Kettle, P. and Whitebread, M. (1995), *Evaluation in the Planning Process*, Oxford, Pergamon.

Lombardi, P. and Curwell, S. (2005), 'Analysis of the INTELCITY Scenarios for the City of the Future from a Southern European Perspective', in D. Miller and D. Patassini (eds), *Beyond Benefit Cost Analysis. Accounting for Non-market Values in Planning Evaluation*, Aldershot, Ashgate., pp. 207–24.

Macchi, S. and Scandurra, E. (1997), 'Evaluating Sustainability: Three paradigms', in D. Borri, A. Khakee and C. Lacirignola (eds), *Evaluating Theory-Practice and Urban-Rural Interplay in Planning*, Dordrecht, Kluwer, pp. 57–66.

Miller, D. (2001), 'Evaluating Environmental Justice for Planning', in H. Voogd (ed.), *Recent Developments in Evaluation*, Groningen, Geo Press, pp. 17–36.

Miller, D. (2005), 'Methods for Evaluating Environmental Justice – Approaches to Implementing US Executive Order 12898', in D. Miller and D. Patassini (eds), *Beyond Benefit Cost Analysis. Accounting for Non-market Values in Planning Evaluation*, Aldershot, Ashgate, pp. 25–44.

Miller, D. and Patassini, D. (2005), 'Introduction – Accounting for Non-market Values in Planning Evaluation', in D. Miller and D. Patassini (eds), *Beyond Benefit Cost Analysis. Accounting for Non-market Values in Planning Evaluation*, Aldershot, Ashgate, pp. 1–12.

O'Neill, J. (1993), *Ecology, Policy and Politics Human Well-being and the Natural World*, London, Routledge.

Rosenhead, J. (2005), 'Problem Structuring Methods as an Aid to Multiple-stakeholder Evaluation', in D. Miller and D. Patassini (eds), *Beyond Benefit Cost Analysis. Accounting for Non-market Values in Planning Evaluation*, Aldershot, Ashgate, pp. 163–72.

Selicato, F., Pace, F. and Torre, C. (2001), 'Evaluation of Environmental Policies and Local Planning Practices in Protected Areas. The Case of Alta Murgia', in H. Voogd (ed.), *Recent Developments in Evaluation*, Groningen, Geo Press, pp. 253–72.

Söderbaum, P. (1998), 'Economics and Ecological Sustainability. An Actor-network Approach to Evaluation', in N. Lichfield, A. Barbanenete, D. Borri, A. Khakee and A. Prat (eds), *Evaluation in Planning. Facing the Challenge of Complexity*, Dordrecht, Kluwer, pp. 51–72.

Woltjer, J. (2001), 'Evaluating Complex Decision Making Processes: An assessment of Dutch communicative planning initiatives', in H. Voogd (ed.), *Recent Developments in Evaluation*, Groningen, Geo Press, pp. 85–110.

Vigar, G., Healey, P., Hull, A.D. and Davoudi, S., (2000), *Planning, Governance and Spatial Strategy in Britain, An Institutionalist Analysis*, London, Macmillan.

Voogd, H. (1983), *Multicriteria Evaluation for Urban and Regional Planning*, London, Pion.

Voogd, H. (1997), 'On the Role of Will-shaping in Planning Evaluation', in D. Borri, A. Khakee and C. Lacirignola (eds), *Evaluating Theory-Practice and Urban-Rural Interplay in Planning*, Dordrecht, Kluwer, pp. 23–34.

Yewlett, C. (1993), 'The Integration of Plan Evaluation with Plan Generation', in A. Khakee and K. Eckerberg (eds), *Process and Policy Evaluation in Structure Planning*, Stockholm, Swedish Council for Building Research.

PART 1
Socio-environmental Principles

Methods for Assessing Environmental Justice in Planning Evaluation – an Approach and an Application

Donald Miller

Introduction

Environmental justice is a recently coined term used to convey the idea of distributional equity in the quality of the setting in which people work and live, and especially that all people should be treated fairly and involved in the process of designing, applying and enforcing environmental regulations (Bullard 1995; Rhodes 2005). Early use of the term in the US emerged from a study that found a high association between race, income, and environmental risk (CEQ 1971). A number of earlier investigations found that a disproportionate number of hazardous facilities were located near low-income and minority populations (Bowman 1997; EJRC 2002; Hofrichter 1993). A church-sponsored study that examined the race and income of communities near several hundred listed hazardous waste sites helped to raise environmental justice as a public issue (UCC 1987).

While concerns about social fairness also consider issues such as income, education, and political empowerment, environmental justice tends to focus on the physical dimensions of equity (Bryant 1995). And since the treatment of environmental justice in the US has taken place in the context of environmental impact statements, it is no surprise that it involves identifying social and economic impacts that are connected with physical or natural environmental effects (Johnston 1994).

In order to accomplish this, it is important to design and apply methodologies useful in accounting for these effects (Meier 1993). The following section provides a brief overview of US legislation requiring that environmental justice be an explicit part of the decision-making process, especially for plans and projects. Then a case study of responding to these requirements illustrates how evaluation of this sort can be performed. Finally this case is critically evaluated and lessons are drawn that can inform future efforts to design methods for assessing social fairness in planning.

Definitions and Treatments of Environmental Justice in the US

Environmental justice requirements are found in several federal and state laws and administrative regulations (Bullard 1994). Until recently, each of these specified that minority and low-income populations should receive an equitable distribution of benefits from governmental programs and regulatory programs, without suffering disproportionately high and adverse impacts. It is useful to define several of these terms.

A thoroughgoing definition of adverse impacts is found, for example, at the Federal Highway Administration website (www.fhwa.dot.gov). Included are air, noise, water pollution and vibration; soil contamination; disrupted natural and man-made resources; degraded aesthetic values; interfering with community cohesion or economic vitality; displacement of businesses, persons or non-profit organizations; traffic congestion; and separation of low-income or minority populations from the broader community.

Environmental justice fails when these adverse impacts are mostly borne by a low-income or minority population, or if these groups could be more severely affected than other groups. Low-income refers to those people whose household income is below the poverty level set by the US Department of Health and Human Services. People who are black, Hispanic, Asian-American, or American Indian or Alaska Native are defined as minorities (CEQ 1997).

One of the earliest national laws to address elements of environmental justice is the Civil Rights Act of 1964. Specifically, Title VI of this non-discrimination law requires fair treatment regardless of race, colour, national origin, sex, disability and age. Complaints can be lodged if there is discrimination in access, benefits, participation, treatment, services, contract or training opportunities, or allocation of funds, in addition to environmental impacts.

Similarly, the National Environmental Policy Act (NEPA), in 42 USC Section 4231, requires that anything that federal agencies undertake, sponsor, fund or permit include planning that makes environmental justice a major consideration in decision making (Bass 1998; CEQ 1996). Additional national legislation such as the Federal Aid Highway Act of 1970 also includes provisions dealing with social fairness: in this case requiring assessing the likely effects of transportation projects on businesses, residences, access to public facilities, and tax base.

Finally, at the national level, Presidential Executive Order 12898 focused on avoiding discrimination in federal programs that have the prospect of affecting human health and the environment, especially for low-income and minority populations (Federal Register 1994). This order also requires that these populations be informed about these likely impacts and be involved in assessing these impacts and designing ways to minimize or to mitigate them.

In addition to specifying low-income and minority populations as the principal focus, this 1994 order explicitly seeks to avoid 'disproportionately high and adverse' human health, economic and social effects of federal actions. These concerns are to be treated in the NEPA process, and so in environmental impact statements. Evidence from these impact statements have provided the basis for legal action, when environmental issues have not been adequately resolved, under Title VI of the Civil Rights Act of 19.64 and the Civil Rights Restoration Act of 1984.

Since the case that will be reviewed and examined in the following sections deals with a major transportation facility in Seattle, regulations of the Federal Highway Administration and the State of Washington are pertinent. Two federal sources for these regulations are the US Department of Transportation Order 5610.2 on Environmental Justice (February 1997), and the FHWA Order 6640.23 on Actions to Address Environmental Justice in Minority and Low-Income Populations (December 1998). Both of these can be viewed at www.fhwa.dot.gov/environment/ej2 and at www.fhwa.dot.gov/environment/guidebook/chapters/v2ch16.

Similarly, the Environmental Services Office of the Washington State Department of Transportation and the State Office of Equal Opportunity have responsibilities for implementing Title VI and EO 12898 on transportation projects in the state. Information on the procedures used to accomplish this are found at www.wsdot.wa.gov/oco/pdffiles/App1.pdf.

Before concluding this discussion of the policy and procedures for addressing the environmental justice implications of plans and projects, it is important to note recent efforts by the Environmental Protection Agency (EPA) to water down the requirements and to eliminate funding for their implementation. The current federal administration and Congress has reduced funding to EPA Offices of Environmental Justice, and staff have been reassigned. Much of the current redirection of national government effort on behalf of social fairness is documented by the report of the EPA Inspector General (2004) that even in its title is sharply critical: *EPA Needs to Consistently Implement the Intent of the Executive Order on Environmental Justice*. This report finds major shortcomings in seven key areas of implementation, ranging from leadership to information and analysis of performance results.

One of several documents included in this report is an August 2001 memorandum from the head of the EPA, which restates the commitment of that agency to environmental justice. This document alters the definition of environmental justice found in EQ 12898 to mean '… the fair treatment of people of all races, cultures, and incomes …' in enforcing environmental laws and in meaningful involvement in the decision-making processes of government. This eliminates special consideration of low-income and minority populations. The EPA response to the draft of the Inspector General's report reaffirmed this redefinition of intent (EPA Inspector General 2004, 39–58).

While the promise of equal protection is desirable, many observers assess this as a major retreat from protecting disadvantaged groups from disproportionate adverse impacts resulting from federal projects and implementation of regulations (see also EPA Inspector General 2004, 59–66). The case that follows uses the original criteria of the executive order and focuses on the impacts affecting disadvantaged groups.

Case Study of an Environmental Justice Evaluation

The city of Seattle is bounded on the east by Lake Washington and on the west by Puget Sound. The downtown area faces the saltwater Sound and a series of elderly shipping piers that have been converted to commercial and public uses. Separating the downtown and the waterfront is the Alaskan Way, a four to six lane surface street, and the Alaskan Way Viaduct that carries state highway 99 on two levels, each with three lanes of traffic. This 52-year-old, four-mile elevated highway was damaged by an earthquake a few years ago, and while it has since been reinforced it will not likely withstand any future, significant earthquake. Additionally, the even older seawall along the central waterfront is failing, and it is likely that a future earthquake would cause it to collapse, causing the viaduct also to fail.

Since the viaduct and seawall are critical infrastructure for the region, several alternatives for their replacement are under consideration, and a draft environmental impact statement has been developed to analyse these. One alternative is to continue to try to maintain the existing viaduct, one side of which has settled by three inches. Replacement alternatives include two designs that would tunnel the highway under the existing surface street and incorporate the new seawall into this structure. Another alternative is to remove the viaduct and accommodate all of the traffic currently carried by it on the ground-level Alaskan Way and other surface streets. A third replacement alternative is to replace the existing structure with a new viaduct of about the same capacity. Since the existing viaduct blocks visual access between the downtown and the waterfront, and is a significant source of traffic noise to adjacent properties, various civic groups have advocated its removal even before it became damaged and a hazard. A tunnel would cost an estimated $4.4 billion, and a replacement elevated highway about $3.4 billion. The federal government recently appropriated $231 million for the project, while more was hoped for. The state has earmarked $2 billion of new gas tax revenues to the project. Thus financing for constructing a new viaduct will be difficult, and replacing the viaduct with a tunnel appears unlikely.

Any of the options for replacing the Alaskan Way Viaduct – by putting the traffic on a surface street, in a tunnel, or on a new elevated highway – will require several years of construction and major disruption of traffic. Of specific interest here is how the construction impacts, and the operational impacts of the new facility, will affect low income and minority populations; how these effects can be accounted for in an operationally effective manner; and how to determine whether these populations are disproportionately adversely affected, thus requiring special effort to minimize or mitigate these impacts. The primary source for this information is the *Environmental Justice Technical Memorandum, Draft Environmental Impact Statement, SR 99: Alaskan Way Viaduct and Seawall Replacement Project* (FHWA 2004), published by the US Federal Highway Administration, Washington State Department of Transportation, and City of Seattle.

The evaluation approach taken to assess the environmental justice implications of this project generally follows a conventional series of several steps:

- identify the impacts of the project and the spatial extent of these;
- identify the demographic characteristics of the people living and working within the area significantly impacted;
- compare this demographic information with that for a baseline population, usually that of a larger area such as the jurisdiction as a whole, to determine if low-income and minority populations are disproportionately disadvantaged by the project;
- engage the populations that are impacted in helping to assess the importance of the impacts and to identify useful ways to mitigate impacts that remain after efforts to minimize them.

A procedure along these lines can be used to meet the requirements of Department of Transportation (DOT) Order 5610.2 (USDOT 1997), DOT Order 6640.23 (FHWA 1998), and Executive Order 12898. In most recent cases, efforts are made to predict and quantify adverse impacts. The results are often then presented in incidence tables and maps that display the degree and spatial extent of the undesirable effects. When this information is combined with demographic data for the affected areas, it is possible to estimate whether low income and minority populations of these areas are disproportionately adversely impacted, as well as the level to which they are affected (Miller 2005a). This approach to forecasting the likely effects of plans and project designs is an increasingly used form of planning evaluation (Khakee 1998; Miller and Patassini 2005, 1–12). The following discussion of this case is organized using the series of procedural steps, just outlined.

Identification of Impacts and Impact Area

The environmental justice study of this environmental impact statement determined that the residents and business that would feel the effects of construction activities involved in rebuilding or replacing the Alaskan Way Viaduct would be located in a corridor consisting of Census block groups adjacent to the waterfront route for all of the design options, shown in Figure 2.1. These block groups consist of several city blocks each, and are smaller than Census Tracts. Income and ethnic group data are reported for block groups in the 2000 Census. This same corridor also is the zone that will be most affected by impacts after construction is completed.

Several of the adverse impacts stemming from construction are obvious and significant, such as noise, dust, and traffic disruption. This assessment of the distributional implications sought information through interviews with organizations providing housing and other services to residents of the impacted area, and residents themselves. These interviews helped to identify several effects that are not as immediately obvious. These include:

- Relocation of housing, social services and other facilities. Since project designs use existing rights-of-way, the major negative impact will be dislocation of a Latino workers' centre located on public land that will be used as a construction staging area.

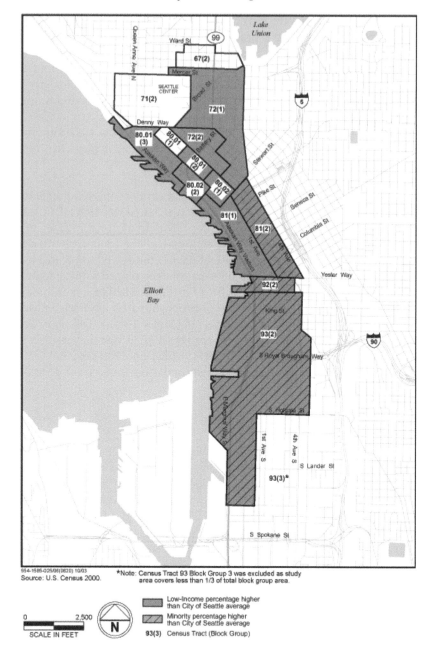

Figure 2.1   Minority and low income populations

- Increased traffic congestion affecting services, deliveries, staff, and volunteers at the social service organizations in the impact area, and increased emergency service response times. This is a major expressed concern, but one that can be minimized by taking care in maintaining access to sites within the affected area.
- Disruption of transit services, including access to ferries. Bus transit is used widely by local residents and by employees in the impact zone, as well as low income and minority people coming to visit the social service agencies located in the affected area. Census data shows that residents of the block groups within the affected area are several times more dependent on public transit than is the population of the city as a whole.
- Construction site hazards. This could affect many of the lower income residents who live in this area, and could pose especial access problems for the physically and mentally handicapped residents.
- Displacement of homeless people who sleep under the viaduct and in parked cars. While this is illegal behaviour, some who do seek shelter in this area would not be able to at least during construction.
- Business and real estate market impacts. The old piers and adjacent warehouse structures on the waterfront have been converted to retail space and to public purposes such as a park and aquarium. Access to these will be difficult at best during the years that construction will take place.
- Elimination of metered parking along the waterfront. This primarily affects shoppers visiting activities along the waterfront, including large numbers of tourists during the summer months.
- Utility disruptions. Since most of these are underground in the project corridor, tunnelling options would have the greatest impact during construction.

None of these negative impacts were measured for this EIS, except the number of social services displaced by the project. Instead, they were identified and described, with the help of personnel at agencies within the corridor and some of the low-income and minority residents. The strategy is to make these impacts part of the public record so that they must be addressed, and then to deal with minimizing and mitigating them as the viaduct replacement design is developed, through continued discussion and negotiations between the affected parties and the governmental agencies that are responsible for these decisions.

It appears that this interactive approach was chosen over developing indicators or metrics, because it would be more sensitive toward the sorts of impacts involved, and to foster a political process that would be more effective in assessing implementation of valid ways to avoid disadvantaged members of the population being negatively impacted disproportionately.

In this case, evaluation is less dependent on analytical methodology and measurement techniques of the sorts discussed earlier, and more on a process that could result in progressive refinement of impacts and their consequences, through directly engaging the parties who would be affected and their advocates.

These impacts of the several-year construction process are likely to be of greater nuisance and hazard than will the operational impacts of the viaduct replacement, once it is completed. In fact, the noise effects of the surface and elevated options are expected to be similar to the current conditions. Pedestrian access to the waterfront would be hindered by the surface street option, but transit access would be equivalent among the several design options once one is built.

Identifying Demographic Characteristics within Impacted Area

Information on characteristics of the residents of the Census block groups identified as the area primarily affected by construction of any of the viaduct replacement options (Figure 2.1) was extracted from the 2000 Census. Of the 15,839 people living in this area, 26 per cent were non-white (versus 30 per cent for Seattle), 8 per cent were Hispanic or Latino (versus 6 per cent for Seattle), and 25 per cent had incomes below the federally designated poverty level (versus 12 per cent for Seattle). Table 2.1 provides this data by block group, showing that ten of these analysis zones had considerably higher rates of low-income population, and three had non-white rates higher than for Seattle as a whole. As shown in Figure 2.1, these are concentrated in the central and southern portions of the corridor, especially around Pioneer Square.

Census information concerning households speaking languages other than English showed higher than usual proportions – 10 and 11 per cent – in the Pioneer Square and International District areas. Federal guidance requires language translation when these rates are greater than 5 per cent or 1,000 persons, whichever is less. This data led to project information being translated into Vietnamese, Spanish, Chinese and Tagalong.

Census date on disabled residents was recognized as important in assessing the environmental justice implications of the viaduct replacement project, since construction could restrict access through the area and to service providers, and also result in higher levels of confusion for some. Eight of the 12 block groups in the corridor exceed the city rate of 6 per cent with disabilities (Table 2.2). Limitations include physical, mental, and employment disabilities.

Transit dependence in ten of the block groups in the affected corridor is two to four times that for Seattle, based on Census counts of dwellings with no vehicle available. However, several of the analysis zones especially in the northern part of the corridor contain high-value condominiums and expensive apartments, where car ownership is lower than average, but separating these from low-income households is not possible. All of the corridor affected by this project is well served by transit.

Finally, it was noted that a number of people of Pacific Island or Asian heritage fish for squid for personal use along the waterfront, especially at Pier 65. Each of the viaduct replacement options could reduce access to these fishing locations during construction.

Comparing the Affected Population with a Baseline Population

In order to determine if low-income and minority populations bear a disproportionate burden from negative impacts, the residential composition of

Table 2.1   Project area demographics

Location	% low-income	% non-white
City of Seattle	12	30
Benefit area	11	18
Affected area		
CT 67, BG 2	1	15
CT 71, BG2	8	17
CT 72, BG 1	**15**	**25**
CT 72, BG 2	**18**	**20**
CT 80.01, BG 1	8	17
CT 80.01, BG 2	**25**	**27**
CT 80.01, BG 3	**23**	**28**
CT 80.02, BG 1	**28**	**27**
CT 80.02, BG 2	**16**	**26**
CT 81, BG 1	**25**	**25**
CT 81, BG 2	**63**	**43**
CT 92, BG 2	**48**	**39**
CT 93, BG 2	49	**35**
Affected area average	25	26

Bold indicates higher percentage than Benefit area; CT: Census tract; BG: block group.

Source: US Census Bureau (2000).

the affected area must be compared to some baseline population. This comparison group is usually a larger population, such as for the city, region, or service area. For example, Figure 2.1 identifies those block groups with low-income and minority percentages that are higher than for the population of Seattle as a whole.

In the case of this EIS, a benefit area was also defined. This benefit area consists of Census tracts that traffic surveys indicate are primary origins of trips using the current Alaskan Way Viaduct. While this viaduct is a major regional facility serving trips originating from three counties in the central Puget Sound region, it was found that five residential areas and two industrial districts in Seattle account for approximately two-thirds of the trips either to or through downtown using this route. The locations of these and their spatial relation to the project area are presented in Figure 2.2.

The 27 Census tracts in this benefit area account for 21 per cent of the total population of the city of Seattle. The benefit area has about the same proportion of low-income residents as the city as a whole, but 18 per cent are non-white, while 30 per cent of Seattle residents are non-white (Table 2.1).

Using the population of the benefit area as the basis for comparison thus amplifies the significance of non-white residents in the affected area in assessing

Table 2.2   Disabled population

Location	Total population	Percent disabled*
City of Seattle	563,374	6
Affected area		
CT 67, BG 2	609	2
CT 71, BG 2	919	3
CT 72, BG1	495	5
CT 72, BG2	2,589	**10**
CT 80.01, BG 1 and 2	2,265	7
CT80.01, BG 3	1,145	5
CT 80.01, BG 1 and 2	2,762	**11**
CT 81, BG 1	2,431	**8**
CT 81, BG 2	1,046	**11**
CT 92, BG 2	911	**17**
CT 93, BG 2	677	**11**
Affected area average	15,839	**9**

* Disabilities include: sensory disabilities (blindness, deafness), physical disabilities, mental disabilities, self-care disabilities, go-outside disabilities, and employment disabilities.

Bold indicates higher percentage than Benefit area; CT: Census tract; BG: block group.

Source: US Census Bureau (2000).

whether this group is disproportionately impacted. It also makes a point of how convenience for some residents and businesses in Seattle may be provided at the expense of residents in the project corridor.

In addition to the higher proportions of low-income and minority residents in the area affected by the construction and operation of this project, 72 low-income housing and social service agencies are located within one city block of the existing viaduct. These agencies serve both nearby residents, and people living outside of this area for whom easy access to this location is provided by public transit, largely by bus. During and after construction, it will be important to minimize and mitigate negative impacts such as increased noise and reduced access on these agencies and their clients.

Involving Affected Parties in the Decision Process

The last step in the process of evaluating environmental justice that was outlined earlier directly involves the affected parties. People living and working in the area impacted by the project are engaged in identifying impacts of concern, in learning about the consequences of these impacts if they adversely affect health or life, and in helping to identify effective means to minimize or to mitigate negative spillovers.

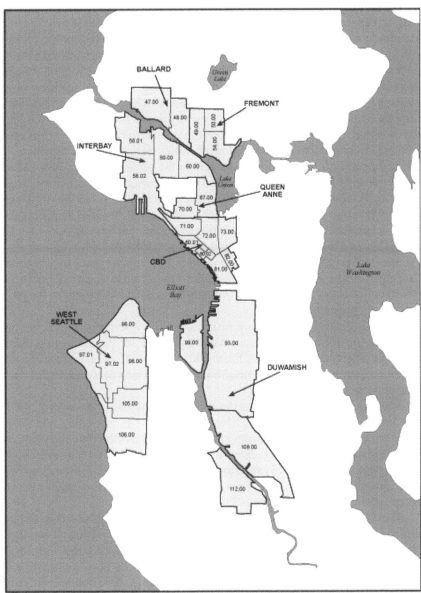

Source: U.S. Census, 2000.

Figure 2.2 Map of benefit area

Public involvement is a major feature of the Alaskan Way Viaduct replacement project. In part, this citizen engagement effort compensates for identifying but not attempting to measure and to forecast adverse impacts, as noted earlier. The draft EIS identifies twelve types of initiatives used in communicating with and obtaining responses from people in the corridor most impacted by construction. In effect, these efforts included interviews with representatives of service agencies, various means of supplying information to interested parties, and means of facilitating responses from the residents. The interviews provide the basis for developing a list of concerns and effects of the project that need to be addressed. The results of these interviews are summarized in an earlier section of this chapter.

Efforts at informing especially people in the project corridor have included advertisements in city and community newspapers, newsletters distributed at community meetings and community centres and mailed to agencies in the area, and information displays at community centres and libraries in the area. In addition, project fact sheets have been translated into the four major non-English languages used by local residents, and these have been distributed through service agencies and posted on a website. Community briefings are made by project staff at regular meetings of several organizations in the area. A phone hotline and e-mail notifications are also used to inform about the development of the project and to invite responses.

The major means of dialogue with residents is a series of public meetings, held in an open house format, to make it possible for people to talk one-on-one with project staff members. From 2001 to 2003, 16 of these meetings were held, in locations accessible to neighbours, property owners and tenants in the downtown area.

The project team working on the environmental justice element of the EIS has concluded that these efforts have been effective in informing and hearing from most of the affected parties in the project corridor. But they have also identified some groups that they need to hear more from, such as disabled and elderly populations. The EIS expresses commitment to engage in even more public involvement activities as the project progresses, in an effort to reach consensus on concerns and solutions (Innes 1996).

The design of this deliberative evaluation approach depends on this continued interaction. And while extensive individual and group meetings have already happened, affected parties will be in a better position to identify and assess project impacts once the decision is made on how the viaduct will be replaced, and as specific staging and design decisions are considered. The public involvement program pursued in the early stages of this project provides a framework for next steps, and an indication of problems that still need to be resolved.

Conclusions

As noted earlier, much of evaluation used in planning and policy analysis involves measuring and forecasting the inputs and outcomes of alternatives under consideration. In practice, benefit cost and revenue cost analysis estimate these effects in physical units that are then converted into monetary units. This requirement is the basis for major criticisms of these methodologies (Miller and Patassini 2005).

Alternative approaches to evaluation, such as multicriteria evaluation and point systems of evaluation, usually also seek to employ metrics in measuring and forecasting the implications of alternatives, but these methodologies summarize this information without converting these effects into market terms. These metrics provide an unambiguous basis for independent critical assessment of the evaluation and thus to help assure creditability.

This tendency toward measuring kinds of effects and presenting estimates in quantitative terms is characteristic of many efforts to deal with the social distributional implications of planning alternatives. An earlier review and assessment of dealing with environmental justice in choosing transit routes illustrates this point (Miller 2005a).

Use of quantitative measures to accomplish this faces several problems. Aside from estimating errors, and difficulty in weighting these performance scores, there is the issue of validity: how well the metric represents the effects. This is complicated by issues such as the prior health and other conditions of disadvantaged people, and consequently how new impacts on air and water quality can affect health, for example. It is also complicated by the possible greater cumulative effect of impacts than that realized by these impacts viewed separately (Miller 2005b).

These are arguments for considering dealing with impacts in qualitative instead of quantitative terms. This is the strategy used in the EIS investigated in this chapter. Members of the organizations providing housing and social services in the project area were interviewed, using open-ended questions, to identify likely problems that would affect local residents. A series of open houses, augmented by an extensive public information program, invited additional ideas and views, which were recorded and provided to project staff. Findings of these activities were presented in terms of descriptions of the different sorts of adverse effects that need to be addressed. No effort was made to estimate most of these effects in quantitative terms. Collaboration between the evaluation staff, local service agency staff, and local residents resulted in what appears to be an exhaustive list of impacts. But more to the point, the governmental agencies involved in the viaduct replacement project have committed to involving the affected parties throughout the design and construction management process in reviewing likely adverse effects as more information becomes available, and in seeding consensus on how these adverse effects can be minimized.

While most of the expected negative impacts of the design and construction alternatives are identified but not measured, the populations most likely to be affected by these negative impacts are assessed in quantitative terms, and compared with counterpart measures of the population of the benefit area, and the city of Seattle as a whole. This data is relatively accurate since there has been little change in these populations between the 2000 Census and the 2004 report. These counts served to satisfy several federal and state regulations that require assessing the extent to which low-income and minority populations are disproportionately negatively impacted. These counts are also useful in making the political case that fairness requires special effort to minimize and mitigate the adverse effects on these populations. The demographic analysis presented in this EIS appears to successfully meet these purposes.

Assessing the environmental justice implications of plans, policies and projects is a new challenge to planning evaluation, for which we have little tradition of strategy or methodology. As a contribution to this task, earlier sections of this chapter reviewed the sorts of adverse impacts that need to be included, and presented a methodological procedure to assess both the spatial extent of these effects and whether the people adversely affected are disproportionately disadvantaged.

Additionally, the case that is presented represents a combination of quantitative demographic analysis and largely descriptive treatment of negative impacts that contrasts with recent quantitative approaches for estimating impacts.

This approach addresses concerns about uncertainty including the several sorts that are noted. However, the approach taken in this EIS will only be adequate and effective if it provides the beginning basis of a promised, on-going deliberative process – engaging affected parties – in finding ways to avoid, minimize and mitigate the negative impacts that the project could generate, including learning by these parties as the project details are developed.

References

Bass, R. (1998), 'Evaluating environmental justice under the National Environmental Policy Act,' *Environmental Impact Assessment Review* 18, pp. 83–92.

Bowman, A. (1997), 'Environmental (In)Equality: Race, class, and the distribution of environmental bads', in S. Kamieniecki, G. Conzalez and R. Vas (eds), *Flashpoints in Environmental Policymaking: Controversies in Achieving Sustainability*, Albany, NY, State University of New York Press, pp. 155–175.

Bryant, B. (ed.) (1995), *Environmental Justice: Issues, Policies and Solutions*, Covelo, CA, Island Press.

Bullard, R. (1995), 'Residential Segregation and Urban Quality of Life', in B. Bryant (ed.), *Environmental Justice: Issues, Policies and Solutions*, Covelo, CA, Island Press, pp. 76–85.

Bullard, R. (ed.) (1994), *Unequal Protection: Environmental Justice and Communities of Color*, New York, Random House.

CEQ (Council on Environmental Quality) (1997), *Guidance for Federal Agencies on Key Terms in Executive Order 12898*, Washington, DC, US Governmental Printing Office.

CEQ (Council on Environmental Quality) (1996), *Draft Guidance for Addressing Environmental Justice under the National Environmental Policy Act*, Washington, DC, Executive Office of the President.

CEQ (Council on Environmental Quality) (1971), *The Second Annual Report of the Council on Environmental Quality*, Washington, DC, US Governmental Printing Office.

EJRC (Environmental Justice Resource Center) (2002), *Environmental Justice Timeline – Milestones*, Washington, DC, EJRC, available at <http://www.summit2.org/>.

EPA Inspector General (2004), *Evaluation Report – EPA Needs to Consistently Implement the Intent of Executive Order 12898*, Report No. 2004–P-00007, Washington, DC, Environmental Protection Agency.

Federal Register (1994), 'Executive Order 12898, Federal Actions to Address Environmental Justice in Minority Populations and Low-Income Populations', Washington, DC ,Federal Register, 59:32, White House Office, p. 7629.

FHWA (US Federal Highway Administration) et al. (2004), 'SR 99: Alaskan Way Viaduct and Seawall Replacement Project, Draft Environmental Impact Statement', available at <http://www.wsdot.wa.gov/projects/viaduct/DEIS/appendices>.

Hofrichter, R. (ed.) (1993), *Toxic Struggles: The Theory and Practice of Environmental Justice*, Philadelphia, PA, New Society Publishers.

Johnston, B.R. (1994), *Who Pays the Price? The Sociocultural Context of Environmental Crisis*, Covelo, CA, Island Press.

Khakee, A. (1998), 'Evaluation and Planning: Inseparable Concepts', *Town Planning Review* 69(4), pp. 359–74.

Mier, R. (1993), *Social Justice and Development Policy*, Newbury Park, CA, Sage Publications.

Miller, D. (2005a), 'Methods for Evaluating Environmental Justice – Approaches to Implementing US Executive Order 12898', in D. Miller and D. Patassini (eds), *Beyond Benefit Cost Analysis – Accounting for Non-Market Values in Planning Evaluation*, Aldershot, Ashgate, pp. 25–44.

Miller, D. (2005b), 'Dutch Integrated Environmental Zoning – A Comprehensive Program Dealing with Urban Environmental Spillovers', in D. Miller and G. de Roo (eds), *Urban Environmental Planning*, Aldershot, Ashgate, pp. 147–68.

Miller, D. and Patassini, D. (eds) (2005), *Beyond Benefit Cost Analysis – Accounting for Non-Market Values in Planning Evaluation*, Aldershot, Ashgate.

Rhodes, E.L. (2005), *Environmental Justice in America – A New Paradigm*, Bloomington, IN, Indiana University Press.

UCC (United Church of Christ, Commission on Racial Justice) (1987), *Toxic Wastes and Race in the United States: A National Report on the Racial and Socioeconomic Characteristics of Communities with Hazardous Waste Sites*, New York, Public Data Access Inc.

Chapter 3

Multidimensional Evaluation for Sustainable Development: Managing the Intermix of Mind, Artefact, Institution and Nature

Jenny Stenberg

Introduction

Implementing the vision of sustainable development implies a need for multidimensional evaluation. The discussion in the present chapter is based on experience from two evaluations of Swedish national funding programmes – the Local Development Agreements and the Local Investment Programmes – and on a study focusing on how urban design and planning are related to changes in 'exposed suburban metropolitan areas' in Sweden (Stenberg 2004; Stenberg et al. forthcoming).

A key challenge in these kinds of evaluation studies is to link environmental and social aspects. The present chapter shows how this relationship can be scrutinized with a special focus on certain environmental and social themes. The evaluation study discussed here involved case study evaluations including ten projects in housing areas from the 1950s, 1960s and 1970s. The empirical material consisted of an environmental matrix, statistical data on the housing areas, interviews with tenants and employees and a smaller mass media study.

The theories applied for analysing the empirical material – the MAIN^{TETRA} (Kain 2003) – have been developed to facilitate our understanding of the complex problems arising from an ambiguous reality. The MAIN^{TETRA} model, thus, enables us to relate different knowledge areas and build up an understanding of how they are related to each other in the perspective of sustainable development. These theories of sustainable development have been put into practice in two pilot studies in Sweden (Kain and Söderberg 2002; Kain et al. 2005).

It is argued throughout the chapter that utilizing the MAIN^{TETRA} as a model for systematizing and analysing empirical material is helpful when evaluating projects from the perspective of linking social and environmental effects. It is helpful, first, because the model implies an awareness of how the empirical material is related to the broad vision of sustainable development. Second, the model implies a focus on how conflicting perspectives presented by different local actors relate to the broad vision of sustainable development. As a result, the multidimensional perspective facilitates an increased understanding of how the different fields of knowledge affect one another.

This chapter also presents some difficulties associated with using the MAIN[TETRA] and discusses the need to include further development of software and multidimensional evaluation tools. However, before further exploring these issues, a brief description of the changing principles for urban planning in Sweden will first be presented as a background.

Sustainable Development and Planning

The prerequisites for urban planning have undergone considerable changes during recent decades, and new theories and approaches have been developed (see Allmendinger 2002; Sager 1994). In Sweden, studies on the collaborative planning approach conducted by planning researcher Patsy Healey (1997) from Britain seem to have had great importance for planning practice and development. The changed prerequisites for urban planning have been closely related to introduction of the concept of sustainable development (WCED 1987). From an overall perspective, the aim of the notion may be considered quite clear, and it was, therefore, easy for all countries to share the vision. Present environmental threats, on the one hand, and material and social poverty, on the other, created the need for a vision of a better world for all people. Thus, the notion of sustainable development began as a marriage between two major societal movements: the economic development movement and the environmental protection movement (Meadowcroft 1999).

However, even if the documents clearly stated the overall goals for sustainable development and how achieving these goals was to be implemented, the notion, on the whole, may still be perceived as rather vague. Yet, it is quite clear that the declaration stressed the importance of amalgamating the social and economic dimensions with the environmental dimension of sustainable development and, further, that such an amalgamation requires an interdisciplinary approach (UNCED 1992). Later, the Commission on Sustainable Development also included a fourth dimension – the institutional – to the framework of sustainable development (CSD 1996).

When relating the vision of sustainable development to urban planning, it is interesting to consider the argumentation in the Agenda 21 document, which maintains that authorities 'should establish innovative procedures, programmes, projects and services that facilitate and encourage the active participation of those affected by the decision-making and implementation process, especially of groups that have, hitherto, often been excluded, such as women, youth, indigenous people and their communities and other local communities' (UNCED 1992, 10,10). In other words, politicians have agreed upon a vision of planning for sustainable development entailing a communicative turn with local partnerships that should include not only local professionals, but also citizens.

In Sweden – a country that had often been considered a forerunner with respect to sustainable development – the effort to implement Agenda 21 at the municipal level progressed at full speed soon after the Rio conference. A National Commission on Agenda 21 was appointed in 1995 to develop, deepen and establish the agenda in Sweden, and in June 2000, a National Commission on Agenda 21 and Habitat was appointed to support and develop implementation of Agenda 21 and the Habitat Agenda. At that point in time, however, the Swedish government mainly stressed the environmental dimension of sustainable development. This was not only obvious in governmental documents; when the Government appointed a new commission to deal with sustainable development, they named it the 'Commission on Ecologically Sustainable Development' (Swedish Government 1997). This understanding of sustainable development as mainly concerning environmental aspects was prevalent during the entire period until the end of the 1990s (Swedish Government 1999). From the point of view of planning, this emphasis on environmental aspects resulted, for instance, in the policy document 'Fifteen Environmental Quality Objectives', objectives to be included in all planning and construction procedures (Swedish Government 1999; Naturvårdsverket 2000). It was not until 2001 that all three aspects of sustainable development were treated as equally important in official documents produced by the Swedish government (2001a) – and even now, in 2006, the fourth institutional dimension has not yet been taken into consideration in official documents.

The Swedish presidency of the EU Council of Ministers in 2001 (Swedish Government 2001b), the national arrangements that took place to prepare for the five year follow-up of the Habitat Agenda in New York 2001 (Swedish Government 2001c), and the ten-year follow-up of Agenda 21 in Johannesburg 2002 (Swedish Government 2002; SOU 2003a; 2003b) seem to have intensified the national focus on sustainable development again, after a weakening trend at the end of the 1990s. One important Swedish contribution to the conference in Johannesburg was a report by the Ministry of the Environment (Azar et al. 2002). This report showed that economic growth cannot solve, at least not on its own, future environmental problems; there is also a need for strong political control. These conclusions may be important to reflect upon in contemporary Swedish society as well, where politicians often present the concept 'economic growth' as synonymous with whatever development policy they are discussing – which increasingly often also pertains to sustainable development.

For the Swedish government, the Johannesburg conference resulted in 12 propositions for further work on sustainable development. The government proposed, e.g., the establishment of an independent national forum for Agenda 21 and Habitat as well as the broadening of physical planning procedures to also include issues of ethnic and economic integration (SOU 2003a).

At the same time as implementation of the vision of sustainable development was initiated, during 1990–1993, Sweden experienced a severe economic crisis – due to overgrown welfare systems or lack of powerful politics, on that point the experts disagree – and since then, the welfare systems have gradually deteriorated.

Sweden, which in the early 1980s was considered one of the most equal countries in the world with regard to income, had, by the late 1990s, income differentiation equivalent to that at the beginning of the 1970s. Additionally, and as a probable consequence of this development, Sweden has shifted, as has e.g. Britain, to policies that stress selective methods rather than general measures to solve social problems (Lindberg 1999).

In fact, during the latter part of the 1990s, Sweden was the most segregated OECD country, insofar as the most exposed housing areas in Sweden had the highest share of immigrants in comparison to all other OECD countries (Swedish Government 1998). The Swedish Integration Board, in their report, stated that the Swedish labour market was not ethnically neutral and that children born in Sweden experienced labour market discrimination if their parents were born abroad. Further, they maintained that it was obvious that people born abroad ended up living in suburban areas in metropolitan cities with increasing frequency, and simultaneously that these areas were also found to be the poorest. Moreover, it was in these areas that pupils most often did not pass the core subjects in school (Swedish Integration Board 2003; 2004).

The United Nations Association of Sweden also highlighted this problematic situation. In one report, they criticized the Swedish government for depicting the situation in Sweden in idealized terms when reporting to the United Nations Human Rights Committees, and concluded in their alternative report that there were serious problems with racism due to hidden structural and institutional discrimination within the Swedish systems (UN Association of Sweden 2004).

One of the governmental propositions described above – to broaden physical planning procedures to also include issues of ethnic and economic integration in order to promote sustainable development – actually highlights a key issue in this complex of problems: How can architects and planners, in their professional roles, contribute to such a development and what are the prerequisites of such a development? Broadening the physical planning procedure implies a need to broaden the evaluation procedure as well –that is to say that the field of planning must further develop tools for multidimensional evaluation of planning processes. We are looking here at a new principle in planning evaluation, implying application of an integrated understanding.

Such enlargement of the field of evaluation has been sought after for some time owing to governmental tendencies to apply selective methods rather than general measures to solve social and environmental problems, because selective methods, as part of the communicative turn in society, often imply local area-based and multidimensional interventions. This chapter will present the results of one attempt to put into operation one tool for multidimensional evaluation of planning processes or national funding programmes.

Undertaking a Multidimensional Evaluation

The MAIN[TETRA] was put into operation as tool in an evaluation of 'multidimensional housing regeneration projects' in Sweden that had been funded by the Local Investment Programme (LIP). The evaluation was initiated by The Swedish Environmental Protection Agency, which was responsible for the programme. The Agency was interested in enhancing knowledge about the interaction between environmental and social effects. Altogether, the evaluation showed clearly that linking social and environmental aspects in the evaluation of the refurbishment projects helped to extend and broaden our knowledge by revealing interesting and complex interconnections between the different aspects. In particular, two themes related to environmental issues emerged in the discussion: social exclusion and organizational learning. These themes stressed caution with regard to the exclusion of certain groups and concern about failure to plan for organizational learning. The results from the evaluation overshadow optimistic confidence in the notion that 'environmental refurbishment projects' contribute to sustainable development (Stenberg et al. forthcoming).

There is also a prehistory to this evaluation that greatly influenced its design – a case study comprising two extensive evaluations of a Swedish area-based funding programme called the Local Development Agreements (*Lokala Utvecklingsavtal* – LDA), aiming at social inclusion and sustainable development (Stenberg 2004). Experiences from this case study research project have facilitated further development of methods for evaluation and research related to the vision of sustainable development, as it focused on a field of knowledge that has largely been neglected: the social dimension of sustainable development.

Turning back to the LIP study, it was also designed as a case study evaluation (Yin 1994; 2000), although the requirement for richness of data was not satisfied, as it was a summative evaluation conducted during a very short time period and we were not able to study events in real-life situations. It may be, therefore, more appropriate to say that the present design was inspired by case study research.

The original set of evaluation questions in the LIP study posed by the Swedish Environmental Protection Agency may be summarized using four main questions at issue:

a) What environmental effects did the projects have and how did the results relate to the project managers' goals?
b) To what extent were the tenants involved and engaged in the process of change?
c) Did the projects lead to the housing areas becoming more attractive?
d) Were the environmental and social effects permanent?

The present evaluation included ten regeneration projects (see Figure 3.1) selected from the total set of 86 multidimensional projects. They were selected on the basis of their focus on both environmental and social aspects, the comprehensiveness of their data, and for their between-project comparability.

	Chosen projects:	Year of con- struction	N:r of flats	LIP- subsidies (EURO)
1	Bergsåker Sundsvall	1972-73	475	470 000
2	Nacksta Sundsvall	1966-72	400	320 000
3	Markbacken Örebro	1958-63	1200	407 000
4	Ringdansen Norrköping	1968-72	1600	27 830 000
5	Östlyckan Alingsås	1959-61	324	963 000
6	Rannebergen Göteborg	1972-75	1600	492 000
7	Norrliden Kalmar	1970	500	300 000
8	Inspektoren Kalmar	1955-57	150	535 000
9	Rådhusrätten Lund	1966-67	470	246 000
10	Augustenborg Malmö	1958-59	1600	4 282 000

Figure 3.1   Ten projects chosen in the LIP study that focused on both environmental and social aspects

Note: The map shows where in Sweden the projects were carried out.

Figure 3.2   Ringdansen in Norrköping before the regeneration project

Note: This was the most extensive project with respect to total cost and level of physical change in the outdoor environment. One measure to increase the attractiveness of the area involved changing its name from Navestad to Ringdansen.

Photo: Jenny Stenberg.

The physical environments of the ten housing areas are quite diverse, although they were constructed in the same time period. Some of them, e.g. the housing area Ringdansen in Norrköping (see Figure 3.2), have tall buildings, while others, e.g. Nacksta in Sundsvall (see Figure 3.3), have a more small-scaled architecture. Most of the projects, however, focused on eco-cycle adaptation, such as changes in the energy, water and sewage distribution systems – and in control and reporting systems – as well as changes in sorting domestic waste and on tenants' participation in implementation of the projects.

The empirical material consisted of four different types of information. First, an environmental matrix was formed using reported environmental data on the projects including the categories: energy, traffic, water, waste water, domestic waste, building material, chemicals and biological diversity. Each of these categories included:

a) information on whether the measure involved efficiency improvement or system conversion;
b) a description of technical measures;
c) a description of management-related measures;
d) a description of tenant involvement;
e) the reported quantity (energy etc.) before and after the project was implemented.

Second, statistical data was compiled on the housing areas covering an eight year period and concerning tenants' age, sex, native country, income, education and occupation (from Statistics Sweden) as well as statistical crime data (from the police). Furthermore, data from the involved housing companies on the number of vacant apartments and annual movements and from the 'satisfied-tenant index' were also used.

Third, an interview study was conducted with tenants and employees. It included 58 qualitative interviews with 78 persons and focused generally on project implementation and specifically on tenant involvement in the process or tenant opinions on project outcomes. When visiting the areas, we also spent time there studying the physical environment and some of the places where people gathered, e.g. a youth recreation centre, to better understand the social sphere.

Fourth, a mass media study was conducted to analyse how the housing areas were described in the daily press before and after the LIP. We studied what thematic categories were discussed in the articles (crime, accidents, social aspects, public sector, physical planning, trade and industry, environment, nature, culture and sport) as well as the tone conveyed in each article (neutral, negative or positive).

Figure 3.3   According to some of the tenants, Nacksta in Sundsvall resembles an English residential area of terraced houses after the regeneration project was completed

Photo: Jenny Stenberg.

Utilizing the MAIN^TETRA

As the assignment entailed a broad range of evaluation questions related to the environmental and social sphere, we needed a tool for multidimensional evaluation. The theories used to analyse the empirical material are based on the idea that sustainable urban development embraces four different main, but overlapping, fields of knowledge about the city, its tenants and its surroundings (see Figure 3.4).

- One area involves the need to study and understand the city's artefacts (a), that is to say man-made things such as works of art, instruments, machines, buildings and physical networks;
- Additionally, there is a need for knowledge of the institutions (i), that is formal and informal relational webs of all sizes and directions, formal and informal norms, and information systems and codified knowledge;

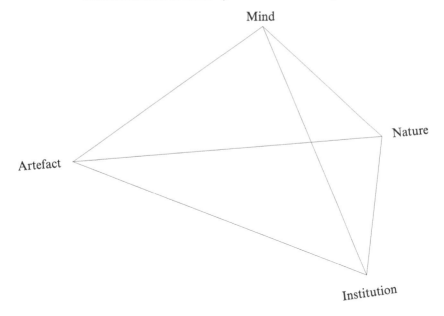

Figure 3.4   The MAIN^{TETRA} **– sustainable urban development embraces
four main, but overlapping, fields of knowledge about the city, its
inhabitants and its surroundings**

Source: Kain (2003).

- Knowledge of nature (n) can then be added, including all kinds of natural elements from the ecosphere as well as the lithosphere;
- Last, but not least, we need knowledge concerning how people, tenants as well as employees, think and feel, as our actions are controlled by our mind (m), that is ethics, worldviews, knowledge, skills and other human attributes.

These fields of knowledge are positioned in a four-dimensional tetrahedron to illustrate how each one of them has a relationship to the other three – that is to say when trying to understand the reality of a given problem, we must deal with a combination of fields of knowledge (see Figure 3.5).

This conceptual model, thus, is intended to facilitate better understanding of knowledge about a complex and confusing reality – that is to say to relate the different fields of knowledge to each other and to increase our understanding of how these fields affect one another. Therefore, it may be more important and interesting to focus on the links between the fields of knowledge than on the fields of knowledge per se.

The starting point for the present analysis was the two evaluation areas – 'environment' and 'social' – that were formed as a result of the many different evaluation questions. These areas were subsequently divided into 15 themes – eight environmental and seven social (see Figure 3.6). The titles of the eight

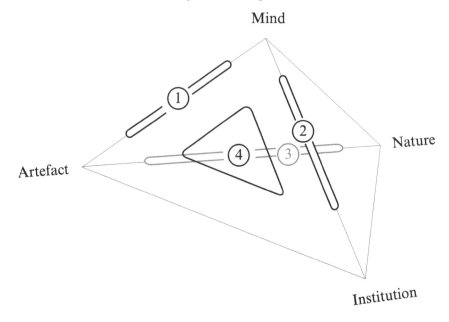

Figure 3.5   The fields of knowledge are positioned in a four-dimensional tetrahedron to illustrate how each one of them has a relationship to the other three – i.e. when trying to understand the reality of a given problem, we must deal with a combination of fields of knowledge. Here, four examples of links are illustrated

Source: Kain (2003).

The evaluation area »environment«	energy
	traffic
	water
	waste water
	domestic waste
	building material
	chemicals
	biological diversity
The evaluation area »social«	information
	dialogue
	participation
	organizational learning
	patterns of behaviour
	social life
	stigmatization

Figure 3.6   The evaluation areas 'environment' and 'social' formed 15 themes

environmental themes follow the previously mentioned matrix and the titles of the seven social themes were based on the evaluation questions as well as on certain theories related to these questions (Andersson 2002; Argyris and Schön 1995; Callon and Latour 1981; Easterby-Smith et al. 1999; Ericsson et al. 2002; Krogstrup 1997; Molina 1997; Putnam 1993; Putnam 2000; Rothstein and Kumlin 2001; Stenberg 2004).

As described above, the empirical material consisted of quantitative data as well as qualitative information concerning both evaluation areas – although the environmental area contained more quantitative data and the social area more qualitative data. In the analysis, quantitative and qualitative material have been integrated, meaning that we related reported environmental effects of different kinds and statistics on the housings areas to the interview statements and to the findings of the mass media study.

The major problem is dealing with the multitude of qualitative material. In practice this has been carried out by entering both the recorded interviews (audio files) and the mass media study (text files) into the software we used (Hyper Research). Interesting parts of the evaluation questions, that is to say specific statements or text segments, were then coded (coupled) to one or more of the above mentioned themes – or more correctly, to one or more of the subgroups (codes) that each theme was divided into (see Figures 3.7 and 3.8 for an illustration of subgroups).

One example of an interesting statement was when a tenant told us why he did not use the kitchen waste disposer that had been installed by the housing company as part of the LIP project. The purpose of the kitchen waste disposer was to enable use of organic waste in a municipal biogas digester to produce biofuel for different kinds of vehicles. However, most of the tenants did not use the kitchen waste disposer, as the system required water, and the housing company simultaneously started charging tenants for water use – as this measure was also part of the LIP project to save energy and water (in Sweden, the cost of energy and water usage in rented apartments is traditionally included in the rent without any specifications). In fact, the annual cost of water for the kitchen waste disposer was not more than the cost of a cinema ticket, but this worked to discourage the users. As a result, the tenants threw the organic waste in the container for mixed waste. Hence, this qualitative information from the tenant, related to a quantitative indicator of the amount of mixed waste, could clarify why the amount of unsorted waste was not reduced, although separation of residential waste was another measure within the LIP project.

Turning back to the theories, these subgroups – coming from the MAIN[TETRA] – increased our knowledge of how interesting specific statements or text segments were related to sustainable development. The number of subgroups increased gradually as the coding progressed. Initially, there were only a few codes, representing the predominantly quantitative empirical material we had early in the process, which consisted of the already reported environmental data, but also the preconception we had formed from reading the local actors' project

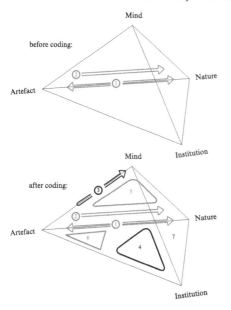

Criterion or indicator in the evaluation for the theme of
ENERGY:

1 A-N: Energy consumption is an indicator that concerns the link between the knowledge area Nature (the environment) and the knowledge area Artefact (the technical system)
2 A: Change of technical system concerns Artefact (the technical system) which influences Nature (the environment)
3 M-A: Change of technical system concerns the link between Artefact (the technical system) and Mind (the experience of the system)
4 M-A-I: Change of technical system concerns the link between Artefact (the technical system), Mind (the indivuduals' understanding), and Institution (organization, decision making, norms, rules)
5 M-A-N: Change of technical system concerns the link between Artefact (the technical system), Nature (the environment, and Mind (the individuals' opinion about how heat shall be charged for – the employees' view as well as the tenants')
6 A-I-N: Change of technical system concerns the link between Artefact (the technical system), Institution (organization, decision making, norms, rules) and Nature (the environment)
7 M-A-I-N: Energy concerns the link between all knowledge areas

Figure 3.7   How the number of subgroups (codes) changed during coding, within the theme of 'energy' in the evaluation area 'environment'

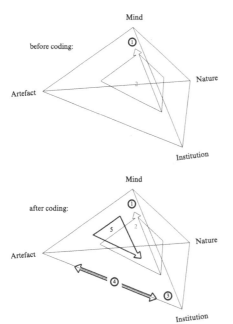

Criterion or indicator in the evaluation for the theme of
STIGMATIZATION:

1 M: Experiences of stigmatization concern Mind (ethics, worldviews, knowledge, etc. of the individual)
2 M-A-I-N: but also the link between Mind (the individual), Artefact (the buildings), Institution (decision-making, norms, organization) and Nature (the environment); a link mainly directed towards Mind (the individual)
3 I: Stigmatization concerns Institution (legislation, rules, norms, social networks, rent control etc.)
4 A-I: stigmatization concerns the link between Institution (the organization of the housing company, decision-making, rules, norms) and Artefact (the housing area as physical environment)
5 M-A-I: and the link between Artefact (the buildings etc.), Institution (decision-making, norms, organizations, institutions) and Mind (the individual)

Figure 3.8   How the number of subgroups (codes) changed during coding, within the theme of 'stigmatization' in the evaluation area 'social'

descriptions. Figure 3.7 shows how the codes in one of the themes grew during the coding process.

Thus, the coding of interviews and the mass media study meant that the entire empirical material was scrutinized and that interesting parts of it, from the point of view of the evaluation questions, were electronically linked to one or more subgroups. Furthermore, the interview and mass media data were coded to both evaluation areas, as the statements or the text segments often concerned both areas. Figure 3.8 shows how the codes developed into another of the themes.

In this way, knowledge about each subgroup – that is information about certain links between two or more fields of knowledge in the perspective of sustainable development – was systematically organized and made ready for analysis. These codings, thus, contain the interviewees' experiences and thoughts about these links, as well as the experiences conveyed in the mass media study. Analysing the material within Hyper Research implied running reports on certain combinations of codings. We made reports on all of the existing overlaps between the evaluation area 'environment' and the evaluation area 'social', that is to say one report on all codings to the theme of energy and to the theme of information, and so on until all of the combinations had been examined. Each report consisted of a list with all codings directly linked to the source – allowing us to immediately listen to, or read, each coding one after the other.

As is obvious in the figures above, it was the breadth of the empirical material that enabled us to discuss, and evaluate, the link between the two disparate evaluation areas. However, these two examples, out of 15, do not reveal how the coding process also gave us the opportunity to understand more about the relationships within the MAIN^{TETRA}. Step by step, during the coding process, we came to understand that the most interesting, and perhaps controversial, information for the evaluation was found when considering what was stated in codes that associated the link Artefact–Nature and the link Mind–Institution. Therefore, when listening to and reading the reports, we particularly focused on information that bridged or broadened this gap.

Simultaneously, when listening to and reading these reports, we also considered the other part of collected empirical material – the environmental matrix and the statistics about the housing areas – and produced texts discussing the evaluation questions. Thus, we did not actually enter quantitative material into Hyper Research in this evaluation, as the time limitations of the project did not give us the opportunity to reflect on the best way of doing this in relation to the analysis model. The present chapter, however, will not present the detailed results from the evaluation itself – or more advanced considerations about its realization. Below, we will instead focus on the analysis model, the MAIN^{TETRA}, used here in combination with the software, and further discuss its possibilities as an evaluation model for urban development processes.

Results and Discussion

Utilizing the MAIN^{TETRA} as a model for systematizing and analysing empirical material is helpful when evaluating projects from the perspective of linking social and environmental effects. It is helpful, first, because the model implies an awareness of how any empirical material may be related to the broad vision of sustainable development. MAIN^{TETRA} helps us establish an increased and more explicit understanding of relevant fields of knowledge, and links between these fields, as reflected by the empirical material. Such awareness may be important not only when policy-makers study evaluation results, but naturally also at the stage of designing a case study evaluation.

Second, the model implies an awareness of how the conflicting perspectives presented by different local actors relate to the broad vision of sustainable development. The model shows how the experiences of various actors relate to the four fields of knowledge and to certain links between these fields. As a consequence, during the coding process, we soon revealed the above-mentioned interesting relation between the link Artefact–Nature and the link Mind–Institution, resulting in a special focus on information that bridged or broadened this gap.

This analysis method may result, for example, in elucidation of a wide range of problem complexes related to the actors' introduction of measurements. In our study, the effects of 'environmental measures' proved to be highly dependent on how the individuals thought about and talked about these new systems: how the initiation of new systems was organized and how the new systems were related to other measures, such as sorting of residential waste.

Included in this complex of problems were also the concept of trust – that is to say in our study, the tenants' trust in society in general and in the housing company in particular – and statistical data showing, for example, a reduction in the share of tenants born abroad in most of the studied housing areas funded by LIP. Although not all the questions in an evaluation such as this can be fully answered, the MAIN^{TETRA}, as a model for systematizing and analysing data, does enable us to integrate aspects from different evaluation fields and reflect on them simultaneously in the same analysis.

For example, thanks to the multidimensional analysis, the present study provided interesting results on organizational learning. The evaluation showed that one reason why changing technical systems and economic incentives gave an effect was that their introduction initiated a learning process. The housing companies, however, had not planned for a learning process to take place and therefore failed to consider the possibility that they could learn – as an organization – from employees and tenants involved in the projects. Nor did the tenants' organizations – families, communities, and other social networks – plan for how they could become involved in organizational learning as part of the LIP projects. Thus, the knowledge developed by local actors – employees as well as tenants – during the projects tended to get dispersed from the area when people moved.

These results on organizational learning overshadow optimistic confidence in the notion that 'environmental refurbishment projects' contribute to sustainable development. If the actors fail to focus on the process perspective, referring to how organizational learning is to take place, knowledge acquired from the projects will not remain for future generations.

We have also experienced some difficulties when utilizing the MAIN[TETRA] as a model for systematizing and analysing empirical material. The model is the product of further development of the perhaps more traditional approach of discussing sustainable development – including the three dimensions social, economic and environmental – a development carried out to avoid confusion about what should be included in each dimension. When coding, we felt that the MAIN[TETRA] with its four fields of knowledge was quite clear and easy to use. A key problem, however, was setting the boundaries between the links between fields of knowledge – here we experienced some confusion and, consequently, coded the material into more subgroups than was required theoretically. Naturally, this became a problem when running the reports and analysing the material, as some statements and text segments showed up too often and perhaps in an inappropriate subgroup. Then again, using the MAIN[TETRA] as an evaluation model was a test, and we were prepared for problems. As multidimensional case study evaluations require a wide range of empirical material, qualitative as well as quantitative, there is obviously a need for appropriate software that facilitates systematic treatment of the material. Hitherto, the market for such software does not seem to be extensive. The software Hyper Research we used may be considered appropriate in many ways. For instance, it was extremely time saving, as it allowed analysis of audio files. The fact that we did not have to transcribe the 58 interviews saved us at least three months. Furthermore, and perhaps the most important feature when discussing multidimensional evaluation, this software handles four types of empirical material simultaneously: text, audio, video and pictures. Thus, it not only facilitates a combined analysis of qualitative and quantitative material, but also use of different types of sources, which enables the triangulation required in case study analysis.

As explained in the present chapter, there is a focus in Swedish society on 'exposed suburban metropolitan housing areas' from the 1950s, 1960s and 1970s. These areas have great potential with respect to energy savings, but also suffer from a number of problems, among them ethnic and economic segregation. It is not unusual for 'environmental projects' in such areas to be put forward as good examples of urban development in general, without considering the fundamental social effects. The idea that these are good examples significantly influences how authorities design interventions and funding systems, e.g. measures intended to help save energy, but which may result in something other than sustainable development. In relation to this, it would seem extremely important to consider and further develop multidimensional evaluation tools for planning processes and national funding programmes.

Further development of multidimensional evaluation tools, including appropriate software, may also create potential for cross-disciplinary learning projects on sustainable development, which would be interesting to carry out in a European context, looking at a variety of social housing in metropolitan areas.

Acknowledgements

A number of researchers have been involved in the LIP study to different extents. The evaluation was assigned to the Department of Built Environment and Sustainable Development, Chalmers Architecture, by The Swedish Environmental Protection Agency. It was carried out during a nine-month period and completed in April 2005. Liane Thuvander, PhD, and Jenny Stenberg conducted the analysis and co-wrote a Swedish report on the results. Jenny Stenberg conducted the interviews and made the visits. Lena Falkheden, PhD, worked with the environmental matrix and the case study descriptions. Paula Femenías, PhD, conducted the mass media study, and Birgit Brunklaus, doctoral candidate at the department of Environmental Systems Analysis, Chalmers Energy and Environment, initially worked with organization of the environmental aspects.

References

Allmendinger, P. (2002), *Planning Theory* (Basingstoke: Palgrave).
Andersson, R. (2002), 'Boendesegregation och etniska hierarkier' [Housing Segregation and Ethnic Hierarchies], in I. Lindberg (ed.), *Det slutna folkhemmet: Om etniska klyftor och blågul självbild* [*The Introvert Swedish Welfare State*], Stockholm, Agora.
Argyris, C. and Schön, D.A. (1995), *Organizational Learning II: Theory, Method, and Practice*, Reading, Addison-Wesley Publishing Company.
Azar, C. (2002), *Decoupling: Past Trends and Prospects for the Future*, Stockholm, Environmental Advisory Council, Ministry of the Environment.
Callon, M. and Latour, B. (1981), 'Unscrewing the Big Leviathan: How Actors Macro Structure Reality and How Sociologists Help Them to Do So', in K. Knorr-Cetina and A.V. Cicourel (eds), *Advances in Social Theory and Methodology: Toward an Integration of Micro- and Macro-Sociologies*, Boston, Routledge and Kegan Paul.
CSD, Commission on Sustainable Development (1996), *Indicators of Sustainable Development: Framework and Methodologies*, United Nations, <http://www.un.org/esa/sustdev/isd.htm 000518>.
Easterby-Smith, M., Burgoyne, J. and Araujo, L. (eds) (1999), *Organizational Learning and the Learning Organization: Developments in Theory and Practice*, London, Sage Publications.
Ericsson, U., Molina, I. and Ristilammi, P.-M. (2002), *Miljonprogram och media: Föreställningar om människor och förorter* [Media and Mass Housing from the 1960s and 1970s], Stockholm, The Swedish Integration and National Heritage Board.
Healey, P. (1997), *Collaborative Planning: Shaping Places in Fragmented Societies*, London, Macmillan.

Holmberg, S. and Weibull, L. (eds) (2001), *Land du välsignade?*, Göteborg, Göteborg University.

Kain, J.H. (2003), *Sociotechnical Knowledge: An Operationalised Approach to Localised Infrastructure Planning and Sustainable Urban Development*, Göteborg, Chalmers Architecture.

Kain, J.H. et al. (2005), *Integrerat beslutsstöd för uthålliga VA-system: Fallstudier inom MIKA-projektet i Surahammar och Södertälje kommun* [*Integrated Decision Support for Sustainable Urban Water Systems: MIKA Project Case Studies in the Surahammar and Södertälje Municipalities*], Göteborg, Urban Water Report, Chalmers Architecture.

Kain, J.H. and Söderberg, H. (2002), *LOKOMOTIV – Motiv för lokalt organiserade kretslopp i Bergsjön: En studie om avfallshantering* [*LOCOMOTIVE – Motives for Locally Organised Cycles in Bergsjön: A Study on Waste Management*], Göteborg, Chalmers Architecture.

Kenny, M. and Meadowcroft, J. (eds) (1999), *Planning Sustainability*, London, Routledge.

Knorr-Cetina, K. and Cicourel, A.V. (eds) (1981), *Advances in Social Theory and Methodology: Toward an Integration of Micro- and Macro-Sociologies*, Boston, Routledge and Kegan Paul.

Krogstrup, H.K. (1997), 'User Participation in Quality Assessment: A Dialogue and Learning Oriented Evaluation Method', *Evaluation* 3(2), pp. 205–24.

Lindberg, I. (1999), *Välfärdens idéer: Globaliseringen, elitismen och välfärdsstatens framtid* [*The Ideas of the Welfare: The Globalization, Elitism and Future of the Welfare State*], Stockholm, Atlas.

Lindberg, I. (ed.) (2002), *Det slutna folkhemmet: Om etniska klyftor och blågul självbild* [*The Introvert Swedish Welfare State*], Stockholm, Agora.

Meadowcroft, J. (1999), 'Planning for Sustainable Development: What Can Be Learnt From the Critiques?', in M. Kenny and J. Meadowcroft (eds), *Planning Sustainability*, London, Routledge.

Molina, I. (1997), *Stadens rasifiering: Etnisk boendesegregation i folkhemmet* [*Racialization of the City: Ethnic Residential Segregation in the Swedish Folkhem*], Uppsala, Uppsala University.

Naturvårdsverket (2000), *Fifteen: Sweden's Objectives for Environmental Quality – The Responsibility of Our Generation*, Stockholm, The Swedish Environmental Protection Agency.

Putnam, R.D. (1993), *Making Democracy Work: Civic Traditions in Modern Italy*, Princeton, Princeton University Press.

Putnam, R.D. (2000), *Bowling Alone: Collapse and Revival of American Community*, New York, Simon and Schuster.

Rothstein, B. and Kumlin, S. (2001), 'Demokrati, socialt kapital och förtroende' (Democracy, Social Capital and Trust), in S. Holmberg and L. Weibull (eds), *Land du välsignade?*, Göteborg, Göteborg University.

Sager, T. (1994), *Communicative Planning Theory*, Aldershot, Ashgate.

SOU (2003a), *En hållbar framtid i sikte* [*A Sustainable Future in Sight*], Stockholm, Fritzes offentliga publikationer, SOU 2003:31.

SOU (2003b), *Johannesburg – FN:s världstoppmöte om hållbar utveckling* [*Johannesburg – United Nations World Conference on Sustainable Development*], Stockholm, Regeringskansliet, SOU 2002/03: 29.

Stenberg, J. (2004), *Planning in Interplace? On Time, Power and Learning in Local Activities Aiming at Social Inclusion and Sustainable Development*, Göteborg, Chalmers Architecture.

Stenberg, J., Thuvander, L. and Femenías, P. (2008 forthcoming), 'Linking Social and Environmental Aspects: A Multidimensional Evaluation of Refurbishment Projects in Sweden', *Building Research and Information*.

Stufflebeam, D.L., Madaus, G.F. and Kellaghan, T. (eds) (2000), *Evaluation Models: Viewpoints on Educational and Human Services Evaluation*, Boston, Kluwer Academic Publishers.

Swedish Government (1997), 'Sustainable Sweden: New Messages to Improve the Environment. The Commission on Ecologically Sustainable Development', <http://miljo.regeringen.se/pressinfo/pdf/m98_en3.pdf>, accessed 2004.

Swedish Government (1998), 'Utveckling och rättvisa – en politik för storstaden på 2000-talet' [Development and Justice – A Policy for Metropolitan Areas in the Twenty-first Century], Regeringens proposition 1997/98:165, <http://www.storstad.gov.se>, accessed 2001.

Swedish Government (1999), 'Sustainable Sweden: Steps Towards Sustainable Development. Ministry of the Environment', <http://miljo.regeringen.se/pressinfo/pdf/ m99_015e.pdf>, accessed 2004.

Swedish Government (2001a), *Fem år efter Istanbul: Erfarenheter, lärdomar och utmaningar [Five Years after Istanbul: Experience*, Learning and Challenge], Stockholm, Swedish National Commission on Agenda 21.

Swedish Government (2001b), 'Sustainable Development: An Important Objective for the EU. Ministry of the Environment', <http://miljo.regeringen.se/pressinfo/pdf/m2001-004eng.pdf>, accessed 2004.

Swedish Government (2001c), 'Towards a Swedish National Strategy for Sustainable Development. Ministry of the Environment', <http://miljo.regeringen.se/pressinfo/ pdf/m2001_012e.pdf>, accessed 2004.

Swedish Government (2002), *Från vision till handling [From Vision to Action]*, Stockholm, Sveriges nationalrapport till världstoppmötet om Agenda 21 och hållbar utveckling i Johannesburg 2002, Stockholm, Nationalkommittén för Agenda 21 och Habitat.

Swedish Integration Board (2003), *Integration 2002*, Norrköping, Integrationsverket.

Swedish Integration Board (2004), *Integration 2003*, Norrköping, Integrationsverket.

UN Association of Sweden (2004), 'Alternative Report to the CERD-Committee with Respect to Sweden's Commitments According to the ICERD', United Nations Association of Sweden, <http://www.sfn.se/svefn/files/CERD2004.pdf>, accessed 25 February 2004.

UNCED, United Nations Conference on Environment and Development (1992), 'Agenda 21', UN Department of Economic and Social Affairs Division for Sustainable Development, <http://www.un.org/esa/sustdev/agenda21text.htm>, accessed 31 March 2001.

UNCED, United Nations Conference on Environment and Development (1992), 'Rio Declaration on Environment and Development', UN Department of Economic and Social Affairs Division for Sustainable Development, <http://www.un.org/documents/ ga/conf151/aconf15126-1annex1.htm>, accessed 1 April 2001.

WCED, The World Commission on Environment and Development (1987), *Our Common Future*, Oxford, University Press.

Yin, R.K. (1994), *Case Study Research: Design and Methods*, Thousand Oaks, CA, Sage Publications.

Yin, R.K. (2000), 'Case Study Evaluations: A Decade of Progress?', in D.L. Stufflebea, G.F. Madaus and T. Kellaghan (eds), *Evaluation Models: Viewpoints on Educational and Human Services Evaluation*, Boston, Kluwer Academic Publishers.

Chapter 4

Is There Room for Equity in European Commission Policy-Making? An Evaluation of Selected Impact Assessment Reports

Tom Bauler, Alessandro Bonifazi and Carmelo M. Torre

Introduction

Reducing inequalities has long been acknowledged as one of the main challenges facing planners and policy-makers,[1] all the more when pursuing the objectives of Sustainable Development (SD) (Lehtonen 2004). In the present chapter, we analyse how equity issues are addressed in European Commission (EC) Impact Assessment (IA). Paying attention to inequalities when policy proposals are framed and alternatives are developed, could be a basic but necessary condition for the European Union (EU) to more effectively tackle these issues.

One can argue that if the Impact Assessments addressed sufficiently inequalities, then they might be better integrated in the constellation of EU normative and administrative measures. This chapter dwells on the first part of this basic assumption, by illustrating critically a coherence analysis we carried out between selected Impact Assessment reports and the set of principles, standards and guidelines they were expected to follow.

The chapter starts with a brief overview of the EC-IA system, the political processes at its roots, and its position in the evaluation culture domain. Section two weaves the role of equity into the general discourses we touch upon in our work, namely sustainable development, evaluation in planning and European policy-making. Then follows an account of our methodological approach, which describes into details the complete workflow and explains how we tapped into the traditions of metaevaluation and multicriteria analysis. In section four we show the most interesting outputs of the empirical part of the research. We end by discussing some salient issues and research perspectives.

For the purpose of this chapter, we see Impact Assessment as a type of evaluation, rather than as a separate construct, irrespective of the fact that the two related bodies of literature have developed largely in isolation.

---

1 As witnessed, inter alia, by some contributions to the very same series of workshops this book originated from, for example Miller (2001) and Clemente et al. (1998).

The European Commission's Impact Assessment

When the European Commission introduced its Impact Assessment scheme in 2002, the intention was twofold (EC 2002a): develop a decision-aiding process to sustain the drive of the institutions for more coherence between their sector policies and an overall better quality of their regulatory outputs; enhance communication with civil society and engage into participatory processes with relevant actors and stakeholders.

Politically, the EC-IA inscribes itself both in the *Göteborg process* (in relation to sustainable development and the ensuing considerations for multi-dimensional impacts) and in the *Lisbon process* (as regards supporting competitiveness in a knowledge-based economy). Impact assessments were also seen (ibid.) as a way to seek a bottom-up integration of both agendas (i.e. the competitiveness agenda and the sustainability agenda) into everyday policy developments. Furthermore, the EC-IA participates in implementing horizontal principles of the 'Better regulation' process (EC 2002b), by operationalizing the ongoing simplification and improvement of the policy-making procedures within the Commission services. Technically, the EC-IA is an inter-service evaluation exercise built around the collaboration of relevant Commission services, specifically configured for each policy proposal.

Historically, the rationale of the EC-IA has developed from a series of parallel evaluation practices. On the one hand, linked to the striving for a general reduction of regulatory interventions in market dynamics, the evaluation scheme leans on the (mostly Anglo-Saxon) experiences with Regulatory Impact Assessments (RIA), which are recently gaining momentum in some member states (Radaelli 2005). On the other hand, the EC-IA taps into the latest developments in the environmental field (e.g. Strategic Environmental Assessment) and into other more specifically oriented, programmatic evaluations (e.g. Sustainability Impact Assessment of Trade agreements at the EC Directorate General [DG] Trade[2]). In effect, the promotion of EC-IA is motivated also with the attempt to integrate existing sectoral or mono-dimensional evaluations (e.g. Business Impact assessments, Environmental Impact Assessments, Gender Assessments) of policy proposals, which tended to spread heterogeneously over different DGs.

The EC-IA scheme can be described (Pope et al. 2004) as a 'sustainability assessment', or defined (Devuyst 2001, p. 9) as 'a tool that can help decision-makers and policy-makers decide what actions they should take and should not take in an attempt to make society more sustainable'. First, EC-IA is an *ex ante*[3] assessment that runs parallel to the ongoing policy-making process by informing European decision-makers of the positive and negative impacts of selected alternatives. The

---

2   See DG Trade official website at <http://ec.europa.eu/trade/issues/global/sia/index_en.htm> (last retrieved on 28 December 2006), or Kirkpatrick and George (2004).

3   We understand the term *ex ante* in the general sense it has in the evaluation literature, rather than in the specific one it takes on in the EC context, where 'ex-ante evaluation' is an evaluation procedure focussing on cost-effectiveness that is mandatory for any programme and action resulting in expenditure from the general budget of the EC (2002c).

principal evaluative questions are posed in terms of impacts mitigation (in case of negative, unwanted effects) and policy strengthening (in case of positive impacts to be amplified). In more general terms, EC-IA contributes to enforce policy learning through evaluation (Scrase and Sheate 2002) by (1) adjusting existing policies, (2) generating new policies and (3) influencing value systems which determine policy development phases such as 'problem identification' and 'policy formulation'.

Second, EC-IA is an *integrated* assessment. While 'integrated assessments' are not univocally defined (ibid.), the integrative character of the EC-IA can be described in our context as twofold: (1) integration in the sense of interdisciplinary assessment exercises, which combine different perspectives and dimensions into a common methodological framework; (2) integration in the sense of a multi-actor process, which assimilates (non-institutional) stakeholders as co-actors of the evaluation process.

The EC-IA system has become operational in 2003,[4] and it concerns all policy proposals inserted in the yearly working program of the EC. After a first screening (the 'preliminary Impact Assessment'), only major policy proposals are selected to undergo a more advanced evaluation process (the 'extended Impact Assessment') which materializes in an *evaluation report* that is eventually made available on a dedicated website.[5] The evaluation report should not only cover the prediction and assessment of impacts, but include as well a rather detailed description and justification of the policy proposal and its context, the evaluation rationale (e.g. from the policy alternatives taken into account to the possible participatory process organized) and its results. The process, use of expertise, participation and consultation are all officially formalized (EC 2002c, e, f, g, h) and so is the structure of the final output (i.e. the Impact Assessment report).

The number of realized evaluations per year is still progressing, but has already increased from 21 extended Impact Assessments (ExIAs) in 2003, to 72 in 2005. The EC-IA is thus a fairly new process at the level of the EC, and its rather experimental (or at least evolutionary) nature is fully acknowledged by the Commission services as well as by independent researchers (Wilkinson et al. 2004). Right from its first implementation year, however, the evaluation scheme raised quite a strong interest among scholars and stakeholder organizations, who intended to evaluate its strengths and weaknesses, notably with regard to its performance in coping with the SD-agenda. Some actors feared that the EC-IA could be ascribed in a family of policy evaluation systems that could undermine, in the long run, the environmental policy agenda by introducing into the policy-making cycle an all-inclusive evaluation process. These 'integrated' evaluations might prove too strongly oriented towards the identification of trade-offs and mitigation measures, and thus not flexible enough to open-up entirely the decision-space and to account, among other flaws, for non-valuable (environmental)

---

4 The procedural and methodological setting of the CEC-IA scheme was updated twice, in June 2005 and March 2006. However, since our review considers only IA reports published in 2004, we refer here to the process applicable at that moment.

5 See <http://europa.eu.int/comm/secretariat_general/impact/index_en.htm> (last consulted on 20 January 2007).

benefits (Scrase and Sheate 2002). Others (COWI et al. 2004) hoped that the EC-IA could 'compensate for the lack of clear definitions and political mandates [by providing] a common language for delineating and scoping the SD discussions, and for targeting policy preparations exactly towards SD issues'. Simultaneously, it appeared that the EC-IA scheme could develop over the years into an effective bottom-up lever for the policy-making cycle to integrate the sustainability agenda, thus being the necessary complement to more programmatic initiatives (such as the EU-Sustainable Development Strategy).

Equity, Sustainable Development, European Policy-making and Evaluation in Planning

The findings in this chapter originated from previous collaborations among the authors in the field of institutionalized evaluations for sustainable development (Bauler et al. 2007; Bonifazi et al. 2007). While discussing possible developments, we moved from the discontent we shared on the focus of much evaluation practice in this context, that is, the centrality of the 'three pillars' logical framework. No matter how strongly the sheer juxtaposition of environmental, social and economic aspects has been criticised, it still lingers in both academics and practitioners' words and deeds. What this loose framework seems to suggest is the need to take at least account of these three dimensions, somehow fostering trade-offs in the management of sustainable development. However, the triple bottom line alone says nothing about how to integrate the three aspects, thus giving free play to largely arbitrary processes and potentially leading to opaque decisions.

It is argued (Pope et al. 2004) that a principles-based approach would emphasise interconnections and interdependencies among the different areas of concerns of sustainability, rather than promoting conflicts and trade-offs. Moreover, specifying principles at the beginning of the process shifts the burden of identifying 'sustainable' solutions onto policy-makers and planners, rather than evaluators, also anticipating the moment in time when sustainability is likely to be taken into account. We therefore looked for some core principles to base our methodology on. We were not aiming at a comprehensive report, yet we wanted to be reasonably sure to grasp the relevance of the evaluations under scrutiny to SD. Dwelling on both law and policy milestones of, and the literature on, sustainability principles, we were attracted by the recurrent call for preventing and redressing inequalities. Equity is the single concept recalled in the most widely accepted definitions of SD, such as the Brundtland Report and the European Union SD Strategy's (WCED 1987; EC 2001 and 2002d). At the same time, poor effectiveness in addressing issues of equity and the social dimension represented so far the most important blind spot in the sustainability discourse, and 'greater equity as a desirable and just social goal is intimately linked to a recognition that, unless society strives for a greater level of economic and social equity, both within and between nations, the long term objective of a more sustainable world is unlikely to be secured' (Agyeman et al. 2003, p. 2).

Notwithstanding the importance of other aspects of SD (notably, the ecological limits on human activities), we would expect any sustainability evaluation to check policy proposals for their links to inter- and intra-generational equity. These two very broad definitions, potentially encompassing all that is relevant to equity, would require a clarification of the philosophical views they embody (Gosseries 2005). However, by choosing equity as the focus of our metaevaluation we did not adopt any particular stance with respect to the different theories enlivening the current debate on justice. In this regard, we derived our evaluation criteria directly from the regulatory and management apparatus set up in the institutional context we were investigating, namely, the European Union. In other words, our criteria are nothing more than a subset of all 'instructions' issued by the relevant EU bodies to define how the Impact Assessments had to be carried out. Of course, the very understanding of issues of equity emerging from criteria all together could be questioned, as they mirror the official EU perspective on the matter, with all its contentions and compromises. Though an interesting endeavour, such an investigation falls beyond the scope of our research, which is rather focussed on triggering a learning process among the actors involved in the IA regime (and primarily, among those EC services that were responsible for the evaluations) about the light and shade of its implementation.

We have chosen to study the EC-IA system because we were fascinated by its potential political influence, which is expanded by the supra-national scope of the policies it addresses. Moreover, many decisions under scrutiny are strategic in nature, as they often shape frameworks all other legislative tiers have to fit into. On a more practical note, the availability of almost all IA reports on the EC Secretariat General public website made it a suitable case. Luckily enough, our equity-centred approach proved relevant for the target of investigation, as all official documents having a bearing on the IA process referred to issues of equity. It was therefore relatively easy to select those elements wherefrom we derived our evaluation criteria. More importantly, such a widespread attention to inequalities made us confident that the yardstick we had picked was very relevant to sustainable development and to European policy-making alike. Indeed, the importance of equity has grown steadily in Europe at the grand level of declarations,[6] treaties,[7] and programmes.[8] The IA methodology has been envisaged consistently to embed this concern. Our work is a preliminary contribution to understanding to what extent real assessment practices turned it into an approach to improve the policy output of the European Commission.

---

6   As witnessed, in particular, by the *Charter of Fundamental Rights of the European Union*: equality is first raised to the status of founding value of the European Union in the preamble, and then spelt out in different issues throughout Chapter III.

7   The *Treaty Establishing the European Community* addresses, for example, gender equality (Articles 2 and 3), discrimination based on nationality (Art. 12), discrimination based on sex, racial or ethnic origin, religion or belief, disability, age or sexual orientation (Art. 13).

8   Though EU structural and cohesion policy *in toto* aims at tackling national and regional disparities, the *EQUAL Community Initiative* deserves a special mention for its focus on combating all forms of discrimination and inequalities in the labour market.

The growing importance of the European dimension in the fields of policy-making and planning has broadened the scope of evaluation theory and practices, which had to follow the emergence of common infrastructural, environmental, and socio-economical development strategies. This evolution can be traced by the slow path towards the European Spatial Development Perspective, which entailed the acknowledgement of the spatial interdependencies triggered by the main lines of intervention (e.g. structural and cohesion policy, common agricultural policy). Thus, European territorial identities in the making are mirrored in the active realizations of the *INTERREG Community Initiative* in general, and of its specific programme *ESPON* (Faludi 2000). There are at least two relevant examples showing that the EU has a bearing on multiple-level planning practices. First, many trans-national and interregional areas (as well as the related governance arrangements) are being defined under the Trans-European Networks. Second, most urban rehabilitation strategies have been deeply influenced by the general philosophy and the management culture of the URBAN Programme.

While this European level kept expanding, workshops on evaluation turned into fora to discuss different approaches to planning (Khakee 1997), pointing to a differentiation of evaluation roles and methods, and witnessing an increase interest for a 'menu for considering the context in which we could be broadening out internationally' (Lichfield 2000, 11). In his opening paper at the *Fourth International Workshop on Evaluation in Planning*, Nathaniel Lichfield (ibid.) drew the attention of participants to the need of linking programme evaluation with the traditions of evaluation in planning. His call went in the direction of both methodological renewal, as the achievements in both fields had remained largely isolated, and the recognition of an increasingly open and multilevel arena where 'segmented' practices push the borders of planning systems beyond their urban and regional boundaries to encompass, *inter alia*, 'European Regional Development Funds, the new public management, and sustainability'. As for us, this call should not remain unheeded, and we tried to contribute to addressing some of the issues it raised.

Carrying out the Metaevaluation of Selected Impact Assessment Reports

Browsing through the early literature on the EC-IA system (Adelle et al. 2006; Wilkinson et al. 2004), we noticed a certain wavering between strengths and weaknesses in the opinions of scholars. Such an ambiguity triggered our resolution to understand better this recently introduced evaluation procedure, and we turned to metaevaluation as the chosen methodological approach. We understood metaevaluation in its original meaning (Scriven 1991) of 'an evaluation of evaluations … indirectly, the evaluation of evaluators' and as a (Widmer 2002) 'systematic assessment of the quality of one or more evaluation studies', where one of the objectives of metaevaluation is to become (Bustelo 2002) 'a tool for the improvement and development of the evaluation function in public administrations'.

On a formal ground, we devised the metaevaluation in a perspective of *internal coherence* of the EC-IA. We identified six core documents (i.e. the 'guidelines') by starting from the Communication (EC 2002a) establishing the EC-IA system, and then checking a web of cross-references to ascertain the relevance of a limited number of sources explaining how the IAs had to be carried out (EC 2002c, e, f, g, h).

In doing so, we were not very much interested in determining *programme integrity*, that is, how precisely the evaluation design had been implemented in practice. We rather aimed at fostering a mutual learning process between the coordination service (the Secretariat General) and the units in charge of the assessment task (at the level of the DGs), trying as well to give any other interested party (Member States, European citizens, NGOs, business, etc.) a critical insight into the process.

It should be clear that our contribution to this rather ambitious goal has been intrinsically limited by taking the IA reports as the sole source of information. No need to explain how understanding the actual influence of each IA on the policy-making process (let alone on the related inequalities) falls beyond the scope of such an approach. We borrow the definitions developed to define different strategies for evaluating Regulatory Impact Assessments (Harrington and Morgenstern 2004) to remember that accomplishing this other task would require shifting from a *content test* to an *outcome test*.

As a first step, we created a database of all elements we found in the aforesaid reference documents that had a bearing on issues of equity. The former exercise yielded 116 items that were subsequently reduced and adapted till we agreed on a checklist including 74 indicators, which we then used to assess each IA report. Table 4.1 shows a sample of 12 indicators out of the total 74 we identified, and the sources we drew them from.

We tried to devise an evaluation methodology that would rely as much as possible on the Commission's own views, objectives, conceptual framework, discourses and wording that define how an extended Impact Assessment should deal with issues of equity. It is important to underline that most criteria we identified are not legally binding obligations on the Commission Services, but rather examples of good practice and practical suggestions on how to carry out an IA. However, we argue that taken altogether they represent a benchmark against which the reports can be compared. We used the indicators to assess each IA report. The underlying assumption behind our simple coherence rules was that the *IAs* had dealt with the issue addressed by each indicator IF this was evident in the *reports*. We are aware of this assumption being a particularly strong one, as the reports might sometimes be part of wider strategic and tactical manoeuvres to discredit (or strengthen) certain policy alternatives. Evidence was scored as follows: zero (no evidence), one (weak evidence), and three (strong evidence). An excerpt of the checklist we used for our evaluation exercise is shown in Figure 4.1.

It was possible to rule out certain indicators that were not always relevant, and to report whether *uncertainty* was affecting significantly the judgment. Though information concerning uncertainty has not yet been processed, it is important to stress that we perceived it as an important factor influencing the output of our investigation. In particular, since the scoring task was executed by three different

Table 4.1 A sample of 12 indicators (out of the total 74) we used for our metaevaluation of EC-IA reports

Indicator	Phase of the IA	Source
Does the report identify conflicts and inconsistencies between the economic, environmental, and social dimensions, potentially leading to trade-offs?	All phases	*Impact Assessment in the Commission: Guidelines* (EC, 2002c, pp. 24–25)
Does the report analyses the issue by referring also to the outcome of previous consultations, for examples the lessons drawn from green papers?	Problem definition	*Communication from the Commission on Impact Assessment* (EC 2002a, p. 13)
Does the report explain if the objectives they set are consistent with other (equity-relevant) EU policies?	Setting the objectives	*A Handbook for Impact Assessment in the Commission: How to do an Impact Assessment* (EC 2002g, p. 9)
Are options presented as opportunities to minimize distributive trade-offs and/or to lead to win–win situations?	Developing policy options	*A Handbook for Impact Assessment in the Commission: how to do an Impact Assessment* (EC 2002g, p. 16)
Was the proposal screened for possible economic impacts, such as might it: -change the sectoral distribution of economic activity; -have specific impacts on SMEs; change the composition of employment?	Identifying, measuring and assessing the impacts	*A Handbook for Impact Assessment in the Commission: How to Do an Impact Assessment – Technical Annexes* (EC 2002h, pp. 16–18)
Among possible social impacts were distributional implications (such as effects on the income of particular sectors, groups of consumers or workers, differential impacts on Member States or social and economic groups, etc.) appropriately checked?	Identifying, measuring and assessing the impacts	*Communication from the Commission on Impact Assessment* (EC 2002a, pp. 15–16) *A Handbook for Impact Assessment in the Commission: How to Do an Impact Assessment* (EC 2002g, pp. 20–21)
Were regional, gender or racial impacts compared with the baseline to see if the proposal was likely to leave existing differences unchanged, aggravate them, or help to reduce them?	Identifying, measuring and assessing the impacts	*A Handbook for Impact Assessment in the Commission: How to Do an Impact Assessment* (EC 2002g, p. 22)

Indicator	Phase of the IA	Source
In cases with environmental impacts, does the report show that the requirements of the Aarhus Convention have been met? Was the public consulted at an early stage of the policy-making process? Was appropriate feedback given? Were interested parties consulted on the results of the detailed analysis of the likely environmental impacts of the proposal?	Consultation	*A Handbook for Impact Assessment in the Commission: How to Do an Impact Assessment* (EC 2002g, p. 31)
Did the Commission allow at least 8 weeks for reception of responses to written public consultations, and 20 working days notice for meetings?	Consultation	*Towards a Reinforced Culture of Consultation and Dialogue – General Principles and Minimum Standards for Consultation of Interested Parties by the Commission* (EC 2002e, p. 21)
Were expert groups composed of at least 40% of each sex?	Consultation	*On the Collection and Use of Expertise by the Commission: Principles and Guidelines – Improving the Knowledge Base for Better Policies* (EC 2002f, pp. 11–12)
Did experts clearly highlight the evidence (e.g. sources, references) upon which they based their advice, as well as any persisting uncertainty and divergent views?	Consultation	*On the Collection and Use of Expertise by the Commission: Principles and Guidelines – Improving the Knowledge Base for Better Policies* (EC 2002f, pp. 11–12)
Is the chosen policy meant to prevent discrimination, in line with the following provisions: 'Discrimination based on nationality, sex, racial or ethnic origin, religion or belief, disability, age or sexual orientation is prohibited under the Treaty on the European Community (Articles 12 and 13)'?	Justification of policy choice	*A Handbook for Impact Assessment in the Commission: How to Do an Impact Assessment – Technical Annexes* (EC 2002h, p. 7)

Note: For each indicator both the phase in the IA process it refers to and the source we drew it from are indicated.

INDICATOR	SCALING	Reference
Does the report analyse more thoroughly proposed measures that are likely to have SERIOUS NEGATIVE SIDE EFFECTS or particularly affect CERTAIN GROUPS in society rather than minor technical changes to regulations?	EVIDENCE: NO ☐ WEAK ☐ STRONG ☐ UNCERTAIN ☐	
If COST-BENEFIT ANALYSIS has been used to assess the impacts, was there a special attention not to overlook distributional issues?	IRRELEVANT ☐ EVIDENCE: NO ☐ WEAK ☐ STRONG ☐ UNCERTAIN ☐	

Figure 4.1   An excerpt from the checklist we used for our metaevaluation of EC-IA reports

Note: For each indicator, evidence was scored as no (0), weak (1), or strong (3). It was also possible to rule out some indicators and to report uncertainty in the value judgment.

persons,[9] the ensuing biases should not be overlooked. In order to fine-tune as much as possible our individual yardsticks, we discussed over and over the interpretation of each indicator in the checklist, and we cross-evaluated few IAs to settle the remaining divergences.

In formulating the judgments, we have been particularly cautious when deciding to opt for 'strong evidence'. Given the qualitative approach we chose, the judgments were translated into scores representing broad classes of values, rather than accurate measures. In fact, the score 'three', corresponding to the semantic judgement 'strong evidence', does not stand, in Saaty's (1980) logic, for a three times higher evidence with respect to the judgement 'weak evidence'. We clearly adopted a conventional ordinal scoring, by which we chose to ignore 'medium evidence' as we had realized that most incoherence among the three reviewers who joined the exercise was likely to concentrate there.

On a more technical note, we processed our value judgments through a multicriteria (MC) approach, which is particularly appropriate whenever researchers are faced with numerous, heterogeneous and quali/quantitative considerations. Analyses result in a priority list, usually without absolute reference points. Therefore, each evaluation is contextual and only partially replicable in different settings. Depending on the particular method one chooses, the *dominant alternative* (in this case, the IA report that followed to a greater extent the guidelines with respect to dealing with issues of equity) can be identified as the winner in the highest number of pairwise comparisons in a concordance analysis (Roy and Bouyssou 1993), as the probable winner of pairwise comparisons in the regime method (Fusco Girard and Voogd 1989; Nijkamp et al. 1990; Fusco Girard and Nijkamp 1997), or as the possible winner in the highest number of pairwise comparison in a fuzzy multicriteria assessment (Munda 1995).

The nature of our evaluation fits into this assumption. It is based on qualitative judgments that allowed pairwise comparisons between different reports, around

---

9   Namely, Tom Bauler, Alessandro Bonifazi and Marco Wäktare (who collaborated with us at the time of the scoring).

each criterion. Once we populated the indicators, we used them as sub-criteria in two separate MC evaluations, both carried out as *concordance analysis* by using the *Electre* method (Roy and Bouyssou 1993). Although the presence of multiple levels would have suggested resorting to the Saaty (1980) approach, this would have required an unmanageable series of 2,664 pairwise comparisons.

For the first analysis we sorted 69 indicators into seven criteria, each representing a different *phase* in the IA process, as they are defined in the guidelines (i.e. problem definition, setting objectives, developing policy options, assessing impacts, consultation, policy choice, monitoring). We calculated the criteria as simple averages of the related value judgments. Weights were introduced only because certain phases had been covered by a much larger amount of indicators, as well as addressed more thoroughly in the reports. However, this size effect of indicators sets on weights was curbed, not to override the basic assumption that due attention to equity issues should have been paid in each step of the IA.

For the second analysis, we aggregated the whole set of indicators into eight equally weighted criteria following the *objectives* set by the Commission in introducing the IA system (i.e. enable informed political judgment, identify trade-offs, improve policy design, analyse economic, social and environmental impacts, allow wider consideration of ethical and political issues, contribute to the implementation of the EU-SD Strategy, involve more interested parties, establish a sound knowledge base).

Whereas in the first case we were interested in stressing how the studies performed in each step, our aim in the second case was to investigate their effectiveness in addressing the self-assigned objectives. Moreover, as a side-effect, we could use measures of centrality obtained from unprocessed value judgments as hints pointing to the attention paid by the programme as a whole to issues of equity.

Sensitivity analyses were done to check the relative stability of the rankings vis-à-vis the influence of weights and the possible concentration of extremely good (or bad) scores in certain criteria, especially with the aim of identifying systematic biases ensuing from the three reviewers' judgment attitudes, rather than testing the methodological robustness of evaluation. Our sample includes only the extended IAs carried out in 2004 because 2003 has to be considered as a phasing–in period (EC 2002a; Wilkinson et al. 2004), full operation being expected only as of 2004. Furthermore, since we could not cover all 29 ExIAs, we arbitrarily decided to exclude from our metaevaluation all nine reports linked to the Financial Perspectives 2007–2013.

Investigating the Role of Impact Assessment to Integrate Equity in EC Policy Proposals

The basic outputs of our metaevaluation are the two rankings[10] shown in Table 4.2. A first observation concerns the absolute stability of the four 'best' and the

---

10 Rankings present descending order of overall score, thus rank 1 corresponds to the relatively 'best' performer.

four 'worst' performers over both multicriteria evaluations. Neither the weights introduced in the first evaluation, nor the different aggregation of indicators into criteria have affected the pairwise comparisons with respect to those IAs that seem definitely above (or below) the average. It should be clear that we never aimed at ascertaining which IA performed better in absolute terms, whatever sense this could have. We were interested in rankings only in so far as they helped us build explanatory constructs to explore the conditions for the integration of equity concerns into the Impact Assessment procedure.

A closer look at the distribution of value judgements over indicators and criteria allows to explore whether those extended-IAs paying more attention to issues of equity, are also more accurate in general terms. In fact, a certain link exists with the most trivial parameter of quality, that is the sheer length (number of pages) of the evaluation report. Longer reports perform better than shorter ones, save some remarkable exceptions. It is likely that too short reports, when not ensuing from a painstaking scoping stage, reveal a weak commitment to the programme and eventually result in poor quality-achievements. Probably, this relates to the evaluation apparatus as enshrined in over-detailed guidelines, entailing that short reports cannot fulfil all requirements.

We also developed a more refined test, building on a study on institutionalisation of the evaluation functions in the EC (Williams et al. 2002). EC Services in charge of carrying out the EC-IA are clearly the leading actors[11] in the assessment process. Therefore, we looked for a correlation between the above rankings and the particular Directorate General responsible for each IA. The authors (ibid.) developed a typology of DGs and other Services by sorting them into three classes (corresponding to high, medium and low level of institutionalisation of evaluation, see Table 4.2, column 3) according to the way they dealt with:

a) organisation of the evaluation function;
b) allocation of human resources;
c) management of the evaluation process;
d) management of the feedback process.

Considering carefully possible biases, it cannot be ascertained that more evaluation-literate DGs deliver better assessments (as for the attention paid to address inequalities). Whereas three reports out of four led by DGs described as having a high (DG EMPL) or medium (DG INFSO) institutionalisation level feature among the first nine places in the ranking by phases, two of them performed significantly worse in the evaluation by objectives (see Table 4.2). Generally speaking, sample size and representativeness need increasing before endorsing any interpretation. However, the difference between our results and the typology might point towards a more detailed account. There seems to be

---

11 In reality, however, the picture is made more complex by the interplay between a central unit (the Secretariat General that developed the guidelines and acts as a watchdog for their implementation) and the different services that actually perform the evaluations (either directly or by hiring external consultants).

Table 4.2 Rankings resulting from the two multicriteria evaluations, focussing respectively on the seven phases of the extended IA process (4th column) and on the eight objectives of the EC-IA programme (5th column)

Extended Impact Assessment	Leading DG	Institutionalisation of evaluation	Ranking by 'Phases'	Ranking by 'Objectives'
SERVICES	MARKT	Low	1	1
WORKING TIME	EMPL	High	2	2
PAEDIATRIC MEDICINES	ENTR	Low	3	3
TIMBER	DEV	–	4	4
DESIGN PROTECTION	MARKT	Low	5	6
REFUGEE FUND	JAI	Low	6	5
PREPACKED PRODUCTS	ENTR	Low	7	9
eEUROPE	INFSO	Medium	8	11
EQUAL TREATMENT	EMPL	High	9	14
CRIMINAL PROCEEDINGS	JAI	Low	10	10
REINSURANCE	MARKT	Low	11	8
INPSIRE	ENV	Low	12	7
VISA	JAI	Low	13	12
ePROCUREMENT	MARKT	Low	14	13
DIGITAL TV	INFSO	Medium	15	16
CREDIT INSTITUTIONS	MARKT	Low	16	15
FISHERIES CONTROL	FISH	–	17	17
RAILWAYS	TREN	Low	18	18
ENVIRONMENTAL STANDARDS	ENV	Low	19	19
ENVIRONMENT&HEALTH	ENV	Low	20	20

a better match between the two analyses in terms of procedural quality (the aforesaid EC-IA reports are well structured around the workflow) rather than on the substantial level of attention to equity.

As far as grouping ExIA reports according to the leading DG is considered acceptable to investigate other phenomena. A second analysis reveals the relations between environmental issues and inequalities in a similar fashion. The three pillars of sustainable development (environment, society, economy) were adopted as organizing principles, and all leading DGs were sorted in three groups according to the orientation of their core policy areas towards the environment (DG ENV and FISH), society (DG JAI, INFSO and EMPL), or economy (DG MARKT, ENTR and TREN).

Looking at the two rankings through these lenses, a negative correlation is revealed between equity-oriented choices and environmental protection (see Table 4.3, column 1 and 2). Indeed, there seem to be clear patterns of clustering for the ExIAs executed by environment-oriented DGs (towards the bottom) vis-à-vis those carried out by society-oriented DGs (towards the top). Furthermore, an internal control was performed by extracting from the 74 item list a sub-set of six indicators, which addressed issues that are directly relevant for the environment (e.g. depletion of non-renewable resources).

The third column (Table 4.3) shows the ranking obtained through this restricted analysis. Interestingly enough, the two clusters (i.e. environment-oriented DGs and society-oriented-DGs) are still spread unevenly over the ranking, though in the opposite way, thus reinforcing the evidence of a negative correlation. Different scholars have put such a case forward. Scrase and Sheate (2002) classified 14 understandings of 'integration' in assessment practices, according to what they mean for the environment. Not surprisingly, they placed the integration of equity concerns into governance among the least environment-supportive policy innovations.

However, one of the analysed ExIA reports could have escaped this rule. We chose not to group the Impact Assessment 'TIMBER' (led by DG DEV) because it addresses an environmental problem (illegal logging), yet dwells on its social consequences (corruption and crime), and tries to cope with it by introducing an economic instrument (voluntary licensing scheme). The fact that the ExIA of a highly 'integrated' policy jumps up when using criteria that are at the same time environment- AND equity-oriented would suggest that sustainability assessments are not doomed to sacrifice one of the policy areas they try to reconcile.

Finally, the EC-IA system is explored further as an evaluation programme, based on an overall appreciation of value judgements. Figure 4.2 shows, for all 20 ExIA reports, the minimum, mean, and maximum values[12] taken on by each criterion used in the first MC evaluation (which revolved around the seven phases of the IA process).

Revealing differential performances across the seven phases foreseen in the IA guidelines inscribes itself in the perspective of assessing programme integrity. Remarkably low scores for the phases of 'assessing impacts' and 'justifying policy choice' are observed, especially when compared respectively to the better achievements in terms of 'consultation' and 'problem definition'. Such pairwise comparisons offer more accurate[13] estimates, since criteria are coupled according to a similar sensitivity to extreme value judgements as a consequence of the number of indicators building up each criterion.

Interestingly enough, the 'assessment of impacts' is among the worst-scoring phases, which is not very promising for an Impact Assessment. On the other hand, 'consultation' performs relatively better, though here it is understood as an aggregated measure of 'stakeholders' involvement' and 'use of external expertise'. Likewise, 'problem definition' scored much better than 'justifying policy choice', meaning that inequalities are addressed quite often in the preliminary, analytic steps of the process, without however driving significantly the ultimate political choice.

---

12 It seems remarkable that mean scores are always below half the maximum possible value, and high scores are exceptional.

13 Indeed, criteria 1 and 6, and criteria 4 and 5 show almost exactly the same dispersion, if computed as the inter-quartile ranges.

Table 4.3 Comparing different rankings to highlight correlations between inequalities and policy areas

Ranking 'Phases'	Ranking 'Objectives'	Ranking 'Environmental indicators'
SERVICES	SERVICES	TIMBER
WORKING TIME	WORKING TIME	PREPACKED PRODUCTS
PAEDIATRIC MEDICINES	PAEDIATRIC MEDICINES	INSPIRE
TIMBER	TIMBER	SERVICES
DESIGN PROTECTION	REFUGEE FUND	DIGITAL TV
REFUGEE FUND	DESIGN PROTECTION	ENVIRONMENTAL STANDARDS
PREPACKED PRODUCTS	INSPIRE	FISHERIES CONTROL
eEUROPE	REINSURANCE	WORKING TIME
EQUAL TREATMENT	PREPACKED PRODUCTS	REFUGEE FUND
CRIMINAL PROCEEDINGS	CRIMINAL PROCEEDINGS	DESIGN PROTECTION
REINSURANCE	eEUROPE	PAEDIATRIC MEDICINES
INSPIRE	VISA	eEUROPE
VISA	ePROCUREMENT	VISA
ePROCUREMENT	EQUAL TREATMENT	RAILWAYS
DIGITAL TV	CREDIT INSTITUTIONS	ENVIRONMENT&HEALTH
CREDIT INSTITUTIONS	DIGITAL TV	REINSURANCE
FISHERIES CONTROL	FISHERIES CONTROL	CREDIT INSTITUTIONS
RAILWAYS	RAILWAYS	ePROCUREMENT
ENVIRONMENTAL STANDARDS	ENVIRONMENTAL STANDARDS	EQUAL TREATMENT
ENVIRONMENT&HEALTH	ENVIRONMENT&HEALTH	CRIMINAL PROCEEDINGS

Legend

SOCIETY-oriented	ECONOMY-oriented	ENVIRONMENT-oriented

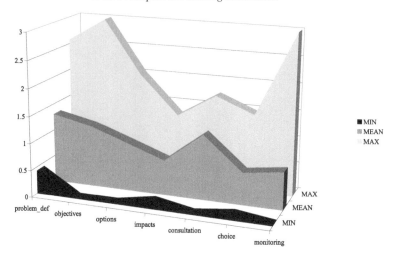

	Problem definition	Setting objectives	Developing policy options	Assessing Impacts	Consultation	Policy choice	Monitoring
MAX	2.571	3	2	1.304	1.739	1.455	3
MEAN	1.3	1.15	0.9	0.64995	1.20865	0.6002	0.7
MIN	0.429	0	0	0.13	0	0.091	0

Figure 4.2   An overview of programme integrity: distribution of the scores with respect to the different phases in the EC-IA process (mean scores of all 20 reports are calculated for each phase)

Figure 4.3 presents a similar analysis, based on value judgments collected for the second MC evaluation, which investigated effectiveness in pursuing the eight objectives set by the European Commission for the IA programme. Relatively satisfactory achievements were attained only on preliminary, formal, and procedural aspects (as indicated by criteria 1, 2, 3 and 7), whereas the programme was neither able to deliver a robust analysis of impacts, nor to allow a wider discussion on political and ethical issues (criteria 4 and 5). The links to sustainable development appear particularly weak (criteria 4 and 6), which is disappointing if one remembers that the Göteborg European Council called for the introduction of IA while adopting the first EU Sustainable Development Strategy.

Following this second analysis, a better insight was gained into the participatory nature of the assessment process. Indeed, what was labelled as 'consultation' in the previous chart (Figure 4.2), is broken down here into 'stakeholders involvement' and 'credibility of expertise', clearly showing that only the former is actually performing well. The process of experts' involvement seems to have remained more opaque and less permeable to due consideration of inequalities.

It was reassuring to realize that equity holds a certain discriminating power and can thus be used to evaluate, under different perspectives, the outputs of the EC Impact Assessment process. It seems like general quality of the evaluation products and attention paid to issues of equity are correlated, at least when the former is relatively high, or extremely low. However, on overall, our metaevaluation reveals an unsatisfactory consideration of inequalities. The most frequent circumstance

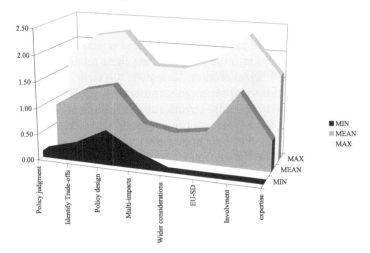

	Enable informed political judgment	Identify trade-of fs	Improve policy design	Analyse social, economic and environmental impacts	Allow wider consideration of ethical and political issues	Contribute to the implementati on of the EU-SDS	Involve more interested parties	Establish a sound knowledge base
MIN	0.13	0.29	0.58	0.23	0.00	0.00	0.00	0.00
MEAN	0.86	1.20	1.34	0.65	0.54	0.64	1.42	0.59
MAX	1.69	2.14	2.25	1.63	1.63	1.83	2.33	1.63

Figure 4.3   An overview of programme effectiveness: distribution of the scores with respect to the EC-IA objectives (mean scores of all 20 reports are calculated for each objective)

we detected was a weak evidence that due attention had been paid to the issues addressed by the different indicators. This observation is in line with what other scholars found. Adelle and co-workers (2006) screened 41 ExIA reports carried out in 2003 and 2004 to discover that the level of consideration of the external dimension of SD[14] was extremely low, whereas Lee and Kirkpatrick (2006) pointed to the poor definition of distributional dimension of problems as a principal weakness in the six 2003 ExIAs they reviewed.

At a closer look, the contribution of the EC-IA regime to SD seems poor in almost all respects. The boost of the coordination function expected under the implementation of the EU-SD Strategy, and more specifically, the ability to put the 'major threats to SD' at the core of every policy process, is far from being recognizable. Even the most down-to-earth understanding of SD, that is, the juxtaposition of economic, social and environmental aspects, in practice proved a hard task to be accomplished. The most favourable trend concerns participation, though understood here merely as consultation. This is not only a remarkable achievement *per se*, but also a promising condition that might eventually trigger further social learning dynamics.

---

14   That is, the impacts of EU policies on third countries, a prominent aspect of intra-generational equity.

As for the environment-society interface of sustainable development, an overall interpretation of our results could read as follows: those departments whose mission is putting forward environmentally-sound policies are the least likely to integrate equity concerns into the ExIAs, unless these relate at the same time to their core policy area. Whereas economy-oriented DGs seem relatively neutral with respect to both inequalities and environment protection, the Directorates focussing on social domains tend to be equity-sensitive and environment-unfriendly. Further investigation should start from adopting different focuses than the relation between leading departments' core policy areas and attention paid to equity. An interesting field might be the interplay between stake and accuracy as a major source of uncertainty (Funtowicz and Ravetz 1993), affecting the contribution of policy evaluation to decision-making. Lee and Kirkpatrick (2006) reasoned along these lines when they called for due consideration of 'perturbations in the programming cycle, which periodically may create pressures for quick decisions [and] result in quality problems for the ExIA process if it has to deliver some reports at relatively short notice'.

Speaking of uncertainty, we must stress that the whole set of conclusions we drew rests on a questionable assumption about the acceptability of the definition of equity as embodied in the guidelines. Indeed, questioning this assumption will be an unavoidable step in the forthcoming research developments, especially when we moved away from the level of internal coherence between the 'instructions' and their 'implementation' to the furthest reaching field of the real influence IAs have on the policies they address.

We acknowledge that the methodological construct we used to evaluate the reports, and in particular the combination of a *content test* (Harrington and Morgenstern 2004) with a multicriteria assessment, shows some interesting perspectives on the potential to perform qualitative assessments with a quantitative instrument. However, particular future research efforts should be dedicated to better cope with the inherent and unavoidable biases introduced when multiple agents assign values to the indicators: the qualitative judgments introduced at this level of the assessment threaten to undermine the derived quantitative rankings considerably.

At the same time, we can envisage how the very same methodological construct could become the target of a lot of criticism, focussing especially on the conditions of comparability of so diverse a set of Impact Assessments. As for us, we tried to keep our investigation within the limits of a credible coherence analysis based only on the IA reports. This shifted the core research question from the ambitious 'how did the IA contribute to a fair policy proposal?' (that is, an *outcome test*), to a lower profile 'how much evidence is there in the reports that the IA followed the guidelines with respect to dealing with inequalities?' (that is, a *content test*).[15] Of course, pushing our assessment beyond the outputs, i.e. the reports, towards covering the entire IA process and its impacts on policy formulation and implementation, would challenge our original approach. However, comparability

15 See Harrington and Morgenstern (2004) for the definitions of *outcome* and *content test*.

will always be an issue, rooted as it is in the EU choice to evaluate all policy proposals through the same system, to develop central guidelines to operate it, and to integrate in that system all sorts of considerations, be that gender, business, or environment related. This homogenization became all the more evident after two successive updates of the guidelines[16] brought about a sudden shift towards the philosophy of Regulatory Impact Assessment and its focus on 'measuring administrative burdens ... imposed by legislation' (EC 2007). Indeed, though assessing costs is not meant to compare different policies, but rather to choose the most efficient option to attain the same policy goal, emphasizing this overarching perspective might in the long run undermine the political legitimacy of those proposals that cannot be framed easily into an economic rationality.

The very nature of the European multiple-level governance system sheds nuances on the political understanding of equity. New rights come into force in the spotlight of high politics and then get systematically infringed in the umbra of delayed implementation and administrative indifference. Growing complexity in terms of scales, fields of intervention, and interrelations among (super) states, (global) markets and (fragmented) societies relegates issues of equity ever more in the realm of those wicked problems (Rittel and Webber 1973) that are bound to lead to 'imperfect compensation' (Glasser 1998).

In this perspective, when even a limited analysis seems to reveal poor internal coherence already at the level of guidelines implementation, is it reasonable to declare that the Impact Assessment system 'identifies the likely positive and negative impacts of proposed policy actions, enabling informed political judgments to be made about the proposal and identify trade-offs in achieving competing objectives' (EC 2002a)? In other words, should the scope of a single evaluation system be stretched enough to accommodate the whole range of complexity ensuing from the supra-national scale (EU-wide), the integrated focus (on social, economic and environmental impacts) and the wide diversity of policy proposals it concerns? Or rather, if no single evaluation can exhaust the debate arising around any proposed action,[17] would not it be better to settle for a limited mandate that could yield more credible outcomes, thus fostering their utilization throughout the argumentation cycle?

Taking advantage of the theoretical constructs introduced by Nooteboom and Teisman (2003), we would like to conclude by praising the 'accountability through transparency' mechanism at work in the EC-IA scheme. Beyond all possible failures,

---

16 In June 2005 and March 2006. A consolidated version of the guidelines is available on-line at <http://ec.europa.eu/governance/impact/key_en.htm> (last retrieved on 15 January 2007).

17 This is very much the case in the context at hand, since the Impact Assessment only covers Commission proposals, which then have to go all the way through the complex decision-making process in place at EU level, before a final decision is taken. This typically rests on an agreement between the Council and the Parliament, which has to consider the opinions of a whole set of other bodies and committees. Many different evaluation practices (be that more or less formalized) might be activated at different stages, and by different actors, during this process.

it is a powerful drive for innovation and change in the European policy-making cycle, as well as an extraordinary chance to carry out evaluation research.

Acknowledgments

We strongly acknowledge the help of Marco Wäktare (formerly based at Université Libre de Bruxelles) for his active participation in the scoring exercise, as well as in the early research design. We are also grateful to the editors and to some informal referees for their very helpful comments on an earlier version of this chapter. This work relies partly on knowledge gained during a research project funded by the Belgian Federal Science Policy. Results at an earlier stage have been presented at the 6th International Conference of the *European Society for Ecological Economics* (14–17 June 2005, Lisbon).

Author Attribution

This chapter is the outcome of a collective research effort, and all authors have contributed to both the study and to writing. Texts have been circulating among the authors, however TB drafted section 1 and contributed mainly to sections 4 and 5, AB drafted section 2, 3 and 5 and contributed to section 4, CMT drafted section 4 and contributed to sections 2, 3 and 5.

References

Adelle, C., Hertin, J. and Jordan, A. (2006), 'Sustainable Development "Outside" the European Union: What Role for Impact Assessment?', *European Environment* 16, pp. 57–72.

Agyeman, J., Bullard, R.D. and Evans, B. (eds) (2003), *Just Sustainabilities: Development in an Unequal World*, Cambridge, MA, MIT Press.

Bauler, T., Wäktare, M. and Bonifazi, A. (2007), 'Exploring the Feasibility of a Sustainability Impact Assessment Procedure for Federal Policies in Belgium', in C. George and C. Kirkpatrick (eds), *Impact Assessment for a New Europe and Beyond*, Cheltenham, Edward Elgar.

Bonifazi, A., Nicoletti, A. and Torre, C.M. (2007), 'Tracing Sustainability Evaluation in Italy', in U. Schubert and E. Stoermer (eds), *Sustainable Development in Europe – Concepts, Evaluation Methods and Applications*, Cheltenham, Edward Elgar.

Bustelo, M. (2002), 'Metaevaluation as a Tool for the Improvement and Development of the Evaluation Function in Public Administrations', paper presented at symposium, 5th European Evaluation Society biennial conference *Learning, Theory and Evidence*, 10–12 October, Seville.

Clemente, M.L., Maciocco, G., Marchi, G., Pace, F., Selicato, F. and Torre, C.M. (1998), 'Evaluation and Equity in Economic Policies for Environmental Planning', in N. Lichfield, A. Barbanente, D. Borri, A. Khakee, and A. Prat (eds), *Evaluation in Planning: Facing the Challenge of Complexity*, Dordrecht, Kluwer Academic, pp. 213–27.

COWI, ECA and Wilson, S. (2004), *Evaluation of Approaches to Integrating Sustainability into Community Policies*, Final Summary Report, Brussels, European Commission-Secretariat General.

Devuyst, D. (ed.) (2001), *How Green is the City? Sustainability Assessment and the Management of Urban Environments*, New York, Columbia Press.

EC (European Commission) (2001), *A Sustainable Europe for a better world: a European Union Strategy for Sustainable Development*, COM(2001)264 final.

EC (European Commission) (2002a), *Communication from the Commission on Impact Assessment*, COM (2002) 276 final.

EC (European Commission) (2002b), *Action Plan: Simplifying and Improving the Regulatory Environment*, COM (2002) 278 final.

EC (European Commission) (2002c), *Impact Assessment in the Commission: Guidelines*, <http://europa.eu.int/comm/secretariat_general/impact/docs/ia_guidelines.pdf>, accessed 17 March 2005.

EC (European Commission) (2002d), *Towards a Global Partnership for Sustainable Development*, COM(2002) 82 final.

EC (European Commission) (2002e), *Towards a Reinforced Culture of Consultation and Dialogue – General Principles and Minimum Standards for Consultation of Interested Parties by the Commission*, Brussels, 11/12/2002; COM (2002) 704 final.

EC (European Commission) (2002f), *On the Collection and Use of Expertise by the Commission: Principles and Guidelines – Improving the Knowledge Base for Better Policies*, Brussels, 11/12/2002; COM 2002 (713) final.

EC (European Commission) (2002g), *A Handbook for Impact Assessment in the Commission: How to Do an Impact Assessment*, <http://europa.eu.int/comm/secretariat_general/impact/docs/imp_ass_how_to_en.pdf>, accessed 17 March 2005.

EC (European Commission) (2002h), *A Handbook for Impact Assessment in the Commission: How to Do an Impact Assessment – Technical Annexes*, <http://europa.eu.int/comm/secretariat_general/impact/docs/imp_ass_tech_anx_en.pdf>, accessed 3 April 2005.

EC (European Commission) (2007), *European Commission-Secretariat General: Impact Assessment News*, <http://ec.europa.eu/governance/impact/new_en.htm>, accessed 15 January 2007

Faludi, A. (2000), 'The European Spatial Development Perspective – What Next?', *European Planning Studies* 8(2), pp. 237–50.

Funtowicz, S. and Ravetz, J. (1993), 'Science for the Post-Normal Age', *Futures* 25(7), pp. 735–55.

Fusco Girard L. and Nijkamp, P. (1997), *Le valutazioni per lo sviluppo sostenibile della città e del territorio*, Milano, Franco Angeli.

Fusco Girard, L. and Voogd, H. (1989), *Conservazione e sviluppo: la valutazione nella pianificazione fisica*, Milano, Franco Angeli.

Glasser, H (1998), 'On the Evaluation of Wicked Problems', in N. Lichfield, A. Barbanente, D. Borri, A. Khakee and A. Prat (eds), *Evaluation in Planning: Facing the Challenge of Complexity*, Dordrecht, Kluwer Academic, pp. 229–49.

Gosseries, A. (2005), 'The Egalitarian Case against Brundtland's Sustainability', *Gaia-Ecological Perspectives for Science and Society* 14(1), pp. 40–46.

Harrington, W. and Morgenstern, R.D. (2004), Evaluating Regulatory Impact Analyses. Resources for the Future', Discussion Paper 04–04, available <http://www.rff.org/rff/Documents/RFF-DP-04-04.pdf>, accessed 20 January 2007.

Khakee, A. (1997), Evaluation and Planning Process. A Methodological Dimension', in P.S. Brandon, P. Lombardi and V. Bentivegna (eds), *Evaluation of the Built Environment for Sustainability*, London, E. and F.N. Spon, pp. 327–44.

Kirkpatrick C. and George C. (2004), 'Trade and Development: Assessing the Impact of Trade Liberalisation on Sustainable Development', *Journal of World Trade* 38(3), pp. 441–69.

Lee, N. and Kirkpatrick, C. (2006), 'Evidence-based Policy-making in Europe: An Evaluation of European Commission Integrated Impact Assessments', *Impact Assessment and Project Appraisal* 24(1), pp. 23–33.

Lehtonen, M. (2004), 'The Environmental-Social Interface of Sustainable Development: Capabilities, Social Capital, Institutions', *Ecological Economics* 49, pp. 199–214.

Lichfield, N. (2001), 'Where Do We Go From Here?', in H. Voogd (ed.), *Recent Developments in Evaluation in Spatial, Infrastructure and Environmental Planning*, Groningen, Geo Press, pp. 7–15.

Miller, D. (2001), 'Evaluating Environmental Justice for Planning', in H. Voogd (ed.), *Recent Developments in Evaluation in Spatial, Infrastructure and Environmental Planning*, Groningen, Geo Press, pp. 17–35.

Munda, G. (1995), *Multicriteria Evaluation in a Fuzzy Environment*, Amsterdam, Physica-Verlag.

Nijkamp, P., Rietveld, P. and Voogd, H. (1990), *Multicriteria Evaluation in Physical Planning*, Amsterdam, Elsevier.

Nooteboom, S. and Teisman, G. (2003), 'Sustainable Development: Impact Assessment in the Age of Networking', *Journal of Environmental Policy and Planning* 5(3), pp. 285–30.

Pope J., Annandale, D. and Morrison-Saunders, A. (2004), 'Conceptualizing Sustainability Assessment', *Environmental Impact Assessment Review* 24, pp. 595–616.

Radaelli, C. (2005), 'Diffusion without Convergence: How Political Context Shapes the Adoption of Regulatory Impact Assessment', *Journal of European Public Policy* 12 (5), pp. 924–943.

Rittel, H. and Webber, M. (1973), 'Dilemmas in a General Theory of Planning', *Policy Sciences*, Vol. 4, Amsterdam, Elsevier Scientific, pp 155–69.

Roy, B. and Bouyssou, D. (1993), *Aide multicritère à la décision: Méthodes et cas, Production et techniques quantitatives appliqués à la gestion*, Paris, Economica.

Saaty, T.L. (1980), *The Analytic Hierarchy Process for decision in a Complex World*, Pittsburgh, RWS Publication.

Scrase, J.I. and Sheate, W.R. (2002), 'Integration and Integrated Approaches to Assessment: What do they Mean for the Environment?', *Journal of Environmental Policy and Planning* 4, pp. 275–294.

Scriven, M. (1991), *Evaluation Thesaurus*, Newbury Park, Sage.

WCED (World Commission on Environment and Development) (1987), *Our Common Future*, Oxford, Oxford University Press.

Widmer, T. (2002), 'Evaluation Standards in the Context of Sustainability', paper presented at symposium, *Easy-Eco 1 – Evaluation of Sustainability* European Conference, 23–25 May, Vienna.

Wilkinson, D., Fergusson, M., Bowyer, C., Brown, J., Ladefoged, A., Monkhouse, C. and Zdanowicz, A. (2004), *Sustainable Development in the European Commission's Integrated Impact Assessment for 2003*, London, Institute for European Environmental Policy.

Williams, K., de Laat, B. and Stern, E. (2002), *The Use of evaluation in the Commission Services*, Paris, Technopolis France.

Chapter 5

Sustainable Development in Regional Development Practice: A Socio-Cultural View of Evaluation

Sylvia Dovlén and Tuija Hilding-Rydevik

Introduction

In 1997, the Swedish government proclaimed that sustainable development (SD) should be the lodestar for all public activities (SOU 1997, 105). The following year sustainable development was explicitly introduced as an overarching aim in Swedish regional development politics (Regeringens proposition 1998). The 21 Swedish regions were thus requested to implement and integrate a new goal and perspective, namely SD, as part of their traditional responsibilities to coordinate and promote regional economic growth and employment.

The introduction of SD is part of the introduction and implementation of the new Swedish national regional development policy. It includes a proclamation that evaluations should be used as tools in the promotion of a learning-working mode (Näringsdepartementet 1998). The evaluations and monitoring undertaken thus far have however not gone into any depth in relation to the specific progress of SD. Indeed, SD has only been a minor part of these monitoring and evaluation processes, which have in the main focussed on traditional regional development issues. As such, it is the content of the regional development programmes that has remained the primary focus. The evaluations undertaken thus far moreover indicate that only slow progress has been achieved in respect of SD implementation (ITPS 2003; Nordregio 1999; Nordregio et al. 2000; Näringsdepartementet 2001; 2002b; 2003; 2004). In addition, those evaluations that have taken place moreover have not been able to shed much light on the specificities surrounding this issue.

The aim of this chapter is to present a study[1] a study that evaluates the effects of introducing SD goals in regional development planning. The analysis of the results is still ongoing; therefore the focus here is on presenting the approach while also, tentatively, highlighting some of the preliminary results. The evaluation study complements previous government initiated evaluations. It does so through its:

---

[1]   See acknowledgements for more details on the project.

- focus on socio-cultural conditions (explained below);
- focus on professional every day programming[2] practice;
- regional organizations that have the responsibility to coordinate regional development work in Swedish regions; and
- explicit focus on promoting critical reflection and learning in these regional organizations and specifically in relation to SD.

The occurrence of learning and especially critical reflection is assumed to be of crucial importance in order to be able to promote change in an organization.

The results of the evaluation study presented here relate more specifically to the prevailing discursive conditions (which form a part of the socio-cultural conditions in the regional organizations studied here) in the regional organizations and to what extent these conditions support reflection, interaction and thinking in new directions, i.e. change. In recent years, research emphasis has gradually come to approach policy and planning practice as communicative and interaction dependent processes (Fischer and Forester 1993; Fischer 2003; Lapintie et al. 2001). In our study we thus assume that investigating and understanding the communication processes among the actors involved is important. How they communicate their knowledge and perspectives is then of critical importance for the practical handling of the issues concerned. The discursive conditions are prerequisites for the adaptation of modes of work and ways of organizing every day programming practice to this relatively new task of implementing SD goals. Through the study of the conceptual frameworks and discourses, different ways of handling the SD issues in regional programming practice can be uncovered. It is thus a challenge to grasp the ways in which actors involved in regional programming work understand e.g. the national task of integrating SD goals with day-to-day programming practice. The current study is itself based on the results of a major on-going study of four regions in Sweden and their efforts to integrate SD goals in their regional growth programming work.

This chapter is divided into five parts. In the two sections that follow, the introduction and development of the Swedish government policy concerning regional development and SD is outlined. Next, the chapter describes the nature and results of previous evaluation studies and provides an introduction to our evaluation approach based on a socio-cultural perspective. The evaluation approach, design and theory are outlined in the next section, which also includes a short description of the four regions included in the study. Preliminary results are then presented relating to how the regional organizations have approached the 'top-down' task of implementing SD as a part of their regional development work. Finally, a short set of conclusions are forwarded.

---

2   'Programming work' is the label we use for the planning process that is connected to partnerships and the formulation of regional development programmes based either on national regional development policy initiatives or on European Union structural funds initiatives.

Swedish National Regional Policy – Contents and Organization

As far back as 1995 the Swedish Parliament included 'sustainable economic growth' in its goal formulation for Swedish regional policy (Regeringens proposition 2001, p. 11). Since then, Swedish regional politics has undergone a number of changes in relation to its contents, its planning processes and planning instruments, its organization and in relation to the role given to SD. During this period 'regional policy' has been reformed and relabelled as 'regional development policy', while new tools (programmes) and working modes (a stronger focus on partnerships) have been introduced (Regeringens proposition 2001, p. 10).

The modern Swedish multi-sector based regional politics, characterized by the goal of coordinating sectoral interests and raising awareness across all sectors and regions in order to promote regional growth and development was, in the main, developed in the 1980s (Regeringens proposition 2001, p. 9). Prior to that point, regional policy primarily focussed on financial support to companies in regions in need of economic and labour market support.

In 1998 the Swedish government created a new policy area – 'regional business development policy' ('regional näringspolitik') – as part of its national business policy. The aim was to better utilize the endogenous economic development potential and characteristics in each of the 21 regions and to 'take as a starting point the specific prerequisites in each region in order to stimulate a sustainable economic growth that can stimulate business activities and thereby increase employment for both women and men' (Regeringens proposition 1998, p. 61). One means to achieving this was to adopt a partnership approach, inspired by the EU structural funds work mode (Statskontoret 2004: 9). Another means was the introduction of the Regional Growth Agreements (RGA). These were programmes that were to be produced by regional partnerships encompassing state, business, regional and municipal actors. The RGAs, which later became the Regional Growth Programmes (RGP) (the term used hereafter), are voluntary agreements between the government and the regional partnerships concerning the amount of funding that each actor will contribute in order to promote regional development measures. The government aim with the RGPs was to better coordinate government initiatives in order to support the development of the regional economy (op cit). Thus the content of each RGP was negotiated between the region in question and the government and then finally accepted. The first round of RGPs ran from 2002 to 2004 with the second round running from 2004 to 2007. The government also pronounced that there was a need to work with Regional Development Strategies (RDS) with the focus on the clarification of long-term regional development strategies (Regeringens proposition 1998: 165). The RDSs were later turned into Regional Development Programmes (RDP), which are described in more detail below.

In 2001 the government introduced another new policy area, namely, 'regional development policy' ('regional utvecklingspoliti'). This was, in effect, a merger between the former policy areas of 'regional policy' and the 'regional business policy' introduced in 1998. It was subsequently pointed out by the government that national economic growth is the sum of the economic growth created locally and regionally and therefore that regional development policy as a whole should embrace all regions in Sweden. The aim of the introduction of this new policy area was therefore to highlight the aims, means, and geographical focus of Swedish regional policy (Regeringens proposition 2001, pp. 100–101). The current goals of Swedish regional development policy then are 'well functioning and sustainable local labour-market regions with a good level of service in all parts of the country' (Regeringens proposition 2001, p. 101). These goals remain the same as those produced in 2001 with the introduction of this new policy field, while the partnership working format and the RGPs were to be the main implementation instruments. Recently the government has however introduced another new regional planning instrument – the Regional Development Programmes (RDP). The aim here is to make the RDP the regional umbrella strategy, for example in the context of the regional economic strategies in the RGP, for infrastructure planning and for municipal land use planning.

In parallel with the introduction of the regional development policy field, the government also opened the way for the establishment of a new regional body with responsibility to coordinate regional partnerships and programming work – these are the so-called regional municipal cooperation bodies ('kommunala samverkansorgan'). At present there are three parallel systems of public agencies with the responsibility of coordinating regional development policy. Ten counties have retained the traditional County Administrative Boards which have been the representatives of the central government at the regional level. They are also responsible for regional development policy. Then there are nine newly started Regional Municipal Cooperation Boards who combine the function of the county boards with that of coordinating municipalities' development efforts. Finally, there are two self-governing regions in west and southwest of Sweden. These two regions have their own regional assemblies which together with the regional administration are responsible for coordinating the RGP-work.

The Introduction of Sustainable Development

As stated previously, the government introduced sustainable economic growth as a goal in regional development policy as far back as 1995. In the government bill 1997/98:62, which introduced the RGPs, it is, moreover, stated that the regional politics and the RGPs are to contribute to SD. SD is described as including the mutual interdependencies between economic development, social development and the protection of the environment (Regeringens proposition 1998, pp. 21–22). In the more specific guidelines accompanying the bill and outlining what the regions should include in their RGPs, written by the Ministry of Industry, Employment and Communication (Näringsdepartementet 1998), the broader goal of sustainable development was interpreted as ecologically sustainable development and gender equity (Näringsdepartementet 1998). In addition, the national decree on regional development (Regeringskansliet, 2003) states that all regional development programmes should take into account the issues of environment, gender equity and ethnic integration. As such, these issues were subsequently elevated to the level of all-encompassing 'horizontal goals' by the national regional development actors (Näringsdepartementet 2002a). The goal of sustainable development was also elevated in such a manner. As such, it should now be integrated with (Näringsdepartementet 1998) and even 'permeate'[3] regional development efforts, e.g. the RGPs.

Since 1998, much effort has been made to promote the implementation of SD in the RGP's. This has been realized in the context of different development and educational projects and in the publication of several guidelines (Naturvårdsverket 2000, 2001, 2002, 2003; Nutek 2004, 2005a, 2005b, 2005c). Another national educational programme is to be launched in 2006/2007.

As a result, the previous government focus on a more restricted ecologically-based view of SD has been significantly broadened such that SD and is now manifested in the most recent national regional development policy strategy (Regeringskansliet 2006). In parallel with the introduction of SD as a goal for regional development policy the government's own SD policy also explicitly highlighted regional development policy as a core area for the implementation of its SD goals (Regeringens Skrivelse 2003/04, p. 129). Thus the inclusion of SD in the then Social Democratic government's regional policy was gradually strengthened up to its electoral defeat in 2006. Since the autumn of 2006 however Sweden has had a new right of centre coalition government. This development may however have a significant impact on the future direction of national regional development policy.

Evaluation of Sustainable Development Integration Progress

Existing Evaluations

As can be seen from the brief outline above, the 21 Swedish regions – their political as well as administrative bodies – face a situation where they are expected to implement a number of top-down changes outlined in the national regional development policy. On top of this they have also been given the task of implementing a totally new issue in Swedish regional development policy, namely SD. As a means of implementing these new tasks the government has recognized that a 'learning mode' is necessary (Regeringens proposition 2001, pp. 111–12). According to this government bill evaluation will be the main tool to ensure government control over education and information about SD at the national level. With respect to regional learning, the bill emphasizes two means namely evaluations of the RGPs and benchmarking.

The regional effort to implement this new regional development policy – its new institutional arrangements and its contents – has been evaluated at least

---

3 The term 'permeate' was used repeatedly by actors from and the Swedish Business Development Agency in a speech at a national education programme aiming to outline exactly how such horizontal goals were to be treated in the Structural Funds programmes.

once a year since 1999 as part of government policy. Evaluations have been performed as:

- ex-ante evaluations of draft RGPs – do the contents comply with the intentions of national regional development policy (based on a review of programme contents and interviews) (e.g. Nordregio 1999; Naturvårdsverket 2000; ITPS 2003);
- yearly follow-ups, monitoring of the progress of the RGP work (based on questionnaires and written reports from government authorities) (Näringsdepartementet 2001, 2002, 2003, 2004; Nutek 2004, 2005);
- mid-term evaluations – for the second round of RGPs (based on a review of programme contents and interviews) (ITPS 2006).

Thus far, both the monitoring and the evaluations undertaken have specified the progress of SD and the integration progress of the 'horizontal goals' but have not however gone into any detail in relation to the empirical or theoretical understanding of this issue. As such, the progress of SD has only been a minor part of the monitoring and evaluation process which has, in the main, focussed on traditional regional development issues. The Swedish Environmental Protection Board and the Swedish Business Development Agency (NUTEK) have however conducted a number of development studies in relation to the question of how to promote the integration of SD (Naturvårdsverket 2001, 2003; Nutek 2005b, 2005c). These studies have, through targeted assignments to certain forerunner regions, gathered practical experience of their efforts to integrate SD. Such experiences have been gathered through written contributions from the regions and regional actors, questionnaires and interviews. The results are however often reported as success stories or best-practice advice applicable under favourable programming and organizational conditions. Both evaluations and development projects have however highlighted the difficulties and challenges inherent in the integration of SD further raising the issue of the continuing need for learning and planning approaches. None are however explicit when it comes to what learning is, how it can be defined, and what the mechanisms are that promote learning. As such, the learning approach seems quite superficial in terms of the guidelines, evaluations and in the development projects. The learning approach adopted here relies to a great extent on the mechanisms of progress information to promote learning (e.g. Nutek 2004: 49).

Evaluations of the national regional development work in relation to SD progress show that SD has been taken on board in the programming work (Näringsdepartementet 2001–2004). In most instances the SD goals have however been made superficial, interpreted as mere environmental goals and/or with little integration into core economic and labour market projects. In the implementation of the RGPs slow progress has been made with regards to necessary changes in the regional administration, in funding of projects, in providing information to applicants of projects and for enhancing SD competence etc. only slow progress has been made. It can thus be stated with some authority that eight years after the introduction of SD, as part of the regional development policy rhetoric, regional SD practice still appears to be in its infancy. Moreover, the specific need for capacity building in regional bodies that coordinate the regional partnerships and the RGP-work, in the regional partnerships and in the sequences of the RGP-work remains a high priority. All in all, the evaluations thus far indicate that a level of inertia has built up in relation to the implementation of SD. They do not provide a deeper insight into the nature of and mechanism behind this inertia. The assumption made here is that the evaluations undertaken thus far need to be complemented by studies focussing on the prevalent socio-cultural conditions in the regional development bodies in focus here. This view is elaborated further below.

Evaluation from a Socio-Cultural Perspective

In Sweden, all sectors of society are expected to contribute to sustainable development (Miljödepartementet 2002). Efforts have been made by national and local actors to promote the integration of environmental considerations, as part of SD, in different planning and programming contexts: environmental and planning legislation has been revised; environmental knowledge has been provided; new instruments and 'tools' such as environmental assessment and sustainability indicators have been introduced; and environmental quality objectives have been formulated etc. These efforts have led to progress being made in some areas (Miljömålsrådet 2004). However, in several areas the ambition to integrate environmental concerns and implement SD continues to make slow progress (Asplund and Hilding-Rydevik 2001; Wood 1999; Hilding-Rydevik 2006; Fudge and Rowe 2001; Owens and Cowell 2002). This under-theorized process of inertia is one of the most crucial, and neglected SD problems today. The point of departure taken here is that this inertia emanates from conditions and attitudes at the micro level in organizations responsible for implementing SD in practice – that is to say, in the every day professional practical work of the implementing bodies.

National directives and policy measures are designed from a macro perspective, neglecting the conditions and difficulties that might appear at the micro level. The government's failure to 'anticipate the resistance' (Sanyal 2005) to national top-down inventions is evident in the national discourse on sustainable regional development (Isaksson et al. 2006). For instance, sector interests are presupposed to be compatible even though it is obvious that society in general and the public administration in particular remains separated into different sectors with different interests, competencies and perspectives. In many planning situations where different sector interests are to be weighed in relation to each other, environmental concerns remain a delicate and conflict-laden issue (Asplund and Hilding-Rydevik 2001; Isaksson 2004; Storbjörk 2004). The implementation of SD is then not simply a task for one single profession, but is instead to be handled jointly in cooperation by practitioners and stakeholders representing different sectors, competencies, interests and cultures. Preparatory work and decision-making are not simply instrumentally rational processes guided solely by explicit 'facts', goals and means.

The success or failure of policy implementation can be assumed to depend to a large extent on the implementation level, the context, the organization, the organization of routines, rules, interaction- and cooperation processes between professionals and the other actors involved. It can be assumed then that for the main 'environmental' problem, namely, the observed inertia in respect of new perspectives, norms, rules, and organization etc, to be overcome in favour of SD at the regional administrative level, the institutional and socio-cultural conditions prevailing in the Swedish regional organizations have to be addressed. This inertia is often ignored, or is perceived (especially in political and public organizations) as a 'black box' in national politics and by the regional actors themselves. Individuals and organizations can be 'trapped' in existing but often unrecognized patterns of thought and action especially in situations where strong and long-institutionalized organizational and professional cultures have been established. This can be assumed to be the case for the Swedish regional development sector and its actors. The patterns referred to are the result of 'invisible' socio-cultural processes and conditions. The theories concerning these processes relate to approaches such as social constructivism, discourse analysis, learning and change i.e. theories relating to humans as social beings and to the creation of 'meaning' in the context of professional practice.

Therefore, a socio-cultural approach to evaluation in its organizational context is appropriate if we want to better understand the mechanisms behind the existing inertia towards change and find ways to deal with this problem. A lack of understanding in respect of these socio-cultural mechanisms and an inability to deal with them is we assume the main reason for the existence of the prevalence of this inertia to change. We also assume the same in relation to the low activity in relation to capacity-building and the low level effectiveness of the capacity-building that is being undertaken in relation to SD in the regional development context. In the following section our evaluation approach and design, based on a socio-cultural perspective, is described.

Evaluation Approach and Design

Evaluation Project

The purpose of this section is to outline the evaluation approach applied throughout the rest of this chapter as well as some of the results from the project on Sustainable Development and Economic Growth – A Socio-Cultural Perspective on Regional Programming Processes.[4] The problem addressed in this project concerns the level of inertia in the capacity-building of regional programming practice in relation to its performing of cross-sectoral work aiming at the integration of environmental concerns with economic and social concerns in order to implement SD. The aim of the project was to:

---

4   See acknowledgements for more details of the project.

- produce profound empirical knowledge and understanding of the capacity of regional development programming processes to integrate environmental with economic and social concerns;
- focus in particular on the socio-cultural conditions and processes currently active in day-to-day professional and organizational practice;
- to communicate and discuss the project results with the regional and national level actors in order to facilitate learning and change at different levels.

The evaluation project consists of five different and interrelated sub-projects of which numbers three and four constitute the main parts of the evaluation study:

1) *Review* of all 21 RGPs in order to be able to choose four regions for closer scrutiny.
2) *Discursive conditions at the national level.* The aim here was to identify the discursive conditions set out by the government in its policy influencing how sustainable development is understood and implemented by the regions in their work.
3) *Studies of the regional organizations and their approach to RGP and SD* – the organization, organizing, discourse etc. The aim here was to explore how the participants in the RGP-process perceive, construct and reconstruct their perspectives and conceptual frameworks in relation to the programming context and SD. A minor part of this project focussed on how private actors in one region perceived their role in relation to RGP and SD (Mobjörk 2006).
4) *Aspects of learning.* The aim here was to open the way for new perspectives to emerge in relation to the programming work using the results from sub-project three. The results describe the factors that are perceived to hinder or facilitate discourses, ways of thinking etc, as well as the contextual conditions e.g. organizational conventions and routines. This will make it possible for us, as researchers, together with the actors in the field to further reflect on and discuss those things that are taken for granted on both sides, and the conditions that seem possible, or not, to change.
5) *International experiences.* The purpose of this study is to further the Nordic-European frame of reference i.e., what experiences have others had in this respect, and in the promotion of SD in the regional development context what other approaches have been adopted? Answering this question will enable us to distinguish the special or specific Swedish features of the current research problem.

This evaluation focuses on the conditions for implementing SD in the regional organizations responsible for coordinating the regional programming processes. This differs from most of the evaluations undertaken in the realms of regional development – be they either focussed on regional development programming initiated from the national point of view or on EU Structural Funds initiated programming work. The evaluation of the programme contents is thus not in focus (compare for example with the programme evaluation approach discussed by Roberts 2006). Instead, we focus on organizational capacity-building. We also focus on the micro-level i.e., on the day-to-day-professional practice in an organization. Other evaluations including SD and/or the regional level often focus on the regional territory (e.g. Davoudi 2005), or on more locally defined places (e.g. Khakee 2002). Valve however, evaluated the social learning potential of EU rural programmes (2003). In most instances the organizations that we have chosen to study have the role of putting together the regional partnership, of managing that partnership and of coordinating the group that gathers the baseline information used as a basis for programme-drafting. Such actors often work in small groups actually writing the programme, also having an important part in implementation while encouraging other organizations to implement the regional programme. Hence our major assumption is that the level of capacity-building in coordinating organizations is of crucial importance for the implementation of SD in the entire programming work – from drafting programme documents to implementation.

The focus in this contribution and in the following sections relates to the design and results of sub-projects three and four above.

Theoretical Approach

Social Constructionism and Discourse

The approach taken throughout the evaluation study but specifically in sub-project three described above, is to view regional programming practice as a being a communicative process, thus viewing it from a social constructionist perspective (Healy 1997; Gergen 2001). It is through communicative processes that we as human beings learn and become active participants in society (Säljö 2000). A social constructionist approach views language as important, but not only in a linguistic sense (Berger and Luckman 1966; Hacking 2000). Words and speech carry meaning and shape our view in the first place (Fischer 2003, p. 41; Richardson 2002). Metaphors and conceptual frameworks that are being used reflect discourses in our society (Lakoff and Johnson 1980; Morgan 1997; 1999).

When an object or a phenomenon, e.g. SD, is talked about, it exists in, or is placed, in a *discourse*. A discourse is a set of concepts, categorizations, ideas and meanings that produce forms for how things are presented and interpreted. A discourse thus produces a particular version of events ('this is how it is', 'this is how it happened') (Hajer 1995; Jacobsson 2000; Burr 1995). A discourse is a way of talking about something, which at the same time presents this something in a certain light. Discourses lead the way we perceive and understand what we see. They can be common on a wide national arena or on a local and narrow one. Discourses are *produced* through practices (how, for example, we as professionals go about our day-to-day business) but, crucially, they also *reproduce* that practice.

A number of writers have therefore stated that discourse concerns not only those descriptions (of, for example, SD) that can be made, but also how, by whom and in what kind of institutional context this can be done more or less successfully (see for example Foucault 1970, p. 26; Foucault 1976, p. 2; Hajer 1995, p. 44; Hedrén 1994, p. 26; Mills 1997). The practical and organizational arrangements that are established for the treatment of certain societal issues – for example SD – can thus manifest, embody and reproduce such discourses. The *discourse* concept alludes to aspects that are fundamental in the communicative process.

By supplying forms for how things are to be presented and interpreted, a discourse is always, to some extent, restrictive. By establishing and reproducing broadly supported comprehensions of what is considered to be taken-for-granted (and not), true (and false), more or less important and relevant, realistic (and unrealistic) etc., discourse serves as a fundamental prerequisite for human action, and is of crucial importance from a power perspective (Foucault 1970; 1976; Hedrén 1994, p. 26). It is also worth noting however that discourse not only restricts what can be said and done at each moment, but also provides possibilities for change. The dynamic between actors (individuals or organizations) and discursive structures is thus a fruitful and important focus of evaluation concerning change and thus is potentially a particularly fruitful approach in relation to SD.

Learning and Change

Though actors are guided by discourses, institutions, routines and conventions, they also have the capacity – if there is an opportunity – to reflect upon and review these, and make more deliberate choices i.e. to realize and abandon what they discover to be dysfunctional, obsolete knowledge and/or perspectives thus opening themselves up to new issues and perspectives i.e. learning. In our day-to-day practice however this capacity is rarely used, often because there is simply no opportunity to use it. The process of 'revealing' and of putting words to processes can then open up old perspectives, rules and norms for reflection, and in this context of opening up to new perspectives – learning and change can take place. This process of reflective learning is termed double-loop learning, by Argyris (1990; 1993a, b). Isaacs (1993) moreover points out that it is possible to achieve 'triple loop' learning in dialogue, as a process of social learning. Within this form of learning, basic views on existential conditions and the relationship between the organization in question and the world at large are raised, reflected upon and discussed, a process which is labelled as a 'learning organization'. By uncovering such discourses and conceptual frameworks the opportunity to find the underlying conceptions and attitudes that function as obstacles to communication and interaction between actors is afforded (Asplund 1979; Maasen and Wingart 2000).

By learning we thus mean the reflecting process in which both the existing and new aspects of a phenomenon are brought to the fore, critically examined, and discussed subsequently leading to new ways of conceiving and handling the task at hand, or the problem to be solved. The task or problem may gradually be perceived in a new light, and thus in a different context than was previously the case. This in turn may result in a deeper understanding of how conflicting views, issues or fields may become more penetrated and clearer or be brought closer to each other, or how new and more integrated perspectives can emerge and be applied. A learning process however does not emerge in the heads of isolated individuals but is created socially, i.e., in the relations between people, and exists within socio-cultural conditions, such as in an interaction context, in power relations, and in local organizational patterns with their cultural features. What then are the favourable conditions demanded by the emergence of a learning context? In order for individual and collective learning to become institutionalized in a long-term perspective new perspectives and knowledge needs to be turned into new modes of organization, communication, cooperation, rules, norms etc. Crossing discursive boundaries, choosing an alternative discourse, or working out new ways of understanding are then dependent on the preconditions for reflection in the organization and programming processes and on obtaining distance from everyday activities through discussion.

The discourses and conceptual frameworks are, from the above point of view, worth further analysis and reflection in relation to regional programming practice aiming at the implementation of SD goals. Evaluations focusing on the scrutiny of conceptual frameworks and discourses within the regional programming work can thus be assumed to be an important tool in initiating learning processes and change.

Four Regions – Skåne, Halland, Dalarna and Västernorrland

Four Swedish regions were chosen for inclusion in sub-projects three and four: Skåne, Halland, Dalarna and Västernorrland (Figure 5.1).

These regions were chosen for two reasons. Primarily we wanted to include regions representing the three different types of organizations that today coordinate the RGP work. It was also important for us to choose regions with clearly stated ambitions to integrate SD goals into their RGP. From the group of regions exhibiting these characteristics we also wanted to include those with different economic and social prerequisites as locations in different parts of Sweden. Before final selection we also conducted interviews with two to three key persons in each of the four regions in order to ensure their commitment and interest in being engaged in the study.

As noted previously, in each of the four regions the broad regional development approach and indeed the RGP-coordinating body itself are organized in different ways. Skåne is a self-governing region, Halland and Dalarna have the new Regional Municipal Cooperation Board and Västernorrland has the traditional County Administrative Board for coordinating their RGP-programming work.

The four regions also differ in size and business focus. Skåne has more than one million inhabitants who live in 34 municipalities. Halland includes six municipalities, Dalarna 15 and Västernorrland seven. Halland, Dalarna and Västernorrland have close to 250,000 inhabitants each. The industrial structures of each region also differ. Skåne corresponds to the national average though an

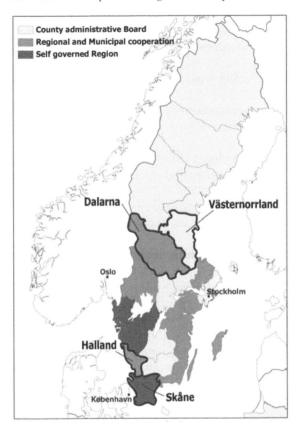

Figure 5.1   Regional governing bodies

Note: Västernorrland – county administrative board; Dalarna and Halland – regional
         and municipal cooperation; Skåne – self governed region.

Source: EuroGeographics Association for the administrative boundaries.

ongoing transformation from manufacturing to knowledge-intensive companies
is currently taking place. Halland's economy is based on many small and medium-
sized companies. In Dalarna the steel, paper and electronics industries dominate.
Västernorrland has many energy-intensive companies producing pulp, paper and
hydroelectricity.

**How the Regions Approach the Top-down Task of Implementing Sustainable
Development**

Since the final analysis of the evaluation project has not yet been concluded we
will restrict the description of the results here to one issue, namely, how each of

the four regions in the study approach the top-down task of implementing SD. In order to understand the empirical basis for the results we will begin by describe in greater detail how sub-projects three and four were conducted.

Aim and Method of the Sub-projects Focusing on Learning Approach

The aim of the closely related sub-projects three and four described above was to:

- explore how the participants in the RGP-process perceive, construct and reconstruct their perspectives and conceptual frameworks in relation to the programming context and SD;
- open the way for new perspectives to emerge in the programming work using the results from sub-project 3 i.e. to promote learning in relation to SD.

The study was designed with the aim of creating interaction between the participants and between participants and the research team in order to stimulate reflection on different relevant issues while providing the possibility to formulate strategies for further programming activities.

In order to attain the stated aims the following steps were conducted:

1) Initial visits to, and meetings with, the four regions chosen (the regions chosen are described in more detail in the next section) in order to gain an overview of how the region organizes its RGP and SD work and to determine the most relevant actors to include in the following focus group sessions. Two to five people from each region participated in the meetings that were held during the autumn of 2004.

2) Preparation of focus group material. Based on the initial visits, review of the RGPs and their experiences, undertaken within the context of our previous (and similar) evaluation approaches and the results gathered in relation to Swedish municipalities (e.g. Asplund and Hilding-Rydevik 2001; Asplund and Skantze 2005; Dovlén 2004) a number of issues worthy of further analysis in the focus group sessions were outlined. An introduction to the project and to the issues in focus for the focus group sessions was produced and sent out to the focus group participants about a week before the focus group sessions.

3) In total, eight focus group sessions in four different regions were arranged. In each region two focus group sessions were held, one with the professionals responsible for coordinating the RGP programming process and one with experts on environmental, equity and ethnic integration issues (professionals responsible for the integration of horizontal goals). The focus group discussion concentrated on the interviewee's experience of integrating SD goals in the regional growth programmes. The reason for dividing the professionals into different focus groups was to ensure an open atmosphere without anyone feeling discomfort at expressing ideas that were felt to be 'not in line' with the mainstream discourse on SD implementation within their own region. The focus groups where held in each region during the period February to June

2005, with between four and eight persons in each meeting. All sessions were tape-recorded.

4) The focus group tape recordings were transcribed and analysed. An analysis of the different emergent discourses in relation to SD implementation in RGP work formed the major part of this phase of the project.

5) The results of the analysis were presented to the regions in feedback seminars held in each region. The participants were the same as in the previously held focus group sessions. In these feedback seminars, which were held in November 2005, the results from the first focus group session were presented. These results describe what we, as researchers, perceive as hindering and facilitating discourses, ways of thinking, seeing and talking about RGP and SD, as well as the contextual conditions e.g. organizational conventions and routines. This made it possible for us, as researchers, together with the actors themselves to reflect on and discuss such things that were 'taken for granted' on both sides, while also identifying those conditions that seemed either possible or impossible to change. The seminars were again tape-recorded.

6) The comments and discussions from the feedback seminars were used to complement the analysis from the focus group sessions.

7) As a final step, a seminar was arranged with all participants from each of the studied regions. The aim here was to present the results from the study and to give the representatives of each of the regions the opportunity to meet each other and to discuss their common experiences. The seminar was held at the request of the participating regions in September 2006.

The main methods used have been focus group discussions (Morgan 1993; 1998; Wibeck 2000) and feedback seminars. All of the focus group meetings with the regions have been arranged with the same moderator while the discussions have focused on exploring the SD goals in the RGP process. The purpose here has been to give the participants the opportunity to discuss matters with each other. Group interaction has been used to generate data. The focus group method is useful in allowing participants to generate their own questions, frames of reference and concepts. This method enables researchers to examine different perspectives and to explore how points of view are constructed and expressed in the interaction (Wibeck 2000; Kitzinger and Barbour 1999). In deciding to use focus groups, we also saw an advantage in the pedagogical possibilities that the method entails. Different ways of organizing the programming process in the regions concerned, the different discourses used and the alternative sets of prevailing power relations it was hoped would then be laid bare, manifesting themselves in the discussion.

Approach to SD

In the analysis of the focus group-sessions four themes were identified as summarizing the contents of the discussions:

1) the stance taken towards the top-down task of implementing SD given by the Swedish government and as part of the RGP-work (this will be described in greater detail below).
2) the interpretation of the contents of SD.
3) the day-to-day organizing of the SD work as part of the RGP drafting and implementation.
4) the role of the region/regional organization in relation to the actors in the field, i.e. those actually creating regional development – business, municipalities etc.

These four themes were further elaborated on and presented at the feedback seminars. The focus of the presentation here was on issue number one above i.e. the stance taken towards the implementation of SD.

The following questions were raised in the description:

• How do professional actors in there day-to-day-practice in the regional bodies, conceptualize, comprehend and discuss the national task of integrating SD goals in the RGP?
• What approaches, vis-à-vis the directives formulated at national level, are used by the regional actors?

Through the interviews and focus group discussions, we identified four different discourses in relation to the national task of integrating the SD goals (horizontal goals; environmental and social goals such as equity and ethnic integration) in regional programming activities:

• an intellectual challenge;
• the goals are accepted;
• a negative stance towards the national task;
• the national task is uninteresting or impossible.

In the regional bodies each of the four discourses could be found to exist in parallel with the others. It was however evident that the actor groups representing the horizontal goals to a larger degree represented a discourse that sees SD as an intellectual challenge and accepts the goals. We did however also uncover a significant point of view that sees the nationally set task as being uninteresting or impossible. In the focus groups representing the traditional core regional development interests we find all of the discourses represented. We cannot. however. determine to what extent which one of the four discourses will become dominant. What we can clearly see is that the representatives of the prevailing regional development discourse were striving practically to take on board SD (discourses 1 and 2) or ignoring in their daily work with regional development (discourses 3 and 4). It is clear that all four regions are in the process of constructing the practice of SD and that, thus far, we cannot firmly state which direction this will take. This particular question is further elaborated in the discussion. The evaluation approach taken thus highlights the different discourses in circulation and the feedback seminars provided opportunities for the regional actors to discuss and reflect upon them. The four discourses distinguished are described in greater detail below.

An intellectual challenge   Implementing the national government's SD goals is considered in this discourse as an intellectual challenge and referred to as a way towards change. The integration of SD goals is a prerequisite for competitiveness and capacity-building in renewing regional governmental practices. SD is a strengthening factor in growth and development work, where growth is not necessary synonymous with increasing production levels. It remains an intellectual challenge to find methods to promote the integration of SD goals in the RGP process and there is a continuing need for innovation. SD is referred to as demanding knowledge and new attitudes; there is a need for education and for the further development of knowledge but this is a time consuming process. One strategy is to be proactive i.e. to think before acting and have carefully prepared strategies ready to promote regional development. Mobilizing resources (competence, economic), and cross-sectoral approaches is one suggestion. The discourse requests that everyone in the organization is responsible for the implementation of the national task irrespective of their ordinary professional responsibility. The professionals responsible for implementing horizontal goals in RGP work commonly use this discourse.

> ... all actors have to move in the same directio n... we believe that growth has to respect sustainable development. (Participant in focus group discussion)

The goals are accepted   Actors work to ensure that the goals are accepted and viewed as being acceptable – a stoic 'we will do it' attitude – but in this discourse the fulfilment of such goals is often viewed as a difficult and complicated task. It is important to show that it is possible to work with SD goals and to trust the power of best practice. One main obstacle here is that some actors within the organization do not accept the task. How the goals are handled is then a matter for the professionals responsible for the programming and is manifest in their communication with the politicians. In other words, competences and interests influence the programming work. Continuity in the programming work is highlighted as important. Developing new views and working conditions is a time-consuming process and an obstacle to goal fulfilment. The lack of cooperation between actors with different interests and competences is one explanation of why the integration of the horizontal goal in RGP is a slow process. One of the participants summarized the discourse as follows:

> We want to demonstrate good practice and ... support good practice. (Participant in focus group discussion)

A negative stance towards the national tasks   This discourse has a negative or critical view towards the national government's perspective and top-down attitude. The national aim of integrating the stated horizontal goals acts as a restriction. National government is often perceived to lack an understanding of the conditions under which regional governing bodies work. National guiding principles and evaluations are seen as not being in accordance with regional and local opinion on priorities.

The concepts used in relation to SD and the horizontal goals imply that they are national 'mantras, mythopoeia, normative or a restricting sidetrack'. The concept of SD used in this discourse is viewed as a magical formula or password that is frequently repeated. The national government seems to rely on SD but in practical programming work it just becomes an empty concept and an abstraction. One strategy promoted here is to redefine the tasks as primary and secondary tasks, where economic growth issues are commonly highlighted as the primary task while the integration of horizontal goals are viewed as secondary. One often proposed 'solution' here is to postpone the issues of SD goal-implementation to future work.

In one of the studied regions where the main discourse (professionals responsible for coordinating RGP) held the negative stance towards national directives, the power position between various interacting professionals became obvious. During the focus group discussion on the integration of social issues in the programming work, one person claimed that the equality aspects were not included in a serious way in the programming process. Another participant then stated that

> ... if you, like you do just now, try to transform equality aspects into a mantra and a normative issue it will be a dangerous path. (Participant at focus group discussion, 18 February 2005)

After this comment, everybody became silent and discussion on this subject ended.

The national task is uninteresting or impossible   The task in this discourse is viewed as being impossible to implement. It is impossible to combine economic growth and the horizontal goals. In addition, how to interpret SD in practice is unclear. Formulations about SD in the RGP work are only made for the purpose of satisfying the national government. SD and the horizontal goals are seen as fancy formulations and as cosmetic devices.

This discourse also implies that the national actors, in written documents and in meetings with regional actors, often have a 'top-down' perspective. Expectations exist at the national level that the national views on SD can be used to tackle the RGP. The national approach will not work at the regional level however it is claimed because each region has its own history, natural and cultural resources and structure of population and business.

It is not possible to unite economic growth and the horizontal goals. What SD is – we can't do it any more – it is a mess. (Participant at focus group discussion, 30 May 2005)

Feedback Seminars

The regional discourses relating to the national task of implementing SD goals in the RGP's presented above highlighted some of the interviewee's core reactions during the feedback seminars. Most of the participants recognized the different discourses and stressed their potentials for discussion but also their governing power. Some of the participants also noted that the discussion had created possibilities for reflection and new thoughts that would probably influence their own programming work in future.

As a general reflection on the feedback seminars it was considered useful that researchers from outside the regional administration helped to identify various perspectives that were prevailing within the organization and the different ways in which SD goals were addressed to in the RGP work. Before the feedback seminars took place the professionals involved were not conscious of the situation where different professional groups used different discourses in the context of the RGP process. They also stressed the lack of possibilities to discuss these kinds of issues in day-to-day practice, as well as the absence of time and organizational routines. During the feedback seminars some of the participants tried to trace the emergence of various problems and circumstances in the discussion. In some situations it became obvious that the participants were conscious of being in a development process and that they had begun a kind of self-reflection process that could eventually deliver new knowledge and working strategies. In other situations the participants remained constrained in traditional roles and routines and displaced the SD implementation task to other actors or other organizations.

Some of the interviewees also felt that they had been let down by national government. One reason for this, they claimed, was that the national goals and instructions were formulated in an abstract language and were therefore difficult to implement. The administratively-driven sectoral division of national policy also inhibits the possibilities for cross-sectoral innovation at the regional level.

Discussion and Conclusions

We have pointed out earlier in this chapter the ambiguity in SD goals and about how to weigh these goals in relation to other regional development goals. We can interpret the national government's SD goals in the context of regional development and RGP work as follows:

- the horizontal goals of environment, social inclusion shall be integrated and 'permeate' RGP work as essential aspects in SD implementation;
- evaluations shall be a learning undertaking at regional and national level. The learning focus therefore implies a view that some kind of change is needed as to

how regional authorities and other actors approach the regional development task and the new issue of SD.

In relation to the results of the evaluations initiated by the national government, our results also show that SD is now clearly on the regional development policy rhetoric agenda in the regional organizations studied here. We have however also shown, for example in the context of the results reviewed here, that the stance taken towards the national SD goals imposed on the RGP work in the regions has not yet, and perhaps may never, permeate the discourse of regional development. We see that the actors representing the horizontal goals are struggling to make their perspective and participation a significant part of the day-to-day regional development work. The integration of SD in the discourse and in the organization of SD and the RGP work is not fully present in any of the regions studied here. Issues and actors representing the horizontal goals and SD perspectives are still mainly seen as representing separate sectoral interests. What the focus group sessions however highlighted was that the government initiative to include SD as a part of regional development work has been the most important factor in placing SD firmly on the regional development agenda in the regions. SD would simply not be on the RGP agenda without the national government's efforts in this regard.

The question to be posed as part of the evaluation is of course whether the identified progress is good enough compared to governmental expectations at the launch of the SD implementation goals? The additional question moreover remains, for what and for whom is the evaluation undertaken in the context of this project valid and useful? Since the analyses of the results from the evaluation thus far are only preliminary we will restrict ourselves to a shorter and indicative discussion in what remains of this chapter.

Our preliminary evaluation results support the initial assumption that a socio-cultural evaluation view and the interaction-focussed methods are useful in understanding the assumed inertia in relation to capacity-building in relation to the implementation of SD goals in the RGP work of the Swedish regions discussed here. Our approach reveals some of the characteristics of the 'black box' that is the implementation conditions, in national government policy, guidelines and evaluations, especially in relation to SD and regional development. In order to evaluate whether SD permeates the RGP work and whether this change is taking place, evaluations of the programme contents are of course valid. If the goal is however to evaluate the changes taking place i.e., whether organizational capacity-building is ongoing and to what extent, then the socio-cultural perspective taken here and the focus on regional organizations seem more appropriate. The focus on capacity-building in the evaluation is also relevant as seen from the point of view that SD was only recently introduced as a goal of regional development work in Sweden. It is therefore important to reveal the progress in capacity-building as a concrete result of the government's policy compared to directly evaluating e.g. the progress of the SD performance of a regional development programme. Once capacity-building is in place the evaluation of SD performance will become more relevant. The first step in achieving SD as a result of regional development work is to ensure that the capacity-building mechanisms are in place.

The capacity-building process has in our study been viewed from a social-cultural perspective – thus highlighting e.g. the discursive conditions for SD integration as a possible evaluation indicator. The evaluation approach was also designed in order to provide opportunities for reflection and learning for the regional actors and thus make the evaluation process a part of the capacity-building process for the region. The preliminary results thus far from the feedback seminars and the final seminar indicate that reflection and learning *did* occur. Further follow-up of this matter is however necessary in order to be able to state this more firmly.

The basic argument for undertaking evaluations from a socio-cultural viewpoint and focussing on the day-to-day professional practice of organizations is thus that the usefulness of the knowledge gained is twofold. If used by the regional organizations the results are useful in terms of the critical reflection of the organization itself in order to promote change. If used by the government bodies it could also be useful in designing government politics and guidelines with a better 'resistance' awareness (as discussed by Sanyal 2005) thus enhancing the possibility for more effective policy implementation.

Another argument for evaluations based on a socio-cultural starting point is the fact that organizations are today 'no longer facing periods of stability in which members of an organization can slowly assimilate and adjust' (Henderson 2002). Thus the capacity to change is an important characteristic needed both at the individual and organizational level. As was indicated in the introduction to this contribution, regional policy in Sweden, as in many of the European Union member countries, has recently undergone a period of manifest change – in its content as well as in its organization. Thus the organizations coordinating regional development work in the 21 regions of Sweden face a situation where the capacity to change is paramount. The question that needs to be answered then is, 'how do individuals adapt, commit, and grow in an environment of discontinuous change?' (Henderson 2002). What Henderson (2002) concludes from his overview and analysis of theories concerning change in organizations, is that critical reflection is essential for 'transformational change' at both the individual and organizational levels. Transformational change in an organization 'involves radical changes in how members perceive, think and behave at work' (op. cit.). Henderson also shows that transformation at the individual level is essential for the outcome of the change process more generally, while yet another characteristic that Henderson (op.cit.) points out as being crucial to the effectiveness of change efforts is the achievement of commitment as opposed to simple conformity:

> Commitment, then, implies a personal decision to participate at an intellectual and emotional level, not a response to a directive from a higher authority or social pressures.

Changing structures, processes, policies, and extrinsic rewards may result merely in conformity rather than in the deeper commitment all the theorists speak of. (Henderson 2002)

Henderson's terminology presented above – commitment and conformity – can be used to label the different discourses presented previously. See Table 5.1 below.

Table 5.1   **The four discourses found, elements in the discourse and labelling the discourses in relation to Henderson's (2002) terminology – conformity and commitment**

Discourse	Elements in the discourse	Approach
An intellectual challenge	Driving force for regional development Prerequisite for regional capacity-building Innovation of cross-sector working methods Demanding creation of new knowledge and new attitudes Everyone is responsible for implementation	Commitment
The goals are accepted	Complicated and time-consuming Continuity is important Showing best practice	Conformity/commitment
A negative stance towards the national task	National guiding principles are restrictions SD a secondary task, an abstraction SD postponed to future work Power play	Opposition/conformity
The national task is uninteresting or impossible	SD in RGP is cosmetic SD and growth cannot be combined	Opposition

If we assume, as Henderson does, that commitment is a necessary prerequisite for transformational change to occur, and if we assume that if SD is to be integrated and to 'permeate' regional development work then this requires a radical change (so-called *transformational change*) in the regional governments selected in our study. Table 5.1 together with the results of our evaluation study show that the stance taken by regional actors towards the national government's assignment of implementing SD cannot be stated in terms of any of the three approaches, namely *conformity*, *commitment* and *opposition* respectively. One of the four discourses can be interpreted in terms of commitment and another one in terms of opposition (Henderson 2002). But the other two discourses balance between conformity or commitment and opposition or conformity respectively. This means that the national government's goal to integrate SD in regional development work is not yet established among all the regional actors. Further analysis may hopefully show whether the capacity-building has reached an acceptable level in the time period stipulated for the implementation of SD goals in regional development in Sweden.

Acknowledgements

The project Sustainable Development and Economic Growth – A Socio-Cultural Perspective on Regional Programming Processes is financed by Mistra. The formal project leaders are Tuija Hilding-Rydevik (Swedish University of Agricultural Sciences) and Karolina Isaksson (Royal Institute of Technology). The design and implementation of the project has however been a joint venture including the project leaders above and Eva Asplund, Maria Håkansson, Sylvia Dovlén, and Ann Skantze (all from the Royal Institute of Technology) and Sofie Storbjörk (University of Linköping). Malin Mobjörk (University of Västerås) has also participated with a separate study. The analysis of the focus group material, concerning the regional approach to the national goals has however been undertaken solely by the current authors Sylvia Dovlén and Tuija Hilding Rydevik, while the results have been discussed and processed by research team as a whole.

References

Argyris, C. (1990), *Overcoming Organizational Defences: Facilitating Organizational Learning*, Boston, Allyn and Bacon.
Argyris, C. (1993a), *On Organizational Learning*, Cambridge, Blackwell.
Argyris, C. (1993b), *Knowledge for Action: A Guide to Overcoming Barriers to Organizational Change*, San Francisco, Jossey-Bass.
Asplund, J. (1979), *Teorier om framtiden*, Stockholm, Liber.
Aspund, E. and Hilding-Rydevik, T. (2001), *Arena för hållbar utveckling – aktörer och processer*, TRITA IP FR 01-88, Stockholm, Gotab.
Asplund, E. and Skantze, A. (2005), *Hållbar utveckling i praktiken – möten, gränser, perspektiv*, TRITA-INFRA 05-011, Stockholm, Gotab.
Berger, P. and Luckmann, T. (1966), *The Social Construction of Reality*, London, Penguin.
Burr, V. (1995), *An Introduction to Social Constructionism*, London, Routledge.
Davoudi, S. (2005), Towards a Conceptual Framework for Evaluating Governance Capacities in European Polycentric Urban Regions', in D. Miller and D. Patassini (eds), *Beyond Benefit Cost Analysis: Accounting for Non-Market Values in Planning Evaluation*, Aldershot, Ashgate.

Dovlén, S. (2004), *Communicating Professional Perspectives – Local Government and Spatial Planning for Sustainability*, TIRTA-INFRA 04-040, Stockholm, KTH.

Fischer, F. and Forester, J. (1993), *The Argumentative Turn in Policy Analysis and Planning*, London, Duke University Press.

Fischer, F. (2003), *Reframing Public Policy, Discursive Politics and Deliberative Practices*, New York, Oxford University Press.

Foucault, M. (1970), *Diskursens ordning. Installationsföreläsning vid Collège de France 2 december 1970* (översättning Mats Rosengren), Brutus Östlings bokförlag.

Foucault, M. (1976), *The History of Sexuality Vol. 1*, London, Penguin Books.

Fudge, C. and Rowe, J. (2001), *Implementing Sustainable Futures in Sweden*, T:19:2000, Stockholm, The Swedish Council for Building Research.

Gergen, K.J. (2001), *Social Construction in Context*, London, Sage.

Hacking, I. (2000), *Social konstruktion av vad?*, Stockholm, Thales.

Hajer, M.A. (1995), *The Politics of Environmental Discourse. Ecological Modernization and the Policy Process*, Oxford, Clarendon Press.

Healey, P. (1997), *Collaborative Planning. Shaping Places in Fragmented Societies*, Hampshire, Macmillan Press.

Hedrén, J. (1994), 'Miljöpolitikens natur', *Linköpings Studies in Art and Science* No. 110, Linköping.

Henderson, G.M. (2002), 'Transformative Learning as a Condition for Transformational Change in Organisations', *Human Resource Development Review* 1(2), pp. 186–214.

Hilding-Rydevik, T. (2006), 'Environmental Assessment – Effectiveness, Quality and Success', in L. Emmelin (ed.), *Effective Environmental Assessment Tools – Critical Reflections on Concepts and Practice*, Blekinge Institute of Technology, Report No 2006:03, Report No 1 from the MiSt-programme, Karlskrona.

Isaksson, K. (2004), *Hållbarhet: Avvägningar prioritet utmaningar*, TRITA-INFRA 04-003, Stockholm, KTH.

Isaksson, K., Håkansson, M. and Hilding-Rydevik, T. (2006), '"Sustainable Growth" – Framing the Swedish Government Discourse on Regional Sustainable Development', in preparation.

Institutet för tillväxtpolitiska studier (2003), *Ex ante-bedömning av de regionala tillväxtprogrammen*, ITPS, Östersund.

Institutet för tillväxtpolitiska studier (2006), *Halvtidsutvärdering av de regionala tillväxtprogrammen*, A2006:013, Östersund.

Isaac, W. (1993), 'Dialogue, Collective Thinking and Organizational Learning', *Organizational Dynamics* 93, pp. 24–39.

Jacobsson, K. (2000), *Retoriska stider. Konkurrerande sanningar i dövvärlden*, Lund, Palmkrons förlag.

Khakee, A. (2002), 'Assessing Institutional Capital Building in a Local Agenda 21 Process in Göteborg', *Planning Theory and Practice* 3(1), pp. 53–68.

Kitzinger, J. and Barbour, R.S. (1999), 'Introduction: The Challenge and Promise of Focus Groups', in R.S. Barbour and J. Kitzinger (eds), *Developing Focus Group Research. Politics, Theory and Practice*, London, Sage.

Lakoff, G. and Johnson, M. (1980), *Metaphors We Live By*, Chicago and London, University of Chicago Press.

Lapintie, K., Maijala, O. and Rajanti, T. (2001), *Governance and Policy Instruments*, Work package 2, GREENSCOM, Helsinki University of Technology, Helsinki.

Maasen, S. and Wingart, P. (2000), *Metaphors and Dynamic Knowledge*, London, Routledge.

Miljödepartementet (2002), *Från vision till handling: Sveriges nationalrapport till världstoppmötet om agenda 21 och hållbar utveckling i Johannesburg 2002*, Nationalkommittén för Agenda 21 och Habitat, Stockholm.

Miljömålsrådet (2004), *Miljömålen – når vi dem?*, Utdrag ur Miljömålsrådets tredje årsrapport till regeringen om uppföljningen av miljökvalitetsmålen, From miljomal. nu 2006-04-21.

Mills, S. (1997), *Discourse*, London, Routledge.

Mobjörk, M. (2006), *Företagsorganisationers roll I regionalt utvecklingsarbete: en studie av RTP-arbetet i Dalarna*, Royal Institute of Technology, Div for Infrastructure, in print.

Morgan, D.L. (1993), *Successful Focus Groups, Advancing the State of the Art*, Thousand Oaks, CA, Sage.

Morgan, D.L. (1998), *Focus Group Kit. Vol 1, The Focus Group Guidebook*, Thousand Oaks, CA, Sage.

Morgan, G. (1997), *Images of Organization*, Thousand Oaks, CA, Sage.

Morgan, G. (1999), *Organisationsmetaforer*, Lund, Studentlitteratur.

Naturvårdsverket (2000), *Utvärdering av hållbar utveckling från ekologisk synpunkt i tillväxtavtalen*, Naturvårdsverket, Stockholm.

Naturvårdsverket (2001), *Bredda perspektiven. Miljöintegration i tillväxtarbetet*, Rapport 5163, Stockholm.

Naturvårdsverket (2002), *Samverkan miljö och regional utveckling: En studie av självstyrelseorgan, samverkansorgan och länsstyrelser*, Rapport 5189, Stockholm.

Naturvårdsverket (2003), *Vägar till hållbar regional utveckling. Processer för hållbarhet i tre län*, Rapport 5324, Stockholm.

Nordregio (1999), *Ex-ante utvärdering av utkasten till regionala tillväxtavtal*, Näringsdepartementet, Nordregio, Stockholm.

Nordregio och Ledningskonsulterna and SIR (2000), *Regionala tillväxtavtal: Utvärdering av förhandlingsprocessen i sju län och på central nivå*, Nordregio, Stockholm.

Nutek (2004), *Regionala tillväxtprogrammen 04. På väg mot hållbar tillväxt?*, Stockholm.

Nutek (2005a), *Regionala tillväxtprogrammen 05. På väg mot starka regioner?*, Stockholm.

Nutek (2005b), *Tre strategier för hållbar utveckling. Miljö, jämställdhet och integration i regional utveckling*, B 2005:1, Stockholm.

Nutek (2005c), *Regionala utvecklingsprogram, RUP – ett metodutvecklingsarbete*, Slutrapport, info No. 062.2005, Stockholm.

Näringsdepartementet (1998), *Regionala tillväxtavtal – näringslivet i fokus*, Promemoria 1998-05-06, Stockholm.

Näringsdepartementet (2001), *Rapport om tillväxtavtalen – Första året*, Ds 2001:15, Stockholm.

Näringsdepartementet (2002a), *Riktlinjer för arbetet med regionala tillväxtprogram*, Bilaga till regeringsbeslut 2002-11-14, No. II 12, Stockholm.

Näringsdepartementet (2002b), *Rapport om tillväxtavtalen – Andra året*, Ds 2002:34, Stockholm.

Näringsdepartementet (2003), *Rapport om tillväxtavtalen – Tredje året*, Ds 2003:43, Stockholm.

Näringsdepartementet (2004), *Rapport om tillväxtavtalen – Fjärde året*, Promemoria 2004-08-09, Stockholm.

Owens, S. and Cowell, R. (2002), *Land and Limits. Interpreting Sustainability in the Planning Process*, London and New York, Routledge.

Regeringskansliet (2003), *Förordningen om regionalt utvecklingsarbete*, Förordning 2003:595, Stockholm.

Regeringskansliet (2006), *En nationell strategi för regional konkurrenskraft och sysselsättning 2007-2013*, Bilaga till regeringsbeslut, ärendenr: N2006/5089/RUT, Stockholm.

Regeringens skrivelse 2003/2004:129, *En svensk strategi för hållbar utveckling – ekonomisk, social och miljömässig*, Stockholm.

Regeringens proposition 1997/98:62, *Regional tillväxt – för arbete och välfärd*, Stockholm.

Regeringens proposition 2001/02:4, *En politik för tillväxt och livskraft i hela landet*, Stockholm.

Richardson, T. (2002), 'Freedom and Control in Planning: Using Discourse in the Pursuit of Reflective Practice', *Planning Theory* and *Practice*, 3(3), pp. 353–361.

Roberts, P. (2006), 'Evaluating Regional Sustainable Development: Approaches, Methods and the Politics of Analysis', *Journal of Environmental Planning and Management* 49(4), pp. 515–532.

Sanyal, B. (2005), 'Planning as Anticipation of Resistance', *Planning Theory* 4(3), pp. 225–245.

SOU (1997), *Agenda 21 I Sverige fem år efter Rio – resultat och framtid*, Miljödepartementet, Stockholm.

Storbjörk, S. (2004), *Att prioritera miljöfrågor? Kommunalpolitiker och det lokala miljöarbetets villkor*, TRITA-INFRA 04-015, Stockholm, KTH.

Statskontoret (2004), *Det regionalpolitiska experimentet, Lärande nätverk för regional utveckling*, Rapport 2004:5, Stockholm.

Säljö, R. (2000), *Lärande i praktiken: Ett sociokulturellt perspektiv*, Stockholm, Prisma.

Valve, H. (2003), *Social Learning Potentials Provided by EU Rural Development Programmes – A Comparative Study on Three Institutionalisation Processes*, Acta Universitatis Tamperensis; 918, Tampereen yliopisto, Tampere.

Vedung, E. (1991), 'Utvärdering och politik', in B. Rothstein (1991), *Politik som organisation*, SNS förlag, Stockholm.

Wibeck, V. (2000), *Focusgrupper. Om fokuserade gruppintervjuer som undersökningsmetod*, Lund, Studentlitteratur.

Wood, C. (1999), 'Comparative Evaluation of EIA Systems', in J. Petts (ed.), *Handbook of EIA. Volume 2. EIA in Practice: Impact and Limitations*, Oxford, Blackwell Science.

Chapter 6

An Ex-Ante Evaluation of an Urban Project through Property Value Increases: An Hedonic Price Approach

Roberto Camagni and Roberta Capello

Introduction[1]

This chapter focuses on the economic assessment of a relevant urban infrastructure project using expected increase in land values as indicators of the potential benefits to the urban population.

The project concerns the underground laying of a part of the urban rail track in the city of Trento, situated on the Alps, in the north-eastern part of Italy, along the Brenner Corridor Verona-Innsbruck-Munich. In fact the rail represents a barrier inside the city in the north-south direction, generating a low-accessibility, low-quality urban land belt running between the rail itself and a second north-south barrier represented by the river Adige (Figure 6.1).

The recent land-use plan of the city, produced under the direction of the Catalan planner Busquets, makes the proposition of laying underground the central part of the rail line, for a length of 2 kilometres, in order to reunite the two parts of the city, create new accessibilities with respect to the city centre (which developed in an eastward semicircle around the central station) and new urban amenities – a boulevard replacing the rails, new high quality buildings and recreation facilities. The policy question regards the economic viability of the project in terms of cost-benefit balance for the local community, given the huge public investment requested.

1   Though the chapter is the result of a common research project, R. Camagni wrote the second section, while the remaining sections were written by R. Capello. A previous version of the chapter was presented at the 6th Regional Science Association International World Conference, held in Port Elisabeth, South Africa, 14–16 April 2004. Authors wish to thank a referee for his useful comments. The chapter is based on the results of a research carried out by the authors for the Municipality of Trento. Authors are grateful to Cristina Lira (University of Milan), Marzia De Tassis, Giuliano Frantoi (University of Trento) for the helpful role in collecting data in Trento, and to Alessia Spairani (Politecnico of Milan) for her support in processing data in the simulation phase. Sincere thanks go to Giuliano Stelzer and Clara Campestrini (Municipality of Trento) for the data on land use in Trento.

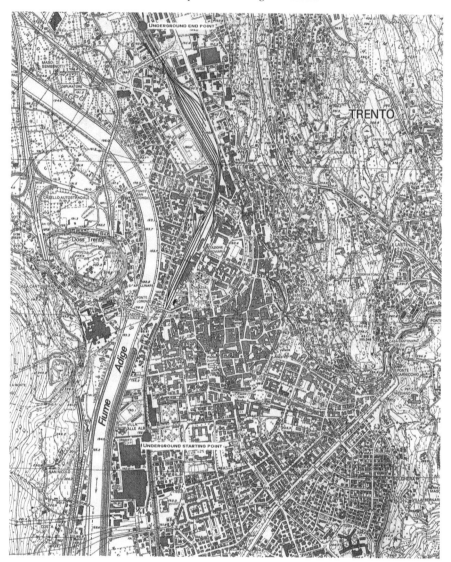

Figure 6.1   The city of Trento

This chapter proposes an assessment of the general social benefits through the evaluation of the expected increase in land and real estate values. In fact, all the potential advantages of the project refer to elements that have a direct impact on these values, namely:

*   increase in accessibility towards the city centre in a wide urban belt;
*   increase in urban quality, as a consequence of the visual disappearance of the rail, of the provision of new public spaces, of the rehabilitation of buildings and of new modern constructions;

- increase in environmental quality, due to noise abatement and provision of new green spaces;
- increase in expected profits on new retail and commercial activities;
- improvement in the general external image and attractiveness of the city.

This main idea is related with well-known theorems in land rent theory, linking net benefits supplied by the city to total differential urban rent, and, from an operational point of view, can be implemented through hedonic prices modelling and econometrics. Therefore, section two is devoted to highlight the underlying theory of land rent, including the distinction between 'differential' land rent *strictu sensu*, operating at the micro-territorial level, and 'absolute' land rent, operating at the scale of the entire city. Sections three and four present the hedonic price model and the results of the estimation and simulation procedures; section five presents some conclusions.

It will be shown that total advantages derived from the project, approximated by the increase in total differential land rent, are likely to be of a comparable size with respect to engineering and adjustment costs, and that very limited (and likely) increases in the general appreciation of the city, approximated by the increase in absolute rents, would easily generate a full cost-benefit equilibrium at the highest level of predicted costs.

The Logics of the Exercise: Differential and Absolute Land Rent

The first logical step of our evaluation exercise consists in linking up with mainstream theory of urban land rent in the Alonso-Fujita tradition (Alonso 1964; Fujita 1989), considering the city of Trento – fairly monocentric – as being in an equilibrium condition, and evaluating the increase in total benefits as a consequence of the implementation of the project.

Except for the last element indicated in the introduction among the expected effects of the project, namely increase in the general image of the city, all other expected effects have a direct impact on differential rent. Accessibility is of course the core of the theoretical model, but also urban quality and urban efficiency are expected to generate an impact on land rent differentials, through the increase in the utility function of households and profitability of firms.

On this subject a well-known theorem of rent theory states that in equilibrium the city's surplus equals, or is absorbed by, the total differential land rents (Fujita 1989, p. 151). The difference between the total output or income generated by the city utilizing its equilibrium population and total population cost – given by the sum of transport costs, costs for purchase of all other goods and opportunity cost of land, viz. agricultural land rent – represent the city's surplus; in equilibrium, it is equal to the total differential land rent, appropriated by landlords.

In case that a city, other things being equal – namely marginal value product of labour and transport conditions – presents a higher level of amenities and urban quality with respect to another city, it will show a greater population supply curve at each per-capita income level and therefore a greater equilibrium population and a greater value of total land rents (ibid.).

Another theorem of land rent theory that could be worth considering and mentioning in our case refers to the relationship between urban size, land rent and urban fixed or infrastructure costs. Taking the case of a utility maximizing community deciding about the optimal city size in presence of infrastructure costs, the theorem shows that optimal population is the one where fixed costs equal urban surplus at its peak, and that it coincides with the equilibrium population of the 'open city' model (ibid., p. 156). In case that urban population is given and fixed, the same theorem defines the level of the maximum investment in order that utility of urban population is maximized (the 'closed city' case); this case proves to be very similar to our policy case, in which policy-makers try to define the maximum infrastructure expenditure that equals urban benefits, or increase in urban rents, given the present level of urban population.

Interestingly enough, this case reminds us the so called 'golden rule of local public finance' or Henry George theorem, which states that, in equilibrium, expenditure in public goods equals differential land rent so that a confiscatory taxation would meet perfectly the budget constraint of the local community. This is not to say that, in case of an expected balance between costs and benefits of the project, the municipality of Trento will automatically find the means to finance the infrastructure cost, but that in economic terms the project is viable and that it could be partially financed through a partial taxation of 'betterments' or 'windfalls'.

The possibility that the project would not just determine a differential advantage in a specific part of the city – in terms of accessibility, urban quality and profitability of economic activities – but also generate a potential upgrading in the general image of the city, and therefore increase its attractiveness with respect to new households or firms, opens an interesting issue in land theory, which in our opinion remains imperfectly explored. We are referring to the concept of 'absolute' land rent.

The issue and the term itself derive from the Marxian tradition, but more recently found new theoretical justifications. The issue regards the possibility of encompassing and interpreting all empirical manifestations of urban land rent with the sole theoretical tool of differential rent.

Main arguments supporting the necessity of adding a new concept, of an aggregate nature, operating simultaneously on the entire urban land, may be listed as follows:

a) The empirical fact that urban land rent at the boundary of the city is higher than agricultural land rent reminds us about the existence of an urban condition and an urban advantage, possibly connected to agglomeration effects or overhead capital accumulation. Similarly, Marx pointed out, referring to the Ricardian theory of agricultural rent, that rent on marginal land is not nil. The Marxist school (Lipiez 1974, 1978; Harvey 1973; Topalov 1984) generally refers this fact to the presence of a class monopoly by landlords, both in rural and in urban areas, or by urban developers, building on some reflections made sometimes by Marx; but Marx himself criticized this theoretical hypothesis, as in this case rent would become a cost element or a tax instead of a distribution element, thus contradicting a pillar of classical economics (Marx 1867, ch. 45).[2]

b) Making an intellectual experiment and supposing a city perfectly accessible in all directions, one would derive from the mainstream model of differential land rent a condition of zero differential rent,[3] while such a city would be highly appreciated by both households and firms (Camagni 1996, para. 9.4).

c) The previous point implicitly refers to scarcity: if all urban sites were of similar quality in terms of accessibility or environmental amenities but not scarce, they would behave like natural assets or commons, and their price would come to zero; but a rent would rise if they were rare compared to the total demand. Along these same lines Piero Sraffa, in his general model elaborated in *Production of Commodities by Means of Commodities*, in his Chapter XI on rent presents two concepts of rent: one linked to differentiated land fertility and decreasing returns, giving rise to *n* production methods and *n* rent levels – what we call differential rent – and one deriving from scarcity of land, in presence of the same fertility in all parcels – what we could call 'absolute', scarcity rent (Sfraffa 1960).

d) An explicit and direct link between absolute land rent and aggregate scarcity in an urban environment was established by Fahri (1973) and Scott (1976) in models of a neo-Ricardian nature. Scarcity introduces a condition on non equilibrium in the abstract general model of land rent, justified by the time length requested in order to implement urbanization processes and supply of urban overhead capital.

e) Still referring to a condition of scarcity and to the rigidity of the urban supply curve in the short run, a relevant case in real life emerges as a consequence of the increase in the aggregate demand for urban locations, due to the expected increase in urban quality or efficiency (something similar to what would happen in the case study considered). In the short run, given the (fixed) size of the existing city, the differential land rent curve would shift upwards, showing a base of extra-rent on all sites and distances from the centre that could be called 'absolute rent by demand of city' (Camagni 1996, par. 9.4). In the longer run the city would expand and at each distance from the centre a similar extra-rent will show up with respect to the initial situation; this extra-rent, which in the logics of the model is technically a 'differential' rent (with respect to the agricultural rent at the new boundary of the city), economically does not depend on differential savings on travel time to the city centre with respect to nearby parcels in a micro-territorial view (differential rent *strictu sensu*) but on aggregate factors depending on the general efficiency/attractiveness of the city, that generate aggregate demand for urban locations and make the city

---

2   Unfortunately, Marx's solution referring to the transformation of 'values' into prices and the permanence of a part of the surplus value into the price of agricultural product is not acceptable, even accepting his theory of value; see Camagni 1996, ch. 9.4.3.

3   And a city of infinite size.

grow.[4] Once again, it could be argued that an absolute land rent adds to the differential rent *structu sensu*.

In the logics of the exercise presented in this chapter, we mainly deal with a differential rent concept (*strictu sensu*), and in particular with its increase due to an upgrading in accessibility, amenities, urban quality and profitability in specific sites, as a consequence of the implementation of the project. But we consider also a potential macro-phenomenon, linked to an absolute rent concept, emerging from the generalized increase in urban attractiveness. This effect is only indirectly quantified, through the calculation of the increase in average land prices that would equal total benefits to the highest level of expected costs.

The empirical evidence on land values and their potential increase is collected through the indicator of prices per square metre of commercial, office or residential floorspace; in correct terms, these prices do not represent rents but their capitalization, even if, for the sake of simplicity, we will refer to them as either prices, values or rents.

Hedonic Price and Urban Project Appraisal

Hedonic Price Methodology: Strengths and Weaknesses

The aim of the appraisal methodology is the quantitative estimate of the property value increase due to the implementation of a large urban project. The methodology is applied to a simulation of the impact that the planned project of laying below ground level the section tracks for 2 kilometres within the city of Trento will have on property value increases. The first step towards such an estimate is to attribute a monetary value to those elements influenced by the project, i.e. accessibility and urban quality.

A well known methodology to find a monetary value to elements, like accessibility and urban quality, which miss a direct market, is hedonic price. The logic on which this methodology rests is that composite and heterogeneous goods exist. The value of each single trait composing the good is capitalized in the value of the good itself; the hedonic price methodology allows the separation of the single values of each trait compositing the good.

Housing is widely recognized as a heterogeneous good, composed not only of internal characteristics of the structure itself (central heating, lift, building quality, number of rooms, number of bathrooms), but also of location-specific characteristics (relative accessibility to the city centre, quality of urban space). While the internal-specific characteristics, related to the structure itself, are

---

4   Less convincing are other arguments proposed by Alan Evans (1988) and David Harvey (1973), linking absolute rent to the need to cover transaction costs or to expectations on future land price increases; in fact costs are included in land prices but are not distributive income shares, and expectations have to be based either on localized or aggregate improvements on the urban scene.

valued by the construction costs, the location-specific characteristics linked to the house location represent the pure land rent, generated by the willingness to pay for more central location or for location in more qualified areas (from both the environmental and the aesthetic point of view). The hedonic price methodology enables us to derive exactly those implicit values that the housing market attributes to each single characteristic, and therefore also to accessibility and urban quality.

Since Rosen's 1974 contribution on hedonic prices,[5] the methodology has been widely applied, especially in the American and Anglo-Saxon world. The interest in the methodology is witnessed by its application in different fields and for different purposes, namely:

- To *estimate environmental externalities* in urban areas, i.e. local traffic, congestion, noise and air pollution, and thus to obtain the willingness to pay in order to avoid such externalities.[6] More recently, hedonic prices have been applied to urban quality of life studies, in order to provide a weight to single environmental indicators when they have to be summed into one synthetic index.[7]
- To *estimate the quality of public goods*. The willingness to pay to have a certain quality in public services (i.e. a school of a certain standard) in an area represents a value of the service, helpful in planning perspectives.[8]
- To *assess ex-post urban policies*. The implementation of a new railway station, of new parks and public areas in general, of a new land use plan are all urban policies which have been assessed by estimating the increase of property values after the implementation of the project.[9]

Despite its attractiveness and wide application, more recently weaknesses of this methodology have been put forward, all related to the issue of consistency and robustness of the estimated parameters. The importance of robust estimations of the parameters is particularly critical in hedonic price methodology for the meaning attributed to such parameters. Assuming the existence of a willingness to pay function, which shows decreasing marginal utility of attribute z at fixed utility index and income ($W(z)$ in Figure 6.2), the estimated parameters represent the marginal increase in the willingness to pay for having a unity of increase in

---

5   The pioneering contributions on hedonic prices are present in Rosen 1974 and Freeman 1979. Recently, Andrew Court has been recognized as the first author dealing with the hedonic price methodology, see Goodman 1998.

6   Among others, see Ridker and Henning 1967; Wilkinson 1973; Freeman 1971; Corielli et al. 1996 in Cheshire and Sheppard 1995, a methodology to estimate distance in non-isotropic space conditions.

7   Pioneering studies are present in Roback 1982; Blomquist et al. 1988; Gyourko and Tracy 1991.

8   See Svitanidou 1996; Cheshire and Sheppard 2002.

9   See Cheshire and Sheppard 1998; Bates and Santerre 2001; Bowes and Ihlanfeldt 2001.

attribute z, and therefore are interpreted as prices of the marginal attribute. A consumer will maximize his utility when the willingness to pay function is tangent to the price function representing what a consumer has to pay for the general attribute z in the market ($p(z)$ in Figure 6.2); this means that the quantity of z chosen will be such that the willingness to pay equals the hedonic price of the attribute. Difficulty arises when errors perturb the consumption level of the attribute away from the optimal level z^*. Each statistical bias in the estimate of the hedonic price in fact perturbs its consumption level away from the optimality point to either a lower level z_2 or a higher level z_2; with a non-linear supply function the choice of a different attribute consumption level enlarges the difference between the estimated and real price of the attribute.[10]

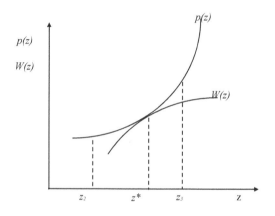

Figure 6.2   The supply function and the willingness to pay function

In the literature, sources of mis-specification of the parameters are identified in:

- *Omitted variables* in the estimate of hedonic prices, generally due to the lack of detailed information. If this is the case, the results obtained for each single characteristic are influenced by the omission of other important characteristics, a problem of mis-specification of the model common to all econometric exercises. As said before, in the case of hedonic prices, the mis-specification is more critical because of the importance attributed to the value of each single coefficient, as equilibrium price of that characteristic.[11]
- *Functional form of the hedonic function*, traditionally exogenously imposed as a linear form. The choice of the functional form influences the results obtained

10 Cf. Cheshire and Sheppard 1998; Deaton and Muellbauer 1980; Brown and Rosen 1982; Linneman 1980.
11 See Cobb 1984; Small 1975; Epple 1987.

for the parameters; therefore, an exogenous choice is somehow critical. In this respect, the literature suggests the use of more advanced and more flexible functional forms, applied also in this study.[12]

- The *geographical extension of the sample*, influencing the variance of cases in urban space, and therefore the estimates. A larger disparity of cases and conditions guarantees a more precise estimate of the hedonic price.
- *Spatial autocorrelation in the estimate of hedonic prices*, both as spatial structural dependence of the dependent variable, and a spatial dependence in error terms.[13] Tests to control for spatial autocorrelation are nowadays available and applied to hedonic price estimates too.

In this work, specific attention has been paid in order to keep these weaknesses under control, by:

- Choosing randomly the sample of 100 residential houses and 50 commercial building units scattered around the whole urban area *in order to guarantee a wide geographical extension of the sample*.
- Collecting via a very wide questionnaire the information concerning both the internal-specific characteristics of houses and flats, and the location-specific characteristics, *in order to limit the number of omitted variables*. The questionnaire was filled in by a group of researchers visiting the houses, the surroundings and covering the distance to the city centre with different modes (by car, by public transport), and in this way collecting one of the main generally missing information in official statistics, i.e. the distance to the city centre.[14]
- Applying spatial autocorrelation tests to the model estimates, *in order to avoid spatial autocorrelation biases*.
- Using flexible forms of hedonic price functions, as discussed in the next section, *in order to avoid exogenous choices of the functional form*.

The Functional Form and the Database

A potentially serious source of bias in hedonic price studies is associated with functional form misspecification. Theory provides no strong restrictions on the functional forms of hedonic functions. In general, empirical studies have specified these forms as either linear or log-linear and have not examined the statistical appropriateness of these forms or the bias they may induce in the marginal characteristic price estimates (Linneman 1980). Recently, a specific functional form has been introduced for the estimate of hedonic prices; Box and Cox (1964) developed a statistical model which determines the functional specification providing (endogenously) the best fit in terms of log likelihood:

---

12 See Bender et al. 1980; Jackson et al. 1984; Box and Cox 1964.

13 On spatial autocorrelation on the stimate of hedonic prices, see Tse 2002. On spatial autocorrelation tests, see Anselin 1988; Anselin and Florax 1995.

14 Cf. Linneman 1980.

$$\frac{y^{\theta} - 1}{\theta} = K + \sum_{i \in D} \beta_i x_i + \sum_{j \in D} \beta_j \frac{x_j^{\lambda} - 1}{\lambda} + \varepsilon$$

(6.1)

where:

y = the housing value per m^2;
x_i = set of indices for dichotomous characteristics;
x_j = set of indices for continuous characteristics;
θ, λ = the standard parameters of the Box-Cox functional form;
K = constant;
β_i, β_j = parameters to be estimated;
ε = error term.

Using the log likelihood function associated with this specification one can solve for the maximum likelihood estimates of the respective sets of coefficients and power transformation factors; if a transformation factor equals one, the associated functional form of the variable is linear. If the power factor approaches zero, the functional form approaches natural logarithmic form.

In order to represent the willingness to pay for each trait, parameters β_i, β_j have to be transformed, given the Box-Cox functional form:[15]

$$\beta'_j = \beta_j y^{1-\theta} x_j^{\lambda-1}$$
$$\beta'_i = y^{1-\theta}$$

(6.2)

β'_j and β'_i represent the willingness to pay for each specific trait; in particular, for the dichotomous traits these trait prices reflect the mean marginal valuations of changing from zero to one: if x_i represents the existence of the railway within 200 meters for the house, coefficient β'_i associated to that trait represents the willingness to pay a higher rent for been located away from the railway. For continuous traits the marginal traits prices describe the change in total valuation associated with a small change in the trait; if x_j represents the accessibility by car to the city centre at a certain distance to the centre, coefficient β'_j represents the willingness to pay a higher rent for being located one minute nearer to the centre.

The elasticity of the housing value to each trait is calculated as follows:

$$e_{yxj} = \beta'_j x_j / y$$

(6.3)

$$e_{yxi} = \beta'_i / y$$

(6.4)

respectively for continuous (6.3) and dichotomous (6.4) traits.

The database used to estimate equation 6.1 is presented in Table 6.1, for both the residential and commercial building markets. For what concerns the residential housing market, a particular attention has been devoted to the collection of

---

15 In Appendix 2 the mathematical solution is presented.

data on house prices. The official source of property agencies is not useful in this respect, since prices are generally overestimated for commercial reasons; the official price registered during the transaction is also not a good source, because it is underestimated for fiscal reasons. The more reliable source in this respect is the value declared for the mortgage contract, a much more realistic value of a house. The collaboration with two local banks has allowed us to obtain such information. Characteristics influencing housing prices can be collected into three groups:

*   *Internal-specific* characteristics: presence of a lift, presence of private heating, restructured building, presence of private parking, type of building (single owned house, flat in a shared owned building).
*   *Neighbourhood-specific* characteristics: *urban space quality* (i.e. width of the street with respect to the height of the buildings; presence of historical buildings), *environmental quality in the street* (both positive environmental elements, like the presence of trees and green areas, and negative, like the presence of local traffic, of air and noise pollution) and *presence of public services* (like schools, university, post office).
*   *Location-specific variables*: distance to the city centre by car or by public transport.

The housing market for commercial buildings is more focused on traits that can influence expected revenues. Data on the prices of buildings for commercial business have been obtained by capitalizing the annual rent value of 1998 provided by the Municipality of Trento.[16] In a strict neoclassical logic à la von Thünen/ Alonso, urban rent is represented by the willingness to pay for a more central location once production costs, transport costs and a certain level of profits have been subtracted to revenue. With expected revenues highly dependent on location-specific traits, equation 6.1 has been estimated by inserting location characteristics that influence business activity, and therefore revenues. Table A6.1a in Appendix 1 presents the variables used and their statistical description.

Hedonic Function Estimates

Accessibility and urban quality in the residential housing market   Table 6.2 presents the results of the estimated hedonic function in equation 6.1. In general, the signs of the hedonic function coefficients are as expected: the presence of private heating, the restructuring of the building and the type of building (civil, exclusive, private house) explain much of the value of the house.

Interestingly enough, the presence of the railway within 200 meters and the urban space quality (expressed by the ratio between building height and width of

---

16 Property agencies estimate a capitalization rate between 4 per cent and 15 per cent, depending on the use destination of the area and of the building; in this work we have chosen a capitalization rate of 8 per cent, representing a mean of the proposed rates. The capitalized value of the building (R) has been obtained by applying formula of perpetual rent: $R=r/i$, where r is the annual rent and i is the capitalization rate.

Table 6.1a   Available data for the residential housing model

Data	Variables
Dependent variable:	
Housing price per m^2	Euro
Internal-specific characteristics	
presence of a lift in the building	yes =1; no=0
presence of a private heating system	yes =1; no=0
presence of garage or private parking	yes =1; no=0
restructured house/flat	yes =1; no=0
type of building * single owned house * flat in a shared owned building	 yes =1; otherwise =0 yes =1; otherwise =0
Urban space-specific characteristics	
ratio of building height/width of the street	If the ratio < ½ =1 If the ratio >½ =0
presence of historical buildings and monuments along the street	yes =1; no=0
typology of buildings in the street	If more than 50% of the buildings in the street are single private houses = 1; otherwise=0 If more than 50% of the buildings in the street are civil buildings with several floors = 1; otherwise=0 If more than 50% of the buildings in the street are exclusive buildings = 1; otherwise=0
Environmental-specific characteristics	
dense traffic street	yes =1; no =0
pedestrian street	yes =1; no =0
presence of parked vehicles on the pavement	yes =1; no =0
presence of green areas (trees, gardens)	yes =1; no =0
presence of the railways within 200 meters	yes =1; no =0
presence of shops	yes =1; no =0
presence of post office	yes =1; no =0
presence of schools	yes =1; no =0
presence of the university within 250 meters	yes =1; no =0
presence of a bus stop within 200 meters	yes =1; no =0
presence of the railway station within 200 meters	yes =1; no =0
Localization-specific characteristics	
distance to the city centre (kilometres)	Kilometers
distance by car to the city centre (minutes)	Minutes
distance by bus to the city centre	Minutes

Table 6.1b   Available data for the model of the commercial building market

Data	Variables
Dependent variable:	
Price of buildings for commercial business per m^2	Euro
Localization characteristics	
distance between the commercial business and the central station	If less than 500 meters =1; otherwise = 0
distance between the commercial business building and the bus stop	Meters
distance to the city centre (meters)	Meters
distance by car to the city centre (minutes)	Minutes
distance by bus to the city centre (minutes)	Minutes
distance by walking to the city centre	Meters
localization on a commercial street	On a commercial street = 1; otherwise=0
localization near a commercial street	If near to a commercial street = 1; otherwise= 0
localization far from a commercial street	If far from a commercial street = 1; otherwise = 0
Characteristic of the area	
central area	If central area = 1; otherwise = 0
medium distance area	If out of the historical centre but not in the periphery (with an easy and fast access to the city centre) = 1; otherwise = 0
peripheral area	If peripheral area = 1; otherwise = 0

the street) are significant in explaining the housing value (at 0.5 per cent only in model four); moreover, even the location of the house with respect to the distance to the city centre (measured in minutes by car), has a central role in determining the value of the market.

The first model of Table 6.2 reports only variables which show a statistically significant value; in the other models of Table 6.2 some potentially interesting variables are inserted, which however have turned out to be non-significant. Table 6.2 also contains the power transformation factor values; the functional form approximates an inverse linear function. Table 6.2 also contains the results of the spatial autocorrelation tests, assuring the absence of spatial autocorrelation.

Table 6.3 contains hedonic prices. The estimated willingness to pay for the sample mean is:

- For private heating, €157 per m^2 (a value that changes from €124 to €167 per m^2 depending on the model chosen).
- For a restructured building, €186 per m^2.
- For the presence of a lift, €228 per m^2.
- For the urban quality, €336 per m^2.

Table 6.2   Estimates of the hedonic function for the residential housing market (*)

Variables	Model 1	Model 2	Model 3	Model 4	Model 5
Constant	1.45	1.42	1.45	1.42	1.42
Presence of a private heating system	0.0004 (0.086)	0.0004 (0.071)	0.0005 (0.102)	0.0004 (0.070)	0.0003 (0.132)
Restructured building	0.005 (0.014)	0.0004 (0.026)	0.00056 (0.016)	0.0005 (0.014)	0.0004 (0.027)
Presence of a lift in the building	0.0007 (0.005)	0.00064 (0.005)	0.0007 (0.006)	0.0006 (0.008)	0.0005 (0.011)
Presence of the railway within 200 meters	-0.0005 (0.034)	-0.0004 (0.051)	-0.0004 (0.038)	-0.00045 (0.046)	-0.004 (0.035)
Quality of the street**	0.001 (0.058)	0.009 (0.067)	0.001 (0.055)	0.0010 (0.045)	
Presence of parked cars on the pavement			-0.00018 (0.773)		
Quality of urban space and enrivonmental quality in the street***					0.0012 (0.091)
Presence of green areas				0.0004 (0.166)	
Presence of trees		0.0001 (0.48)			
Distance to the city centre by car (minutes)	-0.00015 (0.029)	-0.000147 (0.026)	-0.00017 (0.030)	-0.0001 (0.028)	-0.00016 (0.077)
Lambda	0.41 (0.6)	0.38 (0.6)	0.32 (0.7)	0.39 (0.6)	0.24 (0.7)
Theta	-0.68 (0.01)	-0.69 (0.009)	-0.68 (0.01)	-0.69 (0.009)	-0.69 (0.009)
R-square	0.31	0.31	0.31	0.32	0.34
Number of observations	100	100	100	100	100
Theta = Lambda = –1	-711.0**** (0.122)	-710.7**** (0.141)	-710.8**** (0.149)	-709.9**** (0.143)	-711.1**** (0.185)
Theta = Lambda = 0	-713.9**** (0.004)	-713.8**** (0.004)	-713.8**** (0.005)	-713.1**** (0.004)	-714.3**** (0.004)
Theta = Lambda = 1	-741.33**** (0.000)	-741.3**** (0.000)	-741.3**** (0.000)	-740.8**** (0.000)	742.1**** (0.000)
Test: Spatial error:					
Moran's I	0.325 (0.74)	-0.235 (1.186)	0.438 (0.661	0.268 (0.789)	0.983 (0.326)
Lagrange Multiplier	0.000 (0.996)	0.096 (0.757)	0.002 (0.963)	0.001 (0.978)	0.117 (0.732)
Robust Lagrange multiplier	0.000 (0.997)	0.097 (0.756)	0.002 (0.965)	0.001 (0.977)	0.116 (0.733)

Table 6.2 cont'd

Variables	Model 1	Model 2	Model 3	Model 4	Model 5
Test: Spatial lag:					
Lagrange multiplier	1.636	1.965	1.622	1.127	1.947
	(0.201)	(0.161)	(0.203)	(0.288)	(0.163)
Robust Lagrange multiplier	1.636	1.966	1.621	1.127	1.946
	(0.201)	(0.161)	(0.203)	(0.288)	(0.163)

Dependent variable: house price by m^2.
* Beneath each parameter estimate is given the p-value for the parameter in parentheses.
** Ratio between building height of buildings/width of the street < ½.
*** Interaction between the presence of trees and gardens and the ratio between building height/width of the street < ½.
**** Restricted likelihood.

- For not being located within 200 meters from the railways, €168 per m^2.
- Finally, for being located one minute by car nearer to the city centre, €9.4 per m^2, at an average distance of 15.5 minutes by car from the city centre.

Accessibility and Urban Quality in the Market for Commercial Buildings In the market for commercial buildings, accessibility and urban quality are valued differently than in the residential housing market; for this reason equation 6.1 was estimated once again for commercial buildings, and Table 6.4 presents hedonic function estimates for three models fitted to the data for commercial business buildings.

As expected, hedonic price estimates are very different from those obtained in the market for residential houses; different variables influence the price of commercial buildings from those influencing residential houses. In particular, in this case a great importance is attributed to the location of the building near commercial streets, since it influences potential revenues, and therefore the willingness to pay higher rents. Moreover, as expected, the location near the railway does not affect the price of the building, since it has no influence on the expected profits of the business activity.

Hedonic prices are presented in Table 6.5; the willingness to pay for not having the business located far from a commercial street is €1.524 per m^2; for not having it in medium areas is equal to €1.144 per m^2, to have it one minute nearer to the city centre by public transport is €29.6 per m^2 (for an average distance of 13.12 minutes by public transport to the city centre).

For what concerns accessibility, the hedonic price changes according to the relative location; Table 6.6 reports a willingness to pay for being located one minute nearer to the city centre calculated at different distances to the city centre for the observations we have; the results show that the gradient of urban rent is negative, as widely explained by the traditional Von Thünen and Alonso models (Alonson 1964; Von Thünen 1826).

Table 6.3   Hedonic prices for the sample mean in the market for residential housing (values in €s)

	Model 1	Model 2	Model 3	Model 4	Model 5
Presence of private heating	157.2	167.7	124.4	166.2	140.6
Restructured building	186.3	172.9	155.5	184.2	168.0
Presence of the lift in the building	228.3	232.0	217.8	215.7	209.7
Presence of the railway within 200 metres	−168.7	−157.9	−155.5	−159.0	−150.3
Quality of the street*	336.9	326.6	311.1	356.7	
Presence of parked cars on the pavement			−61.1		
Quality of urban space and environmental quality in the street**					402.6
Presence of green areas				150.5	
Presence of trees		47.9			
Distance to the city centre by car (minutes)	−9.4	−9.31	−7.86	−9.19	−7.15

*   Ratio between building height of buildings/width of the street < ½.
** Interaction between the presence of trees and gardens and the ratio between building height/width of the street < ½.

Simulation

Methodology to Measure Differential Rent Increase

As explained in section two, a large urban project generates an impact on urban rent, since it acts on some elements, like accessibility, urban quality, urban attractiveness which in turn have an impact on both components of urban rent, i.e. the differential and the absolute component. In this section and in the next one we present the computational procedure to measure the differential urban rent increase; in the subsequent section we present the results of the methodology. The third section is devoted to the presentation of the methodology and results in terms of the absolute rent increase.

The underground of the urban track section generates a number of urban transformations, namely:

- A *greater accessibility to the city centre* for those areas located in the western part of the town, which at present suffers from the existence of the railway acting as a physical barrier. The boulevard which is planned to replace the actual railway track section is in fact planned to have several crossing points. An *accessibility effect* will therefore take place in these areas, depending on the present relative location of buildings and on the minutes gained.

Table 6.4 **Estimate of the hedonic price function in the market for commercial buildings (*)**

	Model 1	Model 2	Model 3
Constant	90.32	48.56	99.4
Localization of the commercial business far from commercial streets	–26.86 (0.051)		–27.8 (0.06)
Medium distance area***	–20.17 (0.16)	–3.49 (0.61)	–21.1 (0.16)
Proximity to commercial streets		11.73 (0.09)	
Proximity to the railway			–5.28 (0.64)
Distance to the bus stop (meters)	30.04 (0.08)	15.77 (0.063)	28.5 (0.08)
Distance to the city centre with public transport (minutes)	–10.62 (0.037)	–6.04 (0.039)	–11.37 (0.03)
Lambda	–0.17 (0.7)	–0.12 (0.8)	–0.13 (0.7)
Theta	0.51 (0.005)	0.45 (0.01)	0.52 (0.004)
R-square	0.24	0.22	0.25
Number of observations	50	50	50
Theta = Lambda = –1	–474.71*** (0.000)	–473.40*** (0.000)	–474.6*** (0.000)
Theta = Lambda = 0	–447.94*** (0.005)	–447.53*** (0.014)	–447.9*** (0.004)
Theta = Lambda = 1	–449.44*** (0.001)	–451.04*** (0.000)	–449.3*** (0.001)
Test: Spatial error:			
Moran's I	0.462 (0.644)	–0.407 (1.3016)	0.320 (0.749)
Lagrange Multiplier	0.014 (0.907)	0.131 (0.718)	0.002 (0.963)
Robust Lagrange multiplier	0.066 (0.798)	0.056 (0.813)	0.047 (0.828)
Test: Spatial error:			
Lagrange multiplier	2.491 (0.114)	1.247 (0.264)	2.544 (0.111)
Robust Lagrange multiplier	2.543 (0.111)	1.172 (0.279)	2.589 (0.108)

Dependent variable: capitalized per square-meter commercial business rent.
*   Beneath each parameter estimate is given the p-value for the parameter in parentheses.
**   Commercial business localized outside the historical centre but not in periphery.
***   Restricted log likelihood.

Table 6.5   Hedonic prices for the sample mean in the commercial building markets (values in €)

	Model 1	Model 2	Model 3
Localization of the commercial building far from commercial streets	−1524		−1453
Localization of the commercial building on a commercial street		1091.8	
Medium area	−1145	−324	−1105
Presence of the railway within 200 meters			−276
Distance to the bus stop (meters)	20.47	23.12	20.85
Distance to the city centre by public transport	−29	−34.1	−32

- A greater urban space quality for all buildings located on the railway, when the planned 'boulevard' will be constructed, which implies a restructuring both of land at present occupied by tracks and of derelict land located along the railway (*urban quality effect*).
- A greater environmental quality obtained by both a reduction in noise pollution (at present caused by the railway), which impacts on all buildings located within 200 meters from the railway, and by an increase of green areas for those buildings directly located on the railway (*environmental quality effect*).
- An increase in potential revenues in those areas where the commercial activity is expected to increase, especially in the 'boulevard' area (*commercial synergy effect*).

In order to measure the increase of property values caused by each effect, homogenous urban areas in terms of each single trait (accessibility, urban quality, environmental quality and commercial activity) have been envisaged. The present use and future development (as mentioned in the land-use plan 2001 of the Municipality of Trento) of each single lot of the whole area involved in the project (Figure 6.3) have been analysed, and homogenous sub-areas in terms of possible impact on the project, labelled a, b, c ..., envisaged. They differ from one another in terms of urban and environmental quality, in terms of presence of commercial areas, and in terms of accessibility to the centre. In the accessibility case, two elements distinguish the areas: the relative location with respect to the city centre, and the minutes which will be gained thanks to the project (see Table 6.6 above).

As Table 6.7 shows, 18 areas are identified, on which the underground of the track section project impacts differently. For each sub-area volumetry is calculated, and from the latter the total surface, differentiating residential, commercial and office buildings, and between new and already existing buildings.

Table 6.6   Hedonic prices for accessibility

Accessibility to the city centre for residential activities*	Hedonic prices (€/min/sq.m.)	Accessibility to the city centre of commercial activities**	Hedonic prices (€/min/sq. m.)
1	−52,58	3	−178,37
5	−19,46	5	−100,66
7	−15,81	6	−82,07
10	−12,68	7	−69,05
12	−11,33	8	−59,46
13	−10,79	9	−52,11
15	−9,88	10	−46,31
16	−9,49	11	−41,62
17	−9,14	12	−37,76
18	−8,82	13	−34,52
19	−8,53	15	−29,41
20	−8,27	16	−27,36
21	−8,02	18	−23,98
22	−7,80	19	−22,57
23	−7,58	20	−21,31
24	−7,39	23	−18,22
25	−7,20	36	−11,03
26	−7,03	40	−9,80
27	−6,87		
28	−6,72		
30	−6,44		
31	−6,31		
32	−6,19		

*   Distance by car to the city centre (minutes)
** Distance by public transport to the city centre (minutes)

Buildings of the same size have different values according to their use (residential, office, commercial buildings) and to their age (new vs. old buildings). As for buildings used as offices, their value is 7 per cent more with respect to that of houses; for new buildings, the value increases by 30 per cent, as is evident from an analysis of the present property market.[17] These differences in the basic prices have been applied before calculating the increase in property values.

17 Source of property values in the city of Trento, FIMAA 2003.

Table 6.7   Characteristics of each sub-area

Homogeneous subareas	Accessibility effect		Urban quality effect	Environmental quality effect		Commercial synergy effects
	Relative location*	Gained minutes		Noise pollution	Presence of green areas	
a	10	2				Y
b	12	2		Y		
c	10	2	Y	Y	Y	Y
d	5	2	Y	Y	Y	
e	5	2	Y	Y	Y	Y
f	2	0		Y		
g	2	0		Y		Y
h	3	0	Y	Y	Y	
i	3	0	Y	Y	Y	Y
k	12	3	Y	Y	Y	Y
l	10	0	Y	Y	Y	Y
m	8	3		Y		Y
n	12	3				
o	12	1				
q	5	0				Y
r	15	2				
s	12	2	Y	Y	Y	

*   Distance by car to the city centres (minutes)

Once the physical dimension (m^2 of surface) for each homogeneous area has been calculated, the increase of differential urban rent due to the large urban project is easily measurable.

The computation procedure is as follows:

For dichotomic variables (traits)

For continuous variables (accessibility)

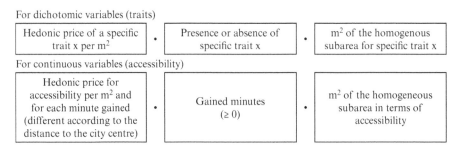

In all models estimating hedonic prices, the results model 2 for both residential and commercial building markets have been used.

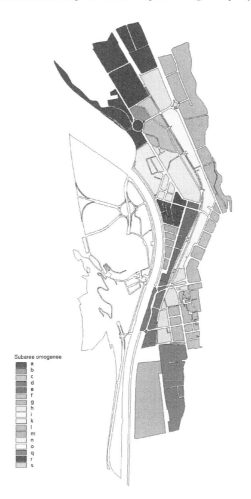

Figure 6.3   Map of the homogeneous sub-areas

Results of the Simulation for the Differential Rent Increase

In this section we report the results obtained by the simulation run on the city of Trento. The total increase in the differential rent obtained for laying underground of the section tracks is equal to €633 million, which represents 15 per cent of the property value of the entire area covered by the project and 3 per cent of the whole Trento area. Of this value, nearly 68 per cent stems from an increase in value of already existing buildings, while 32 per cent stems from new constructions[18] (Table 6.8).

18 Property values have been calculated by considering an average urban rent of 1.800 €/sq. meter.

Most of the increase in the differential rent (nearly 90 per cent) is due to *urban and environmental quality effects*, 43 per cent for the urban quality and 47 per cent for the environmental effects; while for the former the increase is mainly due to the high value attributed by the market to this trait, for the latter, the significant increase is partially due also to the size of the area involved by the project (more than 2,300,000 m^2) (Table 6.9c).

The accessibility effect has a minor impact on the property values; the increase of the differential rent due to increased accessibility is around 4.5 per cent, for both residential and commercial activities. The reason for this result is that the gain in terms of minutes by car is limited, given the limited size of the city as a whole (Table 6.9c).

The commercial synergy effect increases differential rent by only 5.8 per cent, despite the great value attributed by the market to this trait (€1,091 per m^2). The reason is the limited built surface targeted for commercial buildings in relation to that for private houses and offices (Table 6.9a).

Figure 6.4 shows the spatial distribution of the increases in the differential rent with respect to the value of the whole urban area. The main result is that the areas which register the highest percentage increase in property values are the areas along the railway, where the urban quality effect and the environmental quality effect are more prominent; among these areas, higher increases are registered in those areas destined to host commercial activities. The accessibility effect is not so prominent, as witnessed by the small difference between the incremental rates of the areas benefiting from greater accessibility to the city centre and those that do not.

In general, the present derelict land, where large urban requalification projects are planned, represents the areas getting the highest increase of incremental rent; in the areas presently occupied by the track section the increase is around 25 per cent, while in the old industrial areas around 40 per cent (sites where the old industrial firms Michelin and Carbochimica – SLOI and OET were located) (Figure 6.5).

The increases in differential property values caused by the different effects can be easily illustrated on maps. For example, Figure 6.5 shows the accessibility effect on the different sub-areas; in the areas at the west side of the railway the increase of property values is around 1.5 per cent to 3 per cent. The greatest increase is registered by those areas that at present are located far from the crossing bridges on the railway (Figure 6.6).

Urban quality effect has a major impacts on the present derelict land where projects of urban requalification are planned, and along the present railway tracks, which will benefit from the future boulevard. All these areas register an increase of 20 per cent in property values (Figure 6.7).

Absolute Rent Increase: Methodology and Results

As widely explained earlier, an urban project like the one we are analysing has a high impact in terms of attractiveness of the city as a whole and the image of the city:[19] it is therefore likely to anticipate that the demand for an urban location increases and, given the rigidity of the supply at least in the short-medium run, the property value is expected to increase in the whole urban area, without any territorial distinctions within the urban area.

The theoretical reasonings behind such an increase are quite clearly identified and easily acceptable; much more complicated is the formulation of hypotheses on how much the increase in absolute urban rent will be, given the implementation of the above mentioned project. Our methodology to estimate such an increase avoids the direct formulation of the hypotheses on absolute urban rent increase and develops a reverse procedure; on the basis of the increase in absolute rent required by the project to be profitable, we are able to say whether the increase is appropriate or not, given the annual average increase of property values in the last 40 years in the city of Trento.[20]

Considering the engineering cost of the project, estimated by experts around €788 million,[21] we calculated the increase of absolute urban rent necessary to achieve a break-even. Knowing that through the differential urban rent increase social benefits are already €633 million, the required increase in absolute urban rent is equal to nearly €154 million in order to achieve a break-even (Table 6.10). This means that only an *una tantum* 0.74 per cent increase in the property values is necessary for the project to be profitable from the social point of view. By looking at the historical trend in property values in the city of Trento in Figure 6.8, one can understand that the average annual percentage increase in the last nearly 40 years is around 2 per cent. Therefore, the 0.74 per cent required to achieve the breakeven is foreseeable.

Conclusions

The aim of the chapter was to implement an appraisal methodology for large urban projects via a synthetic index of social benefits, i.e. property value increases, and to apply it as an ex-ante methodology to measure social benefits associated to a possible future urban project in the city of Trento.

The methodology enables us to measure the increases in urban rent for both components that constitute rent, i.e. the differential and the absolute component.

---

19 The increase of absolute urban rent impacts without distinction on the whole urban area; in our case, the urban area corresponds to administrative municipalities in the valley area, leaving aside the mountain areas that are part of the Municipality of Trento; the areas affected by the increase are Trento, Villazzano, Ravina, Sardagna, Cognola and Gardolo.

20 The average annual rate of increase has been based on official statistics of property values in the city of Trento, published by Tamborrino (2001), after having deflated the values.

21 Cost evalutaion has been provided by the Department of Structural and Mechanical Engineering of the University of Trento, under the responsibilitiy of M. Bonfioli and R. Mauro (2002).

Table 6.8 Differential urban rent for different types of buildings and traits (€)

Present property value in the urban area of Trento	20.904.464.400
Percentage increase of differential property value with respect to the whole urban area of Trento	3,12
Present property value of the area directly involved by the project	4.189.513.440
Percentage increase of differential property value with respect to the area directly involved by the project	15,57

Types of effects	Accessibility effect		Urban quality effect	Environmental quality		Commercial sinergy effect	Total
Types of buildings	Residential accessibility	Commercial accessibility		Noise	Green areas		
Existing buildings							
Residential houses	17,815,396	–	103,342,141	119,870,498	15,188,067	–	256,216,102
Offices	5,517,754	–	86,610,343	66,158,388	12,729,016	–	171,015,501
Commercial buildings	–	649,897	–	–	–	14,348,242	14,998,139
Total	23,333,150	649,897	189,952,484	186,028,886	27,917,083	14,348,242	442,229,742
% on total	5.28	0.15	42.95	42.07	6.31	3.24	100.00
Projects							
Residential houses	2,307,460	–	50,426,298	38,275,941	7,411,091	–	98,420,789
Offices	2,878,886	–	40,038,939	39,147,600	5,884,474	–	87,949,898
Commercial buildings	–	0	–	–	–	23,789,722	23,789,722
Total	5,186,346	0	90,465,237	77,423,540	13,295,565	23,789,722	210,160,409
% on total	2.47	0.00	43.05	36.84	6.33	11.32	100.00
Total							
Residential houses	20,122,856	–	153,768,439	158,146,439	22,599,158	–	354,636,891
Offices	8,396,640	–	126,649,282	105,305,988	18,613,489	–	258,965,399

Table 6.8 cont'd

Types of effects / Types of buildings	Accessibility effect		Urban quality effect	Environmental quality		Commercial sinergy effect	Total
	Residential accessibility	Commercial accessibility		Noise	Green areas		
Commerc. Buildings	–	649,897	–	–	–	38,137,963	38,787,861
Total	28,519,495	649,897	280,417,721	263,452,427	41,212,647	38,137,963	652,390,151
% on total	4.37	0.10	42.98	40.38	6.32	5.85	100.00

Table 6.9 Computation of differential property value increase

a) *Computation with dicothomous variables*

	Sq. meter residential houses		Sq. meter offices		Sq. meter commercial activities		Hedonic price	Property value increase
	Existing	Projects	Existing	Projects	Existing	Projects	(€/mq)	
Urban quality	316,031	118,622	264,863	94,187	—	—	327	280,417,721
Environmental quality								
noise	758,674	186,348	418,724	190,592	—	—	158	263,452,427
Green	316,418	118,767	265,188	94,302	—	—	48	41,212,648
Commercial synergy	—	—	—	—	13,142	21,789	1091,8	38,137,964

b) *Computation with continuous variables*

	Sq. meter residential houses		Sq. meter offices		Sq. meter commercial activities		Hedonic price	Min. gained	Property value increase
	Existing	Projects	Existing	Projects	Existing	Projects			
Residential accessibility									
5 minutes	91,613	13,991	130,056	40,657	—	—	19	2	15,200,509
8 minutes	47,651	0	0	0	—	—	16	3	2,287,248
10 min	126,071	746	26,427	4,667	—	—	13	2	4,677,416
12 min a	123,945	25,962	2,897	0	—	—	11	1	1,788,826
12 min b	19,953	2,641	9,708	0	—	—	11	2	877,578
12 min c	204,659	4,479	118,619	36,745	—	—	11	3	15,661,817
15min	10,644	36,620	7,405	0	—	—	10	2	1,416,770

Table 6.9 cont'd

	Sq. meter residential houses		Sq. meter offices		Sq. meter commercial activities		Hedonic price	Min. gained	Property value increase
	Existing	Projects	Existing	Projects	Existing	Projects			
Total	624,536	84,439	295,112	82,069	–	–	–	–	41,910,163
Commercial accessibility									
7 min	–	–	–	–	1,028	0	69	2	141,864
10 min a	–	–	–	–	414	0	46	2	38,088
10 min b	–	–	–	–	1,524	0	46	3	210,312
13 min	–	–	–	–	500	0	35	3	52,500
15 min	–	–	–	–	3,067	0	29	2	177,886
18 min	–	–	–	–	1,082	0	24	1	25,968
Total	–	–	–	–	7,615	0	–	–	646,618

c) Total surface involved by the different effects

	m²
Urban quality effect	793,704
Environmental quality effect	2,349,014
Commercial synergy effect	34,931
Accessibility effect	1,093,771

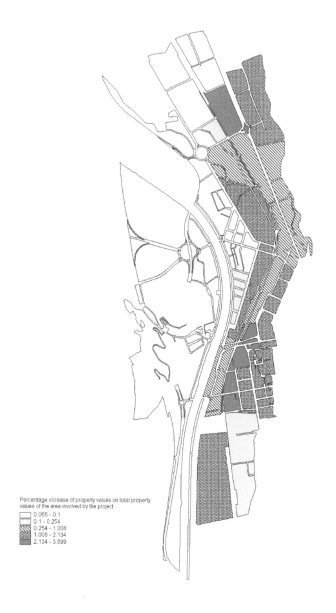

Percentage increase of property values on total property
values of the area involved by the project

   0.055 - 0.1
   0.1 - 0.254
   0.254 - 1.008
   1.008 - 2.134
   2.134 - 3.699

**Figure 6.4   Percentage increase of property values on property value of the
whole city of Trento**

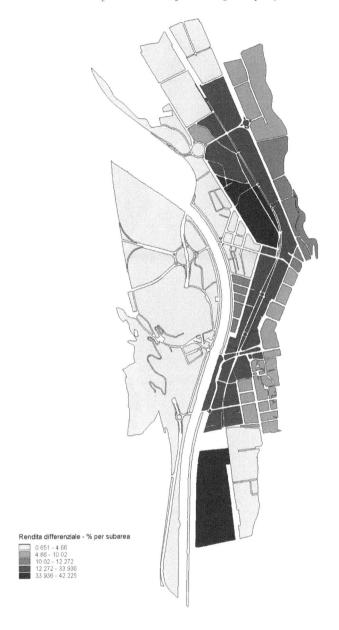

Figure 6.5   Percentage increase of property values on total property values of the area involved by the project

Figure 6.6   Accessibility effect

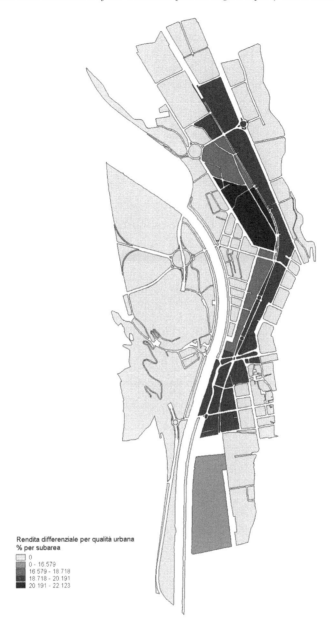

Rendita differenziale per qualità urbana
% per subarea

- 0
- 0 - 16.579
- 16.579 - 18.718
- 18.718 - 20.191
- 20.191 - 22.123

Figure 6.7   Urban quality effect

Elements influencing urban rent, like accessibility, urban space quality, urban
environmental quality, have been attributed a monetary value through hedonic
price methodology. Simulating the effects that a large urban project might generate
on these elements, forecasts of the increase in property values are obtained. The
increase in the absolute urban rent is also investigated, and added to the differential
increase in property values.

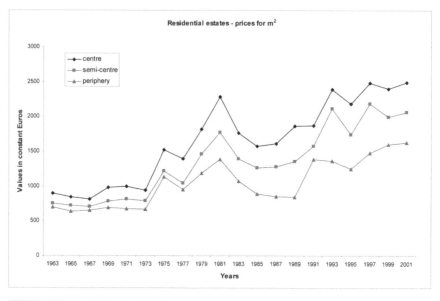

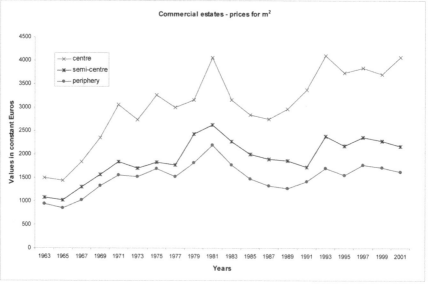

Figure 6.8   Property values trends in Trento between 1963 and 2001

Table 6.10   Increases of urban rent and total property value increase (€)

% increase of property values	Increase in absolute rent	Increase in differential rent	Total property value increase
0.50	104,522,322	652,390,151	756,912,473
0.60	125,426,786	652,390,151	777,816,938
0.74	**154,546,177**	**633,453,823**	**788,000,000**
0.75	156,783,483	652,390,151	809,173,634
1.00	209,044,644	652,390,151	861,434,795
1.50	313,566,966	652,390,151	965,957,117
2.00	418,089,288	652,390,151	1,070,479,439

It has been shown that the total advantages derived from the project, estimated in terms of the increase in total differential land rent, are likely to be of a comparable size with the engineering and adjustment costs, and that very limited (but likely) increases in the general appreciation of the city, approximated by the increase in absolute rents, would easily generate a full cost-benefit equilibrium at the highest level of predicted costs.

The results obtained by the simulation suggest that the project is profitable from the social point of view; this does not mean that a financial-economic equilibrium of the project is achieved, but that the society as a whole achieves benefits which are higher than costs. The financial sources of the project have to be identified by the public sector; a part of the necessary financial resources can be obtained by una-tantum taxes on parts of the increases on property values.

References

Alonso, W. (1964), *Location and Land Use: Towards a General Theory of Land Rent*, Cambridge, MA, Harvard University Press.

Anselin, L. (1988), *Spatial Econometrics: Methods and Models*, London, Kluwer Academic Publishers.

Anselin, L. and Florax, R. (eds.) (1995), *New Directions in Spatial Econometrics*, Berlin, Springer Verlag.

Bates, L.J. and Santerre, R.E. (2001), 'The Public Demand for Open Space: The Case of Connecticut Communities', *Journal of Urban Economics* 50, pp. 97–111.

Bender, B., Gronberg, T.J. and Hwang, H. (1980), 'Choice of Functional Form and the Demand for Air Quality', *Review of Economics and Statistics* 62(4), pp. 638–43.

Blomquist, G.C., Berger, M.C. and Hoen, J.P. (1988), 'New Estimates of Quality of Life in Urban Areas', *The American Economic Review* 78(1), pp. 89–107.

Bonfioli, M. and Mauro, R. (2002), *Studio di fattibilità per l'interramento della ferrovia del Brennero in Trento*, Dipartimento di Ingegneria Meccanica e Strutturale, Università di Trento.

Bowes, D.R. and Ihlanfeldt, K.R. (2001), 'Identifying the Impacts of Rail Transit Stations on Residential Property Values', *Journal of Urban Economics* 50, pp. 1 –25.

Box, G.E.P. and Cox, D.R. (1964), 'An Analysis of Transformations', *Journal of the Royal Statistic Society* 2, pp. 211–52.

Brown, J.N. and Rosen, H.S (1982), 'On the Estimation of Structural Hedonic Price Models', *Econometrica* 50(3), pp. 765–68.

Camagni, R. (1996), *Principes et modèles de l'economie urbaine*, Paris, Economica.

Cheshire, P., and Sheppard S. (1995), 'On the Price Land and the Value of Amenities', *Economica* 62, pp. 247–67.

Cheshire, P. and Sheppard, S. (1997), 'Welfare Economics of Land Use Regulation', Research Papers in Environmental and Spatial Analysis, London School of Economics.

Cheshire, P. and Sheppard, S. (1998), 'Estimating Hedonic Demand Using Single-Market Data: A Pratical Solution Using "Nearby' Instruments", Research Papers in Environmental and Spatial Analysis, London School of Economics.

Cheshire, P. and Sheppard, S. (2002), ' Capitalising the Value of Free Schools: The Impact of Land Supply Costraints', article presented at the Conference on the Analysis of Urban Land Markets and the Impact of Land Market Regulation, Cambridge MA.

Cobb, S. (1984), 'The Impact of Site Characteristics on Housing Cost Estimates', *Journal of Urban Economics* 15, pp. 26–45.

Corielli, F., Frigeri, P., Messori, A. and Tedeschi, P. (1996), 'Applicazione della teoria dei prezzi edonici al mercato immobiliare milanese', R. in Camagni (ed.), *Economia e pianificazione della città sostenibile*, Bologna, Il Mulino.

Deaton, A. and Muellbauer, J. (1980), 'An Almost Ideal Demand System', *American Economic Review* 70(3), pp. 312–26.

Epple, D. (1987), ' Hedonic Prices and Implicit Markets: Estimating Demand and Supply Functions for Differential Products', *Journal of Political Economy* 95(1), pp. 59–80.

Fahri, A. (1973), 'Urban Economic Growth and Conflicts: A Theoretical Approach', *Papers and Proceedings of the RSA* 31, pp. 95–124.

Freeman, M.A. III (1971), 'Air Pollution and Property Values: A Metodological Comment', *Review of Economics and Statistics* 53(4), pp. 415–16.

Freeman, M.A. III (1979), 'The Hedonic Price Approach to Measuring Demand for Heighborhood characteristics', in D. Segal (ed.) *The Economics of Neighborhood*, New York, Academic Press.

Fujita, M. (1989), *Urban Economic Theory*, Cambridge, MA, Harvard University Press.

Goodman, A.C. (1998), 'Andrew Court and the Invention of Hedonic Price Analysis', *Journal of Urban Economics* 44, pp. 291–98.

Gyourko, J. and Tracy, J. (1991), 'The Structure of Local Public Finance and the Quality of Life', *Journal of Political Economy* 99(4), pp. 774–806.

Harvey, D. (1973), *Social Justice and the City*, London, Edward Arnold Publisher.

Jackson, J.R., Johnson R.C. and Kaserman D.L. (1984), ' The Measurement of Land Prices and the Elasticity of Substitution in Housing Production', *Journal of Urban Economics* 16, pp. 1–12.

Linneman, P. (1980), 'Some Empirical Results on the Nature of the Hedonic Price Function for the Urban Housing Market', *Journal of Urban Economics* 8, pp. 47–68.

Lipiez, A. (1974), *Le tribut foncier urbain*, Paris, Maspero.

Lipiez, A. (1978), 'Terre,rente et rareté', *Revue d'economie politique* 5, pp. 746–54.

Marx, K. (1867) *Das Kapital*, Berlin, Dietz Verlag (English version (1977), *Capital: A Critique of Political Economy*, Vintage Books, New York).

Paci, R. and Usai, S. (1999), 'Externalities, Knowledge Spillovers and the Spatial Distribution of Innovation', *GeoJournal* 49, pp. 381–90.

Ridker, R.G., Henning, J.A. (1967), 'The Determinants of Residential Property Values with Special Reference to Air Pollution', *The Review of Economics and Statistics* 49(2), pp. 246–257.

Roback, J. (1982), 'Wages, Rents, and the Quality of Life', *The Journal of Political Economy* 90(6), pp. 1257–78.

Rosen, S. (1974), 'Hedonic Prices and Implicit Markets: Product Differentiation in Pure Competition', *The Journal of Political Economy* 82 (1), pp. 34–55.

Scott, A. (1976), 'Land Use and Commodity Production', *Regional Science and Urban Economics* 6, pp. 147–60.

Sivitanidou, R. (1996), 'Do Office-Commercial Firms Value Access to Service Employment Centers? A Hedonic Value Analysis within Polycentric Los Angeles', *Journal of Urban Economics* 40, pp. 125–49.

Small, K.A. (1975), 'Air Pollution and Property Values: Further Comment', *Review of Economics and Statistics* 57(1), pp. 105–7.

Sraffa, P. (1960), *Produzione di merci a mezzo di merci*, Torino, Einaudi.

Topalov, C. (1984), *Le profit, la rente et la ville*, Paris, Economica.

Tse, R.Y.C. (2002), 'Estimating Neighbourhood Effects in House Prices: Towards a New Hedonic Model Approach', *Urban Studies* 39(7), pp. 1165–80.

Wilkinson, R.K. (1973), 'House Prices and Measurement of Externalities', *The Economic Journal* 83(329), pp. 72–86.

Appendix 6.1

Table 6.A1 Statistical description of variables used

Variables for the residential housing model	Mean value	Standard deviation	Min. value	Max. value
Flat price per m²	1860.5	498.1	1100	5500
Commercial business building price per m²	3799.6	2209.3	642.8	11878.5
Presence of a private heating system	0.74	0.44	0	1
Restructured building	0.65	0.47	0	1
Presence of the lift in buildings of different kinds (old, exclusive)	0.30	0.46	0	1
Presence of the railway within 200 m	0.36	0.48	0	1
Quality of the street*	0.04	0.19	0	1
Presence of parked cars on the pavement	0.03	0.17	0	1
Quality of urban space and environmental quality in the street**	0.02	0.14	0	1
Presence of gardens	0.91	0.28	0	1
Presence of trees	0.57	0.49	0	1
Distance by car to the city centre (minutes)	16.5	8.04	1	32
Variables for the commercial business buildings	**Mean value**	**Standard deviation**	**Min. value**	**Max. value**
Capitalized value of annual rent	3799.6	2209.3	642.8	11878.5
Distance by bus to the city centre (minutes)	13.12	8.14	0.1	40
Distance to the nearest bus stop (minutes)	43.8	28.9	10	160
Location far from shopping streets	0.40	0.49	0	1
Location on a shopping street	0.26	0.44	0	1
Medium distance area***	0.58	0.49	0	1

*   Ratio between building height of buildings/width of the street < 1/2.
** Interaction between the presence of trees and gardens and the ratio between building height/width of the street < 1/2.
***Area outside the historical city centre, but with an easy and fast access to it.

Appendix 6.2   Linear Transformation of Parameters

If we assume the existence of two equations:

$$y = K + \beta x_j$$
$$Y = K + \beta' X_j$$

(6.1a)

where:

$$Y = \frac{y^\theta - 1}{\theta}$$

and

(6.2a)

$$X_j = \frac{x_j^\lambda - 1}{\lambda}$$

The first derivative of Y is equal to:

$$\frac{\partial Y}{\partial X_j} = \frac{\partial Y}{\partial y} \frac{\partial y}{\partial X_j} = \frac{\partial Y}{\partial y} \frac{\partial x_j}{\partial X_j} \frac{\partial y}{\partial x_j}$$

(6.3a)

Being:

$$\frac{\partial Y}{\partial X_j} = \beta$$

(6.4a)

$$\frac{\partial y}{\partial x_j} = \beta$$

equation (3a) can be re-written as follows:.

$$\beta'_j = \beta \frac{\partial y}{\partial Y} \frac{\partial X_j}{\partial x_j}$$

(6.5a)

Since:

$$\frac{\partial y}{\partial Y} = \left(\frac{\partial y}{\partial Y} \right)^{-1} = \left(\frac{\theta \, y^{\theta-1}}{\theta} \right)^{-1} = y^{1-\theta}$$

(6.6a)

and being:

$$\frac{\partial X_j}{\partial x_j} = x_j^{\lambda-1}$$

(6.7a)

equation (5a) becomes:

$$\beta' = \beta y^{1-\theta} x_j^{\lambda-1}$$

(6.8a)

which represents the transformation made on the parameters.

PART 2
Socio-Institutional Principles

Chapter 7

Estimation of a Hedonic Rent Index for the Residential Real Estate Market of Bari: A Contribution to Urban Planning

Maurizio d'Amato and Tom Kauko

Introduction[1]

Traditionally, evaluation of a land-use plan can be performed in many ways. It may be based on ideological grounds such as neoclassical or neo-Marxist principles, or on rational or collaborative action theory. It may also be based on empirical evidence, either in an ex-ante or ex-post framework. Evaluation ex-ante is concerned with the change from present to an unknown future, whereas evaluation ex-post is concerned changes that have already taken place as a result of the implementation of policy measures.

Sager (2003) compares two established *ex-ante* evaluation approaches: the goals achievement matrix (GAM by Hill 1966) and the planning balance sheet (PBS by Lichfield 1956), and makes the following conclusions:

- GAM is about the partial goals of the planning, and distances itself from the unrealistic economic rationality concept, which is an advantage, but at the same time the technique is undermined by the possibility of manipulating specific evaluation results.
- PBS is about welfare economic outcomes for all those affected by a policy measure; but it is somewhat arbitrary because of its emphasis on economic rationality.

We can see, among other things, that these *ex-ante* approaches do not explicitly deal with economic (i.e. monetary) indicators; in other words, they do not test their assumptions at the market place. *Ex-post* based evaluation, on the other hand, is based on already known implications, and may therefore be based on

---

1 This work was carried out jointly by the two authors but the third and fourth sections (and the appendices) are written by Maurizio d'Amato while the first and second are written by Tom Kauko.

actual monetary costs and benefits. Logically, this approach is considered more appropriate than *ex-ante* evaluation insofar as we can rely on valid methodology and data. In *ex-post* evaluation it is common to compare effects categorized as positive (i.e. benefits) and negative (i.e. costs) involved in implementing a certain planning measure. If not, direct costs can be identified. One must isolate them indirectly from observed data. Using *hedonic price modelling* (i.e. the hedonic approach) consumer choices made at the market place may be isolated and quantified as monetary units. The idea is to reveal implicit prices (or shadow prices) for attributes the prices of which cannot be measured directly. The costs and benefits do not only need to be economic (in the strictest sense), as more social, environmental or other issues can be dealt with this way too. While this approach in principle can be applied on any kinds of consumer goods it may be argued that in the context of looking at the built environment, property is more apt consumption good than for instance automobiles. In fact, since the seminal paper by Rosen (1974) the hedonic approach has frequently specifically been applied for house price determination and price index construction.

While the formal theory behind this approach is extensive, for our purposes – i.e. within real estate market analysis and planning evaluation – only a rudimentary account is necessary. According to the basic principle of the hedonic model, as applied within empirical housing market modelling, the market behaviour is influenced by the property prices, the high durability of the property asset and by the fixed geographic location. From this relationship, an econometric relationship between the price and the property characteristics may be derived. If we include the time of transaction among the independent variables, the principle behind hedonic index construction for an urban residential rental market can be formulated in the context of *ex-post* planning evaluation too.

The connection between housing market analysis and planning is indeed important, although not necessary self-evident and, as will be shown in section 2, never straightforward. Following (Oxley 2004, 4), land-use planning instruments may be promoted in order to influence the volume, type, location, allocation and affordability of new housing, but this ought to be examined together with other policies designed to influence both new and existing housing. More or less the same is true for other property and land-use types.

From the point of view of conducting research, real estate and land use planning practice focus on partly different, partly similar objectives and methods. The former deals with the past and present, using mostly quantitative approaches. The latter projects a plausible state of affairs into the future using mainly qualitative but occasionally also quantitative analysis.[2] Nonetheless, both planning

---

2   The quantitative modelling traditions are still going strong despite a generation long era of resistance towards computerized modelling, following the 'Wonderland' trilogy of the special issue of the *Journal of the American Institute of Planners*, May 1973. In particular, Lee's 'Requiem for Large Scale Models' did a lot of damage, not only for modelling, but also for planning. The argument was that the 1960s computer models built largely in the spirit of system dynamics theory were seen as overambitious and downright dangerous for the development of the (urban) planning profession. The anti computer modelling circles

and real estate are preoccupied with spatial and functional allocation of resources within the built environment: these two applications are undoubtedly interrelated, which ought to be recognized in planning evaluation research as well.

In this study we explore the contribution of house rent index to planning decision-making using inputs from more than one discipline. Here the theoretical and methodological basis is provided by microeconomics, specifically by tools of the urban economics and index research subdisciplines. Furthermore, a geographical aspect is included too: the spatial disaggregation of the city area under study is approximated into homogeneous units referred to as 'microzones'. More specifically, we concentrate on isolating certain key housing market effects within the City of Bari, Italy, and for a period of five yearly cross-sections (1999–2003). Besides the spatial and temporal variables, one that determines the heterogeneity of the housing stock is also included in order to produce a more unbiased estimate. The variable selected for this purpose is 'number of rooms'.

The study is organized as follows. Following this brief introduction the second section discusses the linkage between real estate market and land-use planning. In the third section we estimate a residential rent index that enables us to detect a residential 'rent bubble' that, in the light of our evidence, may be seen as a symptom of non–optimal planning decisions. The conclusions and future directions of research are offered in the fourth section of the paper.

Planning and the Real Estate Markets: Possibilities for Cross-fertilization

As already explained, the aim of the paper is to inform planning on the basis of housing market modelling. This becomes possible, if we can produce valid estimates that show potential problems and courses of action in order to solve these problems. Using micro-level datasets with market indicators, our idea is, first, to identify and quantify a price bubble when the Bari housing market is looked at from a spatio-temporal perspective, and, second, based on the resulting estimates, to inform policy decisions in order to improve the social quality of residential areas.

In our view, real estate and planning are not real disciplines but rather sets of applications preoccupied with monitoring of the market development in relation to social-economic indicators. Therefore, the relationship between real estate value and planning is relevant in practice – not academic research. On the other hand, the interface between planning and real estate market can be analysed from different points of view including some helpful theoretical perspectives. In this section we show in which ways planning interacts with the real estate market; we also argue that the market outcome, in the form of prices and rents, in our view, serves as a measure of success for a certain policy decision. The difference

---

could not however foresee any aspect of today's data processing capabilities. A further, more conditional argument in favour of modelling is that, when facilitating the market, rather than working against it, planning is better suited as an objective of mechanistic modelling (cf. Harris 1994; Lee 1994).

in the two types of applications is contributed to how the market outcome is evaluated, and weighted vis-à-vis other kinds of outcome such as social factors, environment and aesthetics. Whereas the real estate paradigm concerns *potentials* and is restricted to monetary evaluation of the past and current circumstances, the planning problem field pays attention to *processes* for improving the state of affairs, often using mostly qualitative criteria (e.g. Sager 2003). The literature reviewed in subsections below mostly concern the connections between planning practice and the housing and housing land market (see also Henneberry et al. 2005).

In our view, planning could benefit pragmatically from incorporating state-of-the-art empirical property price/value modelling applications and automatic valuation methodologies that can 'shed light' on the policy decisions to be made.[3] In doing so, issues of complexity and interaction need to be addressed in an operatively credible way. The reasons, why we think the relationships between planning and real estate markets are such complicated matters that require a responsible treatment are related to the ways in which planning and the market place interact.

When discussing the value of integrating real estate market analysis into planning practice, we provide robust tools for identification of problems and, depending on the assumed causes of these problems, we also suggest some simple courses of action in order to improve the situation. We do not only propagate the hedonic approach applied, but all kinds of sophisticated quantitative empirical property market and value modelling approaches. Unfortunately, in our experience planners tend to play down empirical market evidence here. However, it can be argued that the discussion below picks up issues that are not new for the planning practice;[4] it would then be wrong to claim that indicators for value/price, supply, demand and quality have not been considered for land-use planning purposes. It may be so, but then such work has been carried out an inconsistent and *ad hoc* basis. It is likely that the political climate and governance balance of a certain era or jurisdiction determines the goals to be prioritized; one can, for example, easily assume that in a right wing coalition pro-market issues get the most back up when planning issues are prioritized. The novelty here then is to formalize the relationships and – as clearly as possible – generalize them as principles applicable for further research and practice.

Planning Controls and House Prices

According to Oxley (2004, 14) the impact of planning controls on housing production needs to be examined in the context of the many determinants of residential development; furthermore, the effect of housing production changes on house prices will depend on the relative significance of production changes

---

3   Likewise, we also think that research on property valuation and property markets could benefit from incorporating some more explicit planning elements, but that would be a topic for another paper.
4   This is a point brought up by the referee Abdul Khakee.

compared with all the factors that influence house prices. Nevertheless, planning controls, especially in the form of residential density regulations and green belts constraining the development, may be considered relatively important determinants of land values. This goes for improved (i.e. build-up) and unimproved land (i.e. land without structures) alike (Balchin and Kieve 1977, 36–37; Healey 1998). Applying the simple scheme provided by Kauko (2003), land use regulation and planning controls can basically guarantee building land as (A) pure space without further qualifying the type of land use, and (B) environmental amenities, as well as prevent (C) oligopolistic outcomes that can be seen as harmful in a social-economic sense. The different kinds of relationships, that we call the 'market view', the 'amenity view' and the 'policy view', each shows one aspect of the relationships between property price formation and land use regulation:

A *The 'market view'*: The outcome is determined by the supply and the demand, either on a total market or a submarket level. By restricting the amount of land available for residential construction, the market drives house prices up and vice versa (e.g. Evans, cited in Adair et al. 1991). This relationship between land use restrictions and market clearing pertains to the comparative statics and capitalization framework (of mainstream economics), according to which a land-use constraint is capitalized in reduced supply and, as a consequence, increased land and house prices. This model however should not be applied mechanically without insight to the local circumstances (see Pogodzinski and Sass 1990, for a critical review of the theory models; see also Fischel 1990).

B *The 'amenity view'*: If land use regulations are relaxed in a situation where the supply of land cannot be increased, an increased density (i.e. plot efficiency) reduces unit house prices. This effect is related to the inverse relation between density and price per housing unit: with an additional floor, the price per square meters decreases. The total price of the property increases when the plot efficiency increases, but at the same time its unitary price decreases. Households will in fact, rather live in low-density than high-density areas, and hence a price increasing effect occurs, whenever the new areas are greener and sparser than the old ones (e.g. Evans 1985: pp. 27–32).[5] In most planning contexts the permitted density numbers can be kept low in order to avoid this price reducing effect (see Monk and Whitehead 1999; Pogodzinski and Sass 1990). On the other hand, relaxing land regulations in a situation where supply can be increased would, in principle, lead to building low-density housing, and this way to an overall price increase, which is exactly opposite to the predictions of the view (A) where density is not explicitly considered.

C *The 'policy view'*: If certain economic and institutional pre–conditions are fulfilled, notably, the macroeconomic cycle is in downturn, the landownership during development is in private hands whereas planning powers are retained by public, the project is sufficiently large without substitutability with other

---

5   This perception may be purely optical. What is after all a more efficiently built block: one with high-rise buildings in a green environment, or one with low-density buildings without any public spaces in between them?

plots in the area, no covenants are allowed, and all prospective landowners consider speculation profitable, relaxing the land use regulations may generate a 'perverse' housing supply function where higher land prices reduce the land supply, which raises house prices (see Kauko 2003). In these circumstances the market mechanism is not allowed to work at all, and therefore policy (or political) considerations play a major role in determining land and housing prices. This is more apt in certain very specific situations involving public–private conflict, typically found in European circumstances (e.g. Roca Cladera and Burns 1998).

Views A to C assume that planning affects property prices.[6] According to the market view, a release of building land in feasible locations (i.e. in locations where demand is sufficient and no major negative externalities cause disturbances) leads to an increase in building land supply, and consequently to a reduced land price, which also affects house price levels (see e.g. Adair et al. 1991). This is assumed to be a fairly direct effect and related to how tight market constraints are both temporally and spatially. The amenity view works the opposite way, relaxing of land use regulations may lead to more spacious living, better neighbourhood quality and price increases, unless a demand is spatially differentiated and the land release is allocated to an inappropriate location, for example far from the service centres. This development is conditional upon the location and character of the project. Taken together, the first two views show that land use planning determines the overall supply of land as pure space and the availability of amenities (Cheshire and Sheppard 2002). The C-hypothesis in turn follows a perspective, where the underlying institutional structure in terms of the planning and development apparatuses may constitute extra costs.

The task of analysis is complicated further when we consider the dynamic feedback between the market outcome and the relevant planning decisions being made. As the current trends in housing market, liberalization and deregulation, in many countries show, any of the models above needs to include a feedback between market outcome and the behaviour of various planning actors (landowner, developer, government) as shown in Figure 7.1.

Land value, the supply of land, the demand for land, and institutional parameters are interlinked into a dynamic system illustrated in Figure 7.1. The sequence of events is commencing with a certain land-use planning decision: either land release, or restrictions such as density controls of growth boundaries being set. At a certain point in time market clearing takes place between supply and demand, which determines a value. As a reaction to this value formation, new decisions that either support or overturn the old decisions are made. As a result, a new market clearing takes place and subsequently new value formation too. This we refer to as a feedback process between market outcome and

6   According to (D), the 'macro-structural view', there are no price implications of land-use planning (see e.g. Fischel 1990). Then the price would be entirely determined as a residual value by the market for the end product – housing services. We do not discuss this view (or any of its variations) here any further (see Kauko 2003, for a discussion).

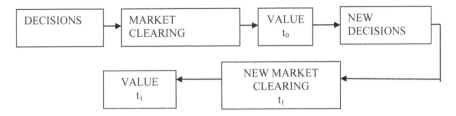

Figure 7.1   The feedback between planning decision and market outcome

decision-making. Exactly how the decisions are made depends on many further parameters such as political climate and governance traditions of the jurisdictions, socio–demographic factors and so forth.

Land-use Allocation according to Attractiveness Potential

We now come to the question: When are the planning decisions of spatial or functional allocation of resources to be considered (in)correct? There are no universal answers to this question. Answering the question requires some form of assessment of a given location or a locationally determined bundle of attributes. Is it desirable that prices rise or fall, given the time-scale of the project and the quality level of its end product? Undoubtedly, an empirical modelling approach to locational value formation becomes essential to enable better policy decisions to be made. When the attractiveness of sites becomes more appropriately managed, a more efficient allocation of resources is achieved.

In this context applications range from the most qualitative urban regeneration issues concerning design to the most quantitative land use modelling based on sophisticated grid technologies. For example, it is sometimes argued that a 'planning lottery' emerges when a new development plan is made and some landowners benefit more than others from the plot specific zoning regulations determining the density and the type of use. Additional negative considerations of arbitrary or even extravagant zoning decisions involve land speculation (see the policy view above) and efficiency problem in the use of infrastructure. Elaborating on the conclusions of Kauko (2002) with regard to the use of automation valuation methodology, our view strongly contradicts with such normative arguments about 'planning lottery' being the problem of the landowner. If the development process is marred by inefficiency, the real problem concerns the planning process, which is carried out without considerations about the real attractiveness of the sites. Rectifying this would lead to a more accurate distribution of profits generated by development rights, supply of housing and residential building land, and a sensible price level, in accordance with the fundamental attributes of the location. Furthermore, if the demand is expected to increase, more land supply and/or higher density is needed in a short time; otherwise a less efficient land use, and a lesser urgency of implementation may be a more sensible strategy.

If full agreement can be reached on the connections between planning and the real estate market, the application of a sophisticated value modelling approach would be easy to break through .In reality the issue is controversial. On the one hand traditional scientific perspective with discipline specific 'gatekeepers' require theory changes to precede changes in practice, on the other hand due to society's constant urge to adopt latest tools imply modus operandi without regard to new theoretical developments. This tension makes it difficult for any ideal approach to get general acceptance.

The three views (A, B and C) presented previously that represent ideal cases, and the more balanced 'dynamic feedback model', can be used for developing a conceptual framework for area assessment with regard to policy measures, quality improvement and price development. The aim of a liberal planning policy goal might be either 'releasing air of a price bubble' caused by housing supply shortages (A), catching up with the environmental quality and land value development (B), or both. However, opponents of such a policy measure may argue for the contrary: that if any price changes are observable, they are the 'unhealthy' outcome of speculation (C). If land development is in the hands of private actors, and the role of the planning system is to ensure the development process to proceed smoothly, then expropriations (i.e. compulsory purchase) of the land to the municipality may be justified, in order to speed up the stock adjustment process and provide substance for the (A) argument.[7]

A variety of examples can be used to demonstrate this framework:

- Monk, Pearce and Whitehead (1991) cited various studies from the UK, where the evidence suggests that the outcome (whether A, B or C is valid) depends on how land price is determined: residually (derived by deducting the costs from the market prices of houses), or through accumulation of costs including those that arise from the planning controls.
- Adair, Berry and McGreal (1991) investigated the connection between land availability and the property market in Northern Ireland, based on four case studies and descriptive statistics (house prices, land values, market shares of different house types, and new build starts). They found evidence of a linkage between *restrictive* planning policy, a shortage of development land, increased property prices and higher density (i.e. they verify the hypotheses A and B).
- In the study by Cheshire and Sheppard (1989) on the effects of land use controls on house prices the issue was, what the impact on house prices were for various market segments, if a more constrained land use regime was loosened substantially (to correspond with the level of a less constrained land use regime). Various scenarios (A)–(C) arose in the simulation.
- Hui and Ho (2003) show, using econometric modelling, that planning indicators (the approval rates of planning applications, the residential floor area, and the area of greenbelt and open space) do have a significant impact on housing prices in Hong Kong, thereby verifying the A hypothesis, as well as accepting the C hypothesis, and the abovementioned feedback model too: the price formation is not only determined by land releases, but by institutionally

---

7   This obviously depends on the legal rights of the local authority and on the local political climate.

specific circumstances that may in itself push prices up, which eventually either reinforces or alters this inertia.

We conclude that, while planning influences land value, a key component of the price of urban housing, the underlying causal mechanism is extremely complex. The outcome may depend on various factors related to market constraints, improved amenities, or policy of particular (land) development-planning regime. Moreover, the tensions between existing and desired outcome underlines the role of planning preferences and decision-making. In all cases a good start to flexible solutions is to look at the current attractiveness of the spatial and functional land use mix and at the future projections of housing demand – also in a qualitative sense.

A Residential Rent Index for the City of Bari

Mass appraisal methodology (i.e. real estate market modelling for valuation purposes using multiple regression analysis or other modelling approaches) comprises a field of growing interest for its application in local tax collection, social rent determination, mortgage market, and property management, to name a few related practices. An Italian reform (431/98) attempted to determine a range of possible rents to be paid by the tenants in the Italian residential property market. According to this reform there are two possible rent determinations. The former is a free agreement between the owners and the tenants based on market rent level. The latter is actually a 'social' rent determination based on agreements among different organizations representing both property owners and tenants. In this second case the municipalities give housing allowance to tenants with low income and tax deductions to the owners. The fiscal arrangement is conditioned by tenants' needs and legal requirements.

As a consequence the level of 'social' rent and the consequent negotiation process may become an important indicator for planning purposes. Using the conceptual reasoning provided in the previous section, and the hedonic index approach presented in this section we now attempt to answer two questions that are of social-economic relevance for the dynamics of the Bari housing market: Is there a spatially defined social-economic problem involved? Is it possible to solve by planning instruments?

The role of indices in the real estate market is strategic to understand the past and future trends of property price or rents. They represent important quantitative tools to provide a representative measure of price movements for housing markets (Maclennan 1977). Palmquist (1980) stressed that 'to assess these trends and proposals, it is necessary to have accurate methods of developing residential real estate price indexes'. There are several different kinds of property indexes.

The hedonic model allows planners to analyse property market rents through econometric modelling. It is possible to define the level of rent per each area of an Italian city. As a consequence it will be possible to locate the area with the highest level of social problems, the area with the highest number of request and those where there is the highest level of rent growth. The analysis of the data may detect rent bubble helping to programme a social housing policy.

Indexes in the Real Estate Market

Two main groups of property index methods are applied to estimate property price and rent (Knight et al. 1995). The former can be defined as *simple price index methods,* while the latter group of methods is also indicated as *constant quality index methods.* A general scheme of different kinds of property indexes is shown in Table 7.1.

Table 7.1 Different groups of property index

Type of house price/rent indices	
Simple price index methods	**Constant quality index methods**
Average price/rent index	Fixed-sample weighted average price/rent
Median price/rent index	Hedonic price/rent index
	Repeat sale index
	Weighted repeat sale index
	Assessed value index
	Hybrid method

In the *simple price index methods* family the index is obtained by comparing different means of property prices (or rents) in the same interval of time. The mean price (or rent) may be calculated in different ways. For example computing an arithmetic sum, an average price (or rent) index will be obtained. Instead of an arithmetic sum it is possible to calculate the median price (or rent) of the distribution for each interval of time considered.

Unfortunately the means obtained in each interval refer to a property or a location whose attributes may change over time. For this reason, several rigorous property index methods based on the hypothesis that the features of a property do not change over the intervals of time have been proposed. This family of methods are indicated as *constant quality index methods.* Within this group, the fixed–sample weighted average price (or rent) method is based on comparing several weighted arithmetic means of rent or prices in different intervals of time. Applying this method the weighted averages are obtained by dividing the properties in homogenous groups and applying the percentage of each specific group of property on the total number of observations as a weight for the mean.

The Hedonic method (Court 1939; Rosen 1974; Gatzlaff and Ling 1995) is based on econometric modelling. In this case the characteristics of a property are regressed on its price or rent. As a consequence a marginal price of these characteristics is defined. Furthermore the model allows the calculation of a rent or price index comparing the price or rent evolution in the considered interval of time. As for potential flaws in this method, Meese and Wallace (1997) have highlighted the risk of omitting important variations of property characteristics: as much information as possible on the heterogeneous composition of the property sample needs to be retained.

There are two different kinds of hedonic methodologies. In the first method the implicit price of housing and the coefficient time dummy variables are calculated using one only equation as in the formula 7.1 below:

$$P_{it} = \sum_{K=1}^{K} \beta_k X_{Kit} + \sum_{T=1}^{T} c_T D_T + \varepsilon_{it} \tag{7.1}$$

The model is described in Gatzlaff and Haurin (1997). Where:

P_{it} is the log of the transacted prices for the ith property
c_T is the estimated coefficient for the dummy variable D_T
β_k is the estimated coefficient for the property characteristics

The other hedonic method (which we apply in the exercise documented in the next subsection) is based on separate regression models for each annual cross section.

Outside the hedonic modelling tradition, the repeat sales method (Bailey et al. 1963) is based on the ratio between two prices or rents concerning the same property sold or rented at two different dates in a specific time. The difficulty of this method is the availability of data in the sample with repeated sales or rents (see Clapp and Giaccotto 1992). Case and Schiller (1989) highlighted an important aspect of repeat sale method. In fact they demonstrate the relationship between the change in price and the length of time between the sales (holding period). For this reason they propose a weighting of the repeat sales method. In the first stage it is necessary to undertake a regression using a normal repeat sale method. Following this, the squared residuals, derived from the initial regression, are regressed further on the length of time between the sales (holding period) with price as dependent variable in order to obtain a predicted value.

An alternative based on assessed value was proposed by Clapp and Giaccotto (1992). In this case the rent or price index is based on repeated valuation by professionals. An integration between hedonic modelling and repeat sale house price index was proposed by Case and Quigley (1991). They proposed a combination between the two; the repeat sale index is specified for the property sold (or rented) more than once, and the hedonic approach controls for the variation in the other property and location characteristics. This is the reason why the method is defined hybrid.

Hedonic Modelling for Residential Rents of Bari

In the exercise documented below a hedonic modelling approach was proposed to define a property rent index. There are several recent studies where hedonic modelling has been applied successfully on the real estate market: inter alia, for constructing constant-quality price indices for apartment buildings and vacant land in Geneva, Switzerland (Hoesli et al. 1997a); for determining rental values of apartments in central Bordeaux, France (Hoesli et al. 1997b); and for explaining the housing market in Tel Aviv, Israel (Gat 1996).

Several applications of rent modelling have also been undertaken for the market for office buildings. Models based on either cross-sectional data (Brennan et al. 1984, Mills 1992) or time series data (Glasscock 1990, Sivitanides 1997, Wheaton 1997) have been used with varying success for this purpose. In some models change in rent is a dependant variable while in other models rent is calculated by making use of information from rent contracts or from statistics on effective rents.

Basically, the studies have provided information about the parameters that affect the office or residential rent at various locations of the world, though none was able to investigate a whole range of parameters. The major difficulties lie within the hedonic regression models are the multicollinearity problem that may exist between the parameters involved and the failure to incorporate the whole range of parameters that may give rise to less accurate modelling. Below a hedonic rent index based on a linear function will be developed by explaining the rent by the number of rooms and the date of the rental contract. Of the two variants of hedonic indices noted above, the one used below is the one based on separate regressions for each annual cross-section.

The data comprises a sample of 2,353 observations obtained from SUNIA, a national organization representing the tenants' interest in negotiations of rental agreements. The properties were rented in the time interval of 1999 to 2003 in the nine urban zones classified as 'microzones' of the municipality of Bari. The data covers several different urban districts; these are grouped together based on spatial contiguity of original districts. These exhibit sufficient internal homogeneity with regard to various social, urban and economic indicators as well as property value and rent level, and are, in principle, well-suited for providing the basis of future property taxation in Italy. Table 7.2 indicates the number and the names of the nine microzones of the city of Bari.

It is worth noting that a microzone in many cases includes more than one administrative zone, like in the microzone 2, which is made of three administrative areas: Murat, Libertà and Madonnella. Furthermore an administrative zone like Madonnella is included in more than one microzone. In this case a part of Madonnella is in the microzone 2 and the rest is in the microzone 3. These observations were collected by a national organization representing the tenants' interest in negotiations. The number of transactions for the years covered in the sample is listed in Table 7.3.

The sample of observations contains information related to the number of rooms, the location of the property and the date of the contract. Starting from these features it was possible to build a linear econometric model measuring the property rent development. In that analysis the number of rooms (NUM. ROOMS), the time of the transaction (DATE), and the microzone where the observation was located (MZ), were regressed on gross rent or gross operate income indicated as GOI. These terms indicate the gross rent paid (passing rent)

Table 7.2   The name and the number of the microzones in the city of Bari

Name of microzone	Number of microzone
San Nicola	MZ1
Murat, Libertà, Madonnella	MZ2
Japigia, Madonnella	MZ3
Carrassi, San Pasquale, Poggiofranco	MZ4
San Paolo, Stanic	MZ5
Libertà, San Girolamo, Fesca	MZ6
San Giorgio, Torre a Mare	MZ7
Carbonara, Ceglie, Loseto	MZ8
Palese, San Spirito	MZ9

Table 7.3   Distribution of observations per years between 1999 and 2003

Years	Number of observations
1999	490
2000	487
2001	510
2002	456
2003	410
Total	2353

by the tenant to the homeowner. The date was measured in months using the date of the transaction. MZ comprises nine dummy variables related to the location of the observation. The general model is indicated in the following formula:

$$\text{GOI=CONST+NUM. ROOMS+DATE+}\lambda MZ_1 + \lambda MZ_2 + \lambda MZ_3 + \lambda MZ_4 + \lambda MZ_5 + \lambda MZ_6 + \lambda MZ_7 + \lambda MZ_8 + \lambda MZ_9 \quad (7.2)$$

Subsequently the property rent index SUNIA–OMING–POLIBA[8] was calculated as follows:

$$PRI = \frac{GOI_t}{GOI_{t-1}} \quad (7.3)$$

---

8   The name of the index is associated to the organizations involved in the index determination: SUNIA is a tenant organization; OMING is the Real Estate Market Observatory of the 1st Faculty of Engineering of the Technical University Politecnico di Bari while POLIBA stands for Politecnico di Bari.

In order to create a hedonic rent index a regression analysis with pooled data was calculated on the sample indicated in Table 7.3 for each year. The results of regression analysis are shown in appendix 7.2. The annual absolute change in hedonic (i.e. quality adjusted) rent is indicated in Table 7.4.

Table 7.4   Absolute variation of rent

Name of microzone	Number of microzone	1999–2000	2000–2001	2001–2002	2002–2003
San Nicola	MZ1	84,73	314,94	212,72	513,18
Murat, Libertà, Madonnella	MZ2	115,93	250,65	286,57	16,38
Japigia, Madonnella	MZ3	490,88	–199,02	705,16	313,96
Carrassi, San Pasquale, Poggiofranco	MZ4	845,32	–626,56	636,33	183,47
San Paolo, Stanic	MZ5	829,23	105,24	440,42	–345,38
Libertà, San Girolamo, Fesca	MZ6	487,30	54,42	398,71	7,77
San Giorgio, Torre a Mare	MZ7	847,58	–258,21	331,60	645,83
Carbonara, Ceglie, Loseto	MZ8	577,18	–136,91	492,24	–415,67
Palese, San Spirito	MZ9	503,67	120,45	246,97	–203,27

The consequent proportionate or relative change is indicate in the following Table 7.5

Table 7.5   The relative change

Name of microzone	Number of microzone	1999–2000	2000–2001	2001–2002	2002–2003
San Nicola	MZ1	0.08	0.26	0.14	0.30
Murat, Libertà, Madonnella	MZ2	0.07	0.14	0.14	0.01
Japigia, Madonnella	MZ3	0.28	–0.09	0.35	0.11
Carrassi, San Pasquale, Poggiofranco	MZ4	0.45	–0.23	0.31	0.07
San Paolo, Stanic	MZ5	0.61	0.05	0.19	–0.13
Libertà, San Girolamo, Fesca	MZ6	0.29	0.02	0.16	0.00
San Giorgio, Torre a Mare	MZ7	0.64	–0.12	0.17	0.29
Carbonara, Ceglie, Loseto	MZ8	0.40	–0.07	0.26	–0.18
Palese, San Spirito	MZ9	0.32	0.06	0.11	–0.08

We see that the average annual variations show the dynamics of rents per each microzone. The general trend is positive with strong growth in MZ1; MZ3; MZ4; MZ5 and MZ7. The data permit the Index to be calculated. In the following Table 7.6 it is possible to observe the SUNIA–OMING–POLIBA index:

Table 7.6   The SUNIA–OMING–POLIBA index

Name of microzone	Number of microzone	1999–2000	2000–2001	2001–2002	2002–2003
San Nicola	MZ1	1.08	1.26	1.14	1.30
Murat, Libertà, Madonnella	MZ2	1.07	1.14	1.14	1.01
Japigia, Madonnella	MZ3	1.28	0.91	1.35	1.11
Carrassi, San Pasquale, Poggiofranco	MZ4	1.45	0.77	1.31	1.07
San Paolo, Stanic	MZ5	1.61	1.05	1.19	0.87
Libertà, San Girolamo, Fesca	MZ6	1.29	1.02	1.16	1.00
San Giorgio, Torre a Mare	MZ7	1.64	0.88	1.17	1.29
Carbonara, Ceglie, Loseto	MZ8	1.40	0.93	1.26	0.82
Palese, San Spirito	MZ9	1.32	1.06	1.11	0.92

The index shows zones with different growth levels in rent. This information is useful indication for social housing policy. In San Nicola, for example, an urban regeneration process resulted in a significant growth of rent. It can be partially considered as a successful effect of the regeneration process. Similarly in San Giorgio, Torre a Mare, a peripheral zone in Bari, for example, the variability in rent index indicates the necessity for social housing investments in order to stabilize the development.

Urban Planning and Real Estate Markets

We have now presented a case of empirical housing modelling that may help us identify problems in an urban planning context. Moreover, the changes in rent level observed with the hedonic index may give important indication of effects caused by market regulation, in which case the various theoretical models discussed earlier potentially become helpful in informing planning decisions.

Urban renewal (in San Nicola) can, for example, imply a justification for the amenity (B) or policy (C) view. In the first case a rent increase is caused by a tangible quality improvement. In the second case (in San Giorgio) a rent increase is attributed to bottlenecks within the microzone. In Bari several microzones have experienced a significant annual rent growth, which may create affordability problems for potential renters. In Bari as a whole the growth of rent has to do with demographic and socio-economic changes. The market view (A) may be relevant in this case. A steep rise in residential rents may require a 'release of air from the rent bubble'. Investments in social housing or other measures may be helpful in order to abate excessive demand.

Concluding Reflections

In this chapter we explore the relationship between real estate market and urban planning with the help of hedonic index analysis. In the case of Bari residential index shows a significant increase in rents. In fact we can describe the development in terms of a 'rent bubble', a phenomenon that is harmful for real market in particular and urban development more generally. The three models that we have described in this chapter namely the *market*, the *amenity* and the *policy* view enable us to suggest different courses of action in order to contain the 'bubble'. These models suggest measures which extend from stimulating transactions in the housing market to new production of housing. We can also conclude that the application of the SUNIA-OMING-POLIBA index in Bari provides an opportunity to extend its application to other Italian cities as well.

Real estate market analysis offers useful information to urban planners. Planning decisions can abate price and rent changes and avoid 'rent bubbles' when these are considered harmful for an urban society. Our case study of Bari provides modest knowledge in this direction but further applications of the index developed here or of other hedonic price models may shade more light on the complex relationship between property market and urban planning.

References

Adair, A.S., Berry, J.N. and McGreal, W.S. (1991), 'Land Availability, Housing Demand and the Property Market', *Journal of Property Research* 8, pp. 59–69.

Bailey, M.J., Muth, R.F. and Nourse, H.O.(1963), 'A Regression Method for Real Estate Price Index Construction', *Journal of the American Statistical Association* 58, pp. 933–94.

Balchin, P. and Kieve, J. (1977), *Urban Land Economics*, London, Macmillan.

Brennan, T., Cannaday, R. and Codwell, P. (1984), 'Office Rent in the Chicago CBD', *AREUEA Journal* 12(3), pp. 243–60.

Case, B. and Quigley, J. (1991), 'The Dynamic of Real Estate Price', *Review of Economics and Statistics* 73, pp. 50–58

Case, B. and Shiller, R. (1989), 'The Efficiency of the Market for Single Family Homes', *American Economic Review* 79(1), pp. 45–56

Cheshire, P. and Sheppard, S. (1989), 'British Planning Policy and Access to Housing: Some Empirical Estimates', *Urban Studies* 26(5), pp. 469–85.

Cheshire, P. and Sheppard, S. (2002), 'The Welfare Economics of Land Use Planning', *Journal of Urban Economics* 52, pp. 242–69.

Clapp, J.M. and Giaccotto, C. (1992), 'Estimating Price Trends for Residential Property: A Comparison of Repeat Sale and Assessed Value Methods', *Journal of Real Estate and Financial Economics* 5, pp. 357–74

Court, A. (1939), *Hedonic Price Indexes with Automotive Examples in Dynamics of Automobile Demand*, New York, General Motors.

Evans, A.W. (1985), *Urban Economics – An Introduction*, Oxford, Blackwell.

Fischel, W.A. (1990), 'Introduction: Four Maxims for Research on Land-use Controls', *Land Economics* 66(3), pp. 229–36.

Gat, D. (1996), 'A Compact Hedonic Model of the Greater Tel Aviv Housing Market', *Journal of Real Estate Literature* 4, pp. 163–72.

Gatzlaff, D.H. and Ling, D.C. (1995), 'Measuring Changes in Local House Prices: An Empirical Investigation of Alternative Methodologies', *Journal of Urban Economics* 35, pp. 221–44.

Gatzlaff, D.H. and Haurin, D.R. (1997), 'Sample Selection Bias and Repeat-sales Index Estimates', *Journal of Real Estate Finance and Economics* 14, pp. 33–50.

Glasscock, J.L., Johanian, S. and Sirmans, C.F. (1990), 'An Analysis of Office Market Rents', *AREUEA Journal* 18, pp. 105–119.

Harris, B. (1994), 'The Real Issues Concerning Lee's "Requiem"', *Journal of the American Planning Association* 60(1) (Winter), pp. 31–34.

Healey, P. (1998), 'Regulating Property Development and the Capacity of the Development Industry', *Journal of Property Research* 15(3), pp. 211–27.

Henneberry, J., McGough, T. and Mouzakis, F. (2005), 'The Impact of Planning on Local Business Rents', *Urban Studies* 42(3), pp. 471–502.

Hoesli, M., Giacotto, C. and Favarger, P. (1997a), 'Three New Real Estate Price Indices for Geneva, Switzerland', *Journal of Real Estate Finance and Economics* 15(1), pp. 93–109.

Hoesli, M., Thion, B. and Watkins, C. (1997b), 'A Hedonic Investigation of the Rental Value of Apartments in Central Bordeaux', *Journal of Property Research* 14, pp. 15–26.

Hui, E.C.-M. and Ho V.S-M (2003), 'Does the Planning System Affect Housing Prices? Theory and with Evidence from Hong Kong', *Habitat International* 27, pp. 339–59.

Kauko, T. (2002), 'Modelling the location Determinants of House Prices: Neural Network and Value Tree Approaches', PhD thesis, Utrecht: Labor Grafimedia, <http://www.library.uu.nl/decollectie/proefschriften/11688main.html>.

Kauko, T. (2003), 'Planning Processes, Development Potential and House Prices: Contesting Positive and Normative Argumentation', *Housing, Theory and Society* 20(3), pp. 113–26.

Knight, J.R., Dombrow, J. and Sirmans, C.F. (1995), 'A Varying Parameters Approach to Constructing House Price Indexes', *Real Estate Economics* 23, pp. 187–205

Lee, D.B. Jr (1973), 'Requiem for Large-scale Models', *Journal of the American Institute of Planners*, 39, pp. 163–78.

Lee, D.B. Jr (1994), 'Retrospective on Large-Scale Urban Models', *Journal of the American Planning Association* 60(1) (Winter), pp. 35–40

Mclennan, D. (1977), 'Some Thoughts on the Nature and Purposes of House Price Studies', *Urban Studies* 14, pp. 59–71

Meese, R.A. and Wallace, N.E. (1997), 'The Construction of Residential Housing Price Indices: a Comparison of Repeat-Sales, Hedonic Regression and Hybrid Approaches', *Journal of Real Estate Finance and Economics* 14 pp. 51–73.

Mills, E.S. (1992), 'Office Rent Determinants in the Chicago Area', *AREUEA Journal* 20, pp. 273–89.

Monk, S., Pearce, B. and Whitehead, C.M.E. (1991), 'Planning, Land Supply and House Prices – A Literature Review', *Land Economy Monograph* 21, University of Cambridge.

Monk, S. and Whitehead, C.M.E. (1999), 'Evaluating the Economic Impact of Planning Controls in the United Kingdom: Some Implications for Housing', *Land Economics* 75(1), pp. 74–93.

Oxley, M. (2004), *Economics, Planning and Housing*, Basingstoke, Palgrave Macmillan.

Palmquist, R.B.(1980), 'Alternative Techniques for Developing Real Estate Price Indexes', *Review of Economics and Statistics* 62(3), pp. 442–48.

Pogodzinsky, J.M. and Sass, T.R. (1990), 'The Economic Theory of Zoning: A Critical Review', *Land Economics* 66(3), pp. 294–314.

Roca Cladera, J. and Burns, M. (1998), 'The Liberalisation of the Land Market in Spain', ERES and AREUEA International Real Estate Conference, Maastricht, 10–13 June.

Rosen, S. (1974), 'Hedonic Price and Implicit Markets: Product Differentiation in Pure Competition', *Journal of Political Economy* 82, pp. 34–55.

Sager, T. (2003), 'Rationality types in Evaluation Techniques: The Planning Balance Sheet and the Goals Achievement Matrix', *European Journal of Spatial Development* 2 (January), <http://www.nordregio.se/EJSD/>.

Sivitanides, P. (1997), 'The Rent Adjustment Process and the Structural Vacancy Rate in the Commercial Real Estate Market', *Journal of Real Estate Research* 13(2), pp. 195–209.

Wheaton, W.C. and Torto, R. (1994), 'Office Rent Indices and Their Behaviour Over Time', *Journal of Urban Economics* 35, pp. 121–39.

Appendix 7.1

Results of MRA carried out on the sample of 2353 observations

1999 (490 observations)

Model

R	R square	Adjusted R square	Std error of the estimate
.789(a)	.623	.615	591.52

Analysis of variance

	Sum of squares	df	Mean square	F	Sig.
Regression	277287998.471	10	27728799.847	79.248	.000(a)
Residual	167601190.582	479	349898.101		
Total	444889189.053	489			

Coefficients

	Unstandardized coefficients		Standardized coefficients	t	Sig.
	B	Std error	Beta		
(Constant)	4711.284	635.200		7.417	.000
VANI	942.167	37.550	.722	25.091	.000
DATA	−1.685	.273	−.175	−6.176	.000
MZ1	−485.989	156.769	−.091	−3.100	.002
MZ2	83.562	81.382	.033	1.027	.305
MZ3	135.024	116.719	.035	1.157	.248
MZ4	248.882	77.953	.104	3.193	.002
MZ5	−248.367	149.092	−.049	−1.666	.096
MZ6	83.562	127.715	−0.045	2.325	.001
MZ7	−289.692	229.663	−.036	−1.261	.208
MZ8	−168.707	95.126	−.055	−1.774	.077
MZ9	−17.006	107.011	−.005	−.159	.874

2000 (487 observations)

Model

R	R square	Adjusted R square	Std error of the estimate
.789(a)	.623	.615	552.9108

Analysis of variance

	Sum of squares	df	Mean square	F	Sig.
Regression	240595846.281	10	24059584.628	78.701	.000(a)
Residual	145518140.482	476	305710.379		
Total	386113986.764	486			

Coefficients

	Unstandardized coefficients		Standardized coefficients	t	Sig.
	B	Std error	Beta		
(Constant)	5463.217	478.475		11.418	.000
VANI	693.805	32.903	.605	21.086	.000
DATA	−2.006	.249	−.229	−8.042	.000
MZ1	−961.745	133.449	−.214	−7.207	.000
MZ2	−260.314	80.397	−.105	−3.238	.001
MZ3	142.396	110.560	.039	1.288	.198
MZ4	615.236	74.194	.277	8.292	.000
MZ5	92.811	125.738	.022	.738	.461
MZ6	87.306	125.716	−0.027	2.295	.02
MZ7	77.460	280.770	.008	.276	.783
MZ8	−50.953	86.066	−.019	−.592	.554
MZ9	−4.441	96.315	−.001	−.046	.963

2001 (510 observations)

Model

R	R square	Adjusted R square	Std error of the estimate
.793(a)	.629	.622	541.66930

Analysis of variance

	Sum of squares	df	Mean square	F	Sig.
Regression	248600449.314	10	24860044.931	84.729	.000(a)
Residual	146409412.301	499	293405.636		
Total	395009861.616	509			

Coefficients

	Unstandardized coefficients		Standardized coefficients	t	Sig.
	B	Std error	Beta		
(Constant)	2736.395	377.759		7.244	.000
VAN	775.454	29.128	.751	26.622	.000
DATI	−.813	.239	−.093	−3.395	.001
MZ1	−622.633	112.724	−.161	−5.524	.000
MZ2	−86.168	73.177	−.038	−1.178	.240
MZ3	−109.429	96.624	−.034	−1.133	.258
MZ4	−68.668	74.784	−.029	−.918	.359
MZ5	149.789	125.317	.035	1.195	.233
MZ6	88.965	125.716	−0.027	2.295	.02
MZ7	−236.643	275.758	−.024	−.858	.391
MZ8	−264.752	84.599	−.096	−3.130	.002
MZ9	70.805	98.380	.021	.720	.472

2002 (456 observations)

Model

R	R square	Adjusted R square	Std error of the estimate
.791(a)	.626	.618	391.2433

Analysis of variance

	Sum of squares	df	Mean square	F	Sig.
Regression	113978857.678	10	11397885.768	74.461	.000(a)
Residual	68116752.840	445	153071.355		
Total	182095610.518	455			

Coefficients

	Unstandardized coefficients		Standardized coefficients	t	Sig.
	B	Std error	Beta		
(Constant)	4278.051	228.750		18.702	.000
VANI	509.373	25.043	.622	20.340	.000
DAT	−1.838	.186	−.295	−9.883	.000
MZ1	−662.889	93.306	−.220	−7.104	.000
MZ2	−129.353	55.377	−.081	−2.336	.020
MZ3	277.205	74.899	.120	3.701	.000
MZ4	294.922	61.326	.164	4.809	.000
MZ5	229.897	99.918	.073	2.301	.022
MZ6	140.312	125.716	−0.027	2.295	0.02
MZ7	−228.689	199.978	−.034	−1.144	.253
MZ8	−92.927	63.271	−.050	−1.469	.143
MZ9	−20.872	68.980	−.010	−.303	.762

2003 (410 observations)

Model

R	R square	Adjusted R square	Std error of the estimate
.801(a)	.642	.633	461.9355

Analysis of variance

	Sum of squares	df	Mean square	F	Sig.
Regression	152419065.949	10	15241906.595	71.429	.000(a)
Residual	85140369.700	399	213384.385		
Total	237559435.649	409			

Coefficients

	Unstandardized coefficients		Standardized coefficients	t	Sig.
	B	Std error	Beta		
(Constant)	2908.301	228.646		12.720	.000
VANI	718.072	33.325	.703	21.548	.000
DAT	−1.368	.256	−.313	−5.336	.000
MZ1	−118.802	117.004	−.032	−1.015	.311
MZ2	−106.830	68.870	−.054	−1.551	.122
MZ3	556.267	107.798	.165	5.160	.000
MZ4	716.375	120.771	.364	5.932	.000
MZ5	−141.959	135.641	−.035	−1.047	.296
MZ6	133.327	125.716	−0.027	2.295	.022
MZ7	589.095	180.274	.100	3.268	.001
MZ8	−491.657	81.708	−.200	−6.017	.000
MZ9	−218.812	86.525	−.083	−2.529	.012

Chapter 8

Looking Inside the Plausibility of Contact in Aid Programmes and Partnerships

Domenico Patassini

A civilization is recognised by what it gives and what it researches ..., and also by what it has the power to reject. (F. Braudel, quoted in Aymard 2004)

Introduction

This chapter emphasizes the principle of plausibility to examine critical difficulties encountered by cooperation and aid programmes. These difficulties have been generally underestimated or ignored in evaluation. In contrast to cooperation theory, the success or failure of cooperation policies depends largely on the way in which contact between communities of donors and recipients (in a broad sense) develops and gains significance amongst the parties involved. The start up of an aid programme is a construction that takes shape in the negotiation phase and which is then followed by a system for resource mobilization. Contacts among parties may be new or tested in time; one-off or continuative; fertile and lasting, innovative, sporadic, a righteous duty, or even a dead weight to be shed as soon as possible. Contact evokes various reactions of assimilation, aloofness or rejection; it may encourage partnership, which is a key feature of effective development programmes. Assimilation, aloofness and rejection are themes of unquestionable interest for evaluation for the very reason that they indicate the plausibility of programmed actions.

A programme is plausible only if it responds to a discourse ethic; if it clarifies any possible compromise amongst varying cultures. Looking above and beyond the programme in itself (its cycle), enables us to enrich concepts such as feasibility and sustainability and thereby change perspective, without simply observing the effects created, or the process, with reference to pre-existing models.[1] The efforts of

---

1   Although the literature and practice may not correspond exactly, the main models are experimental, based on theory – realistic, pragmatic (goal free or utilization focused), and constructivist. Each model proposes its own design. The best known references are: Pawson (1997); Tilley (1997); Weiss (1998); Scriven (1973; 1991); Guba (1989); Lincoln (1989). Two useful Italian references are Stame (1998) and Moro (2005).

participants in aid programming should be focused on the building of a common language (however provisory and difficult that may be) before setting-up the actual partnership (Liebenthal et al. 2004). The assumption should be made that the programme is based on discourse ethics.

The chapter is organized in four sections. The first part takes into consideration and highlights the importance of the notion of contact within cooperation theories and programmes. The second part defines the concept of plausibility within the domain of discourse ethics, and highlights the entailing limitations. The third part considers partnership as a peculiar form of organizational ecology, and the fourth and last part discusses a number of evaluation models, with brief references to decentralized cooperation programmes as experienced by the University Iuav in Ethiopia between 1983 and 2005.[2]

Contact and Culture

A cooperation programme is an opportunity for contact between cultures. That opportunity may be created by diverse configuration factors, but often it undervalues or mystifies cultural factors. It may be requested by one of the parties involved, by a third actor, or derive from unforeseen opportunities or necessities. It may even be the fruit of the idea of an 'inverted world', wherein those on the receiving end see themselves as losers, or failures, and therefore attempt to rid themselves of the label of 'inferior culture'.[3]

For operational reasons (schedule, previous understandings, reference frameworks and alike), the programme takes for granted asymmetries, the most important of which being the unlikely 'rejection' of aid. Aid is at the centre of the cooperation relationship, seen as a net benefit at no or very little cost and as an opportunity to be grasped and exploited as best as possible – at times according to little apparent logic.

Distance between cultures in the design stage leads to a sort of marginality that is temporary but that conceals another form of marginality, which is more substantial. Since it is difficult to reduce culture to a character, whether it would be legal, economic, social, religious, environmental, linguistic and so on, and since none of these considerations constitutes a dominant function, 'culture' cannot be a form of loan to the cooperation agencies (Bateson 1972). This has been made very clear at the negotiating tables concerning the Country Programmes,[4] where various agencies discussed and interpreted (according to their own logics) the motivation behind contact, and laid out and proposed their own cultures of origin. This also happens in decentralized cooperation programmes, which are

---

2   A comment on this experience can be found in Patassini 2004. The title of this chapter is a throwback to Levine 1992.

3   A frequent attitude in Africa, that still conditions relations with the West in general – and not only with the former colonial powers. Globalization, with the discriminatory effects it produces, tends to enhance rather than diminish this attitude.

4   The programmes may be either bi- or multilateral.

usually anchored to simpler negotiating procedures and which expect less from the mediation process. Native cultures can influence contact and its outcomes decisively, but often are ignored, left behind, depriving thus the ex-ante evaluation of crucial demands and documentation.

The marginal position of cultures with regard to contact is therefore of primary importance for reflection. The second most important point is that of the modality and dynamics of contact – and it is here that cultures may have a great influence. These dynamics are an integral part of the design of a programme; they mould the beginnings, and together with specific components, define the theories and mechanisms.[5] In these circumstances the actions and allocations of resources are discussed, alongside the agreements over the timings of the programme, and whether interpretations of its significance are shared (be it explorative, experimental, cooperative, humanitarian, instrumental, political or so on). The various opinions expressed may create disturbance factors, relations which are symmetrical, supplementary, reciprocal, or altogether asymmetrical. These relations, though they may not influence the actual components of the programme, may, however, have some bearing on its realization, sustainability, and on future relationships among the parties. Outcomes of this type are present in all forms of cooperation: multilateral, bilateral, or decentralized; the south-south[6] or north-south models.

Only rarely is there convergence on the meaningfulness of the programme, even when there is formal agreement. The outcomes of contact require interpretation of the cooperation features, of the predominance of one perspective over another, mediation (dynamic balance) with or without 'detachment' from an original stance, assimilation, aloofness, or rejection.

The map of these outcomes serves to understand whether from within a system of meaningfulness, a communicative structure (language) and an ethic may emerge, and whether the resources are sufficient for evaluating the programme and comprehending the underlying logics. To give an example, the meanings of a capacity building programme cannot be assumed as given; they rather come into the arena while developing any action during which all the concerned parties learn how to acknowledge themselves in the provisionality of the outcomes. Looking at the accomplished, they discover logics and resources to reflect (to think over) and be reflected.

---

5   Theory-driven evaluation aids in assimilating the theoretical motivations of the programme, while realistic evaluation studies the outcomes as interaction between the mechanisms and the context. Motives and forms of contact could therefore guide the design of a programme and merit the attention of evaluation.

6   'South-south cooperation started to influence the field of development studies in the late 1990s, fuelled by a growing realization that poor nations might find appropriate, low-cost and sustainable solutions to their problems in other developing countries rather than in the rich north. It drew on clear examples of existing waste and alternative opportunities …' (Stigliz, *Capital*, May 2006). Globalization is, however, modifying this situation.

Plausibility and Evaluation

The particular focus of this chapter is not on the declared objectives of a programme, but on the discursive logic from which they derive, and which indeed they generate.

A programme may have as its declared objective the fulfilment of a specific demand, but take for granted the reasons and meanings behind the demand, thus leaving the problems of reciprocity to others. For other reasons the programme may be utilized as an opportunity, with the deliberate goal of improving the conditions of reciprocity in various fields – cultural, judiciary, economic social and environmental. The programme could fail however, to pay adequate attention to the modalities of contact, or the language that makes it possible, resulting in the need for tiresome preliminaries.

This provisional classification already supplies us with three different objects of evaluation: an action as a response to a demand; an action as an improvement in the conditions underlying reciprocity, or an action as the development of a language, of a discursive ethic. It is the latter that we take into examination here, thereby appealing to the concept of plausibility.

A programme of cooperation is plausible if it respects and develops the conditions of reciprocity between the parties – donors and beneficiaries – throughout the process, and its actual effects. This is realized according to the capacity of the programme to give sense to the contact in terms of language and ethics, thereby laying the foundations for partnership.

Three definitions of plausibility can be defined in this perspective. First of all, *plausible* is that which deserves or seeks praise, merit, approval, consensus (lat. *plaudo*). This is the case when the programme is constructed so as to respect and develop conditions of reciprocity, and although it may be limited, is approved by all. The parties would be equipped with a minimal dictionary that would enable all concerned to work together without excessive interference and maintaining, where necessary, appropriate distances.

In the second instance, *plausible* is that which may be accepted because it is logical, just, rational – even without plaudits; the programme may respond partially to the general objective, pushing for legislation or policies on one or more subjects involved. In this case, the dictionary could be reduced to a minimum or be enriched to some degree, but it would include 'faction words' – deployed for the meaning attributed to them without necessitating difficult interpretation. The lack of a 'translation' would thus make the dictionary hybrid, and the partnership asymmetrical.

The third definition considers *plausible* an action that is persuasive, but illusory, fallacious or even deceitful. This type of action could gain consensus, even if it is unjust, and has been perhaps the most common concept of plausibility in the history of north-south cooperation, enhanced by the myth of the 'inverted world'. This approach is spreading rapidly also in south-south cooperation, resulting in implications leading to the archiving of post-colonial cooperation policies.

These three declinations of plausibility are not necessarily disjunctive, so far as they recognize in plausibility the conditions wherewith to improve the dialogue between the parties, but characterize in a different way the logic of the discourse and pose specific questions to evaluation.

The Communicative Process and Dialogue Ethics

A cooperation policy that upholds the moral principle of discourse ethics should rely on collective dialogue.[7] But how should these dialogues be set up, amongst whom, and when? How should the 'contact community' in which to develop the dialogue be defined? And should this community be anchored institutionally to the programme, or should it have its own life within a logic of permanent contact amongst those who are anyway involved in some sort of dialogue, with or without the formal policies of international cooperation?[8]

If by cooperation we mean a communicative, rather than an instrumental activity, the programme must not limit itself to offering the best means for realizing established ends, but should, whereas, make a relevant contribution[9] to the building of a communicative context. This is why evaluating the actions of the programme requires a critical theory of cooperation able to highlight the communicative problems which arise in both the single-faction and common decision-making processes.

In this regard, Habermas (1983) offers us a point of reference when he says that the theory should be empirical, interpretational, lawful, and practical. He states that at the base are the fundamental rules of universal pragmatism: to communicate in an understandable, sincere, legitimate and truthful way; but above all, that in the relationship between communication and power, certain structural, organizational and political barriers are forcefully erected.

Given the pragmatic rules and having recognized the limitations, it is important to understand to what extent cooperation is oriented towards a critical participation, understanding, a genuine agreement or else towards a mere compromise or even abuse. In either case, an aggregation of given preferences cannot be proposed (with the ensuing well-known problems of trade-off) in that the programme may contribute in modifying them.

It is only through genuine dialogue that there may be a tendency towards (or even a lead towards) a discursive, sustainable situation. Asymmetries could be considered educational, rather than distortions: indeed, where necessary they may be cancelled, or else expanded whenever this may aid reciprocity. A critical and conscious participation (including political limits and legitimization) may help in defining a form of plausibility which would then be subjected to evaluation.

Set aside from the needs of refinement of the means and ends (consequently of secondary importance), the 'evaluation of the threats to the conditions of *free*

---

7 Reflections from Moroni 1997, 177–182.

8 The so-called 'minor policy' could rely on various forms of networks, with many and diverse subjects within differing or similar problems; differently 'free': a mobile political marketplace for the formation of capacities, potentialities, and new meanings.

9 According to Forester (1985), cited in Moroni 1997, 178.

dialogue and of discussion free of dominion' becomes of increasing importance and utility. This underlines how communicative rationality is closely linked to questions of liberty and justice (Forester 1985).

Discourse ethics is not a stranger to criticism. Those who support 'rational communication'[10] highlight the importance of transparency, awareness, criticism, but above all, the fact that there is a deontological ethic wherein the just, and not the good, is the ethic of paramount importance. This is fundamental in the relations between different cultures, wherein there are scenarios of varying encounters from which to extract, when needed, the protocols of cooperation.

Critics and fault-finders are concerned with the 'perfecting of the decision-making techniques' (see Amin and Thrift 2001), and start their case by citing difficulties due to manipulation, invested interest, frauds, wrongful attribution of preferences, false identities, institutionalized disparity, information and capacity asymmetries. It is clear that suggestions made via erroneous assumptions regarding communicative rationality are of a somewhat dubious contribution to the discussion on cooperation and partnership. Therefore we need to move from the 'best practices'[11] to the building of plausible contacts – listening to the 'voices from the frontiers' and giving heed to the difficulties mentioned by the critics.

These pathologies emphasize the procedural limit that could overshadow the substantial values to be pursued and which would lead to a sort of idealism of difficult understanding. Communicative action would be oriented towards understanding, and an acceptance of not always choosing the most obvious public decisions.

So as to guarantee the correct conditions for dialogue, the parties should have equal opportunities to intervene and to listen. This is valid both for those directly involved in the negotiations, and also for those behind the negotiators.

Interim responses to these criticisms are to be found within radical democracy and in approaches that consider the development of capacities a fundamental right. Both attitudes reject the simplistic notions of tolerance and acceptance of differences; they embrace distribution of power, active citizenry, and do not exclude the idea that the building of conditions may be based on dialogue which is explorative, constructive and inter-generational (the meta-discourse ethic) (see Elster 1986). Indeed they go beyond. Radical democracy 'requires the democratization of institutions and the rise to power of weaker groups within a policy of vigorous but fair competition between parties' (Amin and Thrift 2001), a kind of competitive pluralism. Those theories that place capacities (see Sen 1992, 1999; Unger 1998). at the heart of the debate hold that rather than the fair distribution of resources, priorities should lie in the creation of conditions whereby parity of human liberties is established.

---

10 Reference is to deliberative democracy based on Rawls's concept of justice-through-association (neo-contractualism) and to Habermas's thesis that public moral agreement is possible when there is rational communication.

11 Generally speaking, the 'best practices' are those that respond to demands of feasibility, efficiency and sustainability that however die out before the conclusion of the full programme cycle.

Partnership as Peculiar Form of Community of Practices

When considered as a peculiar form of community of practices, the partnership model might become a foundation of the programme in the sense that it can be (either tactically or strategically designed) a valuable source of hybrid institutions, of incremental social capital and, in a broader sense, of value added.

The evaluation literature on partnership (its theoretical foundations, like prisoner dilemma, Coase's theorem, but also related to the assessments on a variety of practices) is far reaching, and a useful publication is that edited by Andres Liebental, Osvaldo Feinstein and Gregory Ingram in 2004 for the World Bank series on evaluation and development. But, few attention has been given so far to the way partnerships emerge from 'contacts communities', to what extent partnerships are able to interpret community discourses, develop social capital and discursive ethics. It is about contact outcomes, a translation of its features, which can then produce significant effects on programme efficacy, efficiency and relevance.

Though our present discussion focuses on what happens just before structuring the partnership, it is a fundamental component of evaluation design and deserves great attention.

As Ingram says, 'partnership has become much more important in development work in recent year. Awareness of limitation of government activity is growing and partnerships among state, private business, and civil society organizations are increasingly used to deliver the goods and services required for balanced, broadly based growth and poverty reduction' (Ingram 2004, xi). Moreover:

> ... partnerships are often essential to deal with the added complexity and the larger number of agencies, groups and stakeholders involved. Another factor encouraging assistance agencies to seek partnerships has been their broad move away from imposing conditions and toward a concern with development effectiveness. The focus on results calls for less hierarchical relationships between providers and recipients of aid, and a move from 'You do as I say' to 'We do what we agreed'. (Ingram 2004, p. xi)

So, a certain degree of equity might be guaranteed at least in theory.

Partnerships do help in reaching a number of results. 'When done well, partnership...helps to ensure that things are done right, improving the efficiency of resource use ... that the right things are done[12] ... it constitutes a means of capacity building[13] and institutional development ... provides accountability for activity promised and for resource used[14]' (Ingram 2004, p. xii). It is a programme resource by itself and outcomes can change for the role it plays.

Stern (2004, quoted in Ingram 2004, xiii) found empirical evidence from evaluation research related to partnership capacity to reduce costs of implementation, to make effective knowledge transfer, and to reflect public-goods dimensions in addition to private ones. It has even been shown that benefit

---

12 Need of local ownership of development programmes.

13 Whenever participants do not enjoy organization and skill available to others.

14 That is mutual accountability of one partner to another or to the population the partnership is intended to serve. It helps also fighting against corruption.

obtained through partnership can outweigh any detrimental effect of the more protracted decision-making process.

We said that partnership is an operational result of contact communities, but it is also a source of new social capital that might foster cooperation policy. As Stern underlines

> ... a new organizational ecology ... raises new questions for evaluators, centring on justification and rational of partnership. For example: 'In what ways do the goals pursued require partnership working? What are the goals and do they determine the scope of specific partnerships? How inclusive are partnerships? Who was involved in the planning and initial definition of partnership tasks? How important is knowledge transfer in a particular partnership? (Quoted in Ingram 2004, p. xvi)

Additional questions can be added to the previous ones: who is acting as leading partner, is the leadership changing over time? How is the network expanding or shrinking, how is its internal structure changing, and what links are there to external environments? Which are the shared values and processes, how are they changing during the programme cycle (if any cycle can be considered relevant)? What remains at the end? Valuable contributions in this respect come from network analysis that has been applied, for example, to assess social capital produced during participatory planning experiences.

In short, a variety of languages and discursive ethics emerge within contacts communities. Languages and ethics do depend upon the way partnerships are able to understand the richness of contacts. Unfortunately, very often, partnerships adopt a poor language, i.e. the procedural languages of protocols to build confidence, reliance and trust in a short time. In this way they certainly help to make manageable the negotiation domain, but at a high risk of being stuck to formulations and terminology of the programme.

Despite the approach followed (theory-driven, process or outcome oriented), evaluation easily reflects a mere accountability to beneficiaries and people affected by the programmes. Languages and discursive ethics go hand in hand while creating a policy community to negotiate the meaning given to the partnerships and programmes, but also to appreciate the meaning of negotiation.

Evaluation Designs: Directions and Results

To use a geometric metaphor, evaluation could be seen as an 'intersection': it intersects an action (that can be represented by a function) at various points and there are as many designs as there are plausible combinations of the intersecting angles. Let us suppose that the function is the definitive itinerary and that it is 'examined' at varying points. The designs therefore respond to differing mandates, be it general or partial. As will be discussed, at least six mandates or designs can be identified. Each of these represents an example of professional and university experiences of urban planning in Ethiopia from 1983 to 2004.

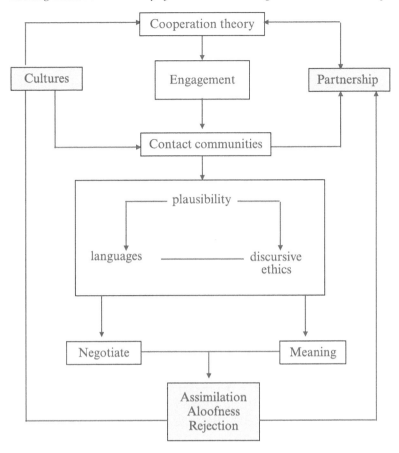

Figure 8.1   Evaluation design: cooperation theory and programme plausibility

The first design, which could be defined *general criticism*, attempts to identify the theories and logic systems behind cooperation that trigger the actual process. The reasons may vary, as may the start up of the programme. One of the most relevant themes at this stage is the financing method that might enhance the donor's 'visibility'. The experience of the Italian Trust Fund for urban capacity building in Ethiopia (2000–2004) demonstrates how a programme, which is well-built, planned, and sensitive to reciprocity may fail due to bad choices in financing (in that case, arranged by the World Bank and its Public Service Capacity Building Programme, Ministry of Capacity Building 2002, 2003). The deadlock that ensues, apart from dating the programme itself, has a very negative bearing on the contact communities, and may often lead to misunderstanding and even hostility.

A second evaluation design may be directed towards the *building of contact communities*, whether that is despite of, or in contrast with current cooperation policies. It is well-known that these policies may influence either positively or negatively the contact communities; in the outcome either encouraging or hindering their growth. This may certainly be a valid topic of evaluation, but perhaps we should ask ourselves how should we build these communities, and which rules of pragmatism[15] should be employed (mindful of limitations, barriers, asymmetries and pathologies) so as to build what sort of capacity. The Addis Ababa Master Plan of many years ago (1983–1986) provides us with an interesting history (see Bruschi 1986; Ceccarelli 1987; Diamantini and Patassini 1993). An offer made by the University Iuav – three million dollars from the Italian government in 1983 – was appreciated for its *technical-organizational model* (presented as *capacity building*), which, together with the Master Plan, foresaw the development of a planning office divided in five units. One of these, the implementation and follow-up unit, whose role was to monitor the city and attempt to create communication between ministries and corporations, was supposed to evolve into a tool of planning governance. We were well aware of the difficulties entailed in the exporting of Western models, and above all of the limit pointed out in a recent publication by Amy Chua (2004). In a review of her text, Umberto Galimberti says 'the version of capitalism as exported to the non-Western world is of a fundamentally liberalist kind, and only on rare occasions does it contemplate the existence of mechanisms of redistribution' (Galimberti 2004, 43). In the Ethiopia of the 1980s there was neither a market nor democracy, which when specifically defined constitute the two cornerstones of Western culture and are essential conditions for urban, territorial and environmental planning. The Iuav waived these conditions, but not for reasons of constraint, laziness or servility; there was simply too much uncertainty regarding their legitimacy and effectiveness. In her book, Chua demonstrates the perverse outcomes of the combination of the opening up of the market and democracy in African countries without even mentioning the tragic breakdown in identity mechanisms and the symbolic circuit so well described by Basil Davidson in *The Africans. The Entry to Cultural History.* In most of the African way of thinking a dynamic co-existence of opposite principles endures time, that of Fate and Justice (that is Oedipus and Job) due to the motionless aims of ancestral laws that cannot be changed, but also to the irresistible forces of life and human nature that try to change such a system of laws. Had Chua taken this into account, her conclusions would have been even more critical.

Under these conditions of obvious non-exportability (and danger) of the model, every action (including the *capacity building* programmes) runs the risk of being absorbed by and thus strengthening bureaucracy, and further stimulating conflicts. It was not at all easy to be conscious of this and translate it into an experience of planning and of cultural and scientific cooperation.

At the outset, certain precautions were taken. The inter-institutional *Steering Committee* was provided with a tool for the co-ordination and monitoring of what was happening in the city. This tool was the *implementation and follow-up unit*. It had never happened before that ministries, corporations and authorities sit round a table to discuss public projects or even merely the integrated programming of expenditure. This was an opportunity that had not been foreseen even by the

---

15 Procedural features of contact include entry, mediation, institutional design, the role of the beneficiaries, learning, reciprocity, and sustainability planning.

Planning Commission, which was the overseeing authority of the five-year Soviet-style programming. This debating table triggered a mechanism that although centrist, managed to link up sectorial programming to public spending and to an idea of the growing city.

Particular attention was paid to the *kebeles* – 287 administrative units similar to Anglo-Saxon style 'neighbourhoods' with an average population of 5,000 in each – where, according to the regime, the redistribution mechanisms (welfare) went hand in hand with political loyalty. A survey conducted on a stratified sampling of the *kebeles* revealed how the link between welfare and political loyalty varied according to living conditions. In the wealthier neighbourhoods, where there was less demand for welfare, there was less political control; in the poorer areas, conversely, the greater demand for welfare created tighter control, and the denial of citizenship rights. What occurred in the illegal neighbourhoods, whereas, was a combination of traditional solidarity and collusion. Engaging with these acquisitive dynamics means not only carrying out continuous testing of the land management policies and of the residential assets already under much pressure, but also taking the side of those deprived of citizenship rights.

AAMPO (Addis Ababa Master Plan Office) sought to identify possible courses of action despite working under a dictatorial regime. These were opportunities which were legitimized, it is worth noting, by a multi-ethnic planning office that was not driven by material incentives, and, at the same time, by a sense of the social utility of the mission: a real contact community.

Paradoxically, the positive results actually caused an unexpected failure, not due to failings of design and implementation of the project, but to the underestimating of the possible success of the initiative and of its implications for the national planning system. What was supposed to become the new municipal planning office was transformed in 1987 into the National Urban Planning Institute (NUPI), the creation of which brought about the first national urban and regional planning law that led to the current planning system[16].

Problems of underestimation are frequently highlighted by Albert Hirschman (1975). He says:

> Since we tend necessarily to underestimate the capacity that we have to give a creative response to any possible difficulty, we need also to underestimate those very difficulties by more or less the same extent, so that we are led by the converging action of the underestimates, which tend to compensate each other, to carry out actions that are within our reach, and that otherwise we would not have the courage to attempt. This principle is of sufficient importance to warrant a name, and since we are theorising on the need for an invisible hand that conceals difficulties to our benefit, I suggest we call this principle 'the hiding hand'. (Hirschman 1975, p. 24)

This 'hand' implies that we have overestimated the difficulties posed by a bloodthirsty regime and have been rendered short-sighted as regards the possibility of modifying planning practices.

---

16 A second national planning law is currently under discussion.

Fifteen years ago NUPI was an innovation not foreseen as an outcome of AAMPO, but endogenous – a small crack in the monolith of the regime. The fact that today it is little more than a wreck is due to a number of factors, the discussion of which would require very close examination. Despite this and being chained to a terrible system of command and control, Aampo moved light years ahead of other institutions while at the same time involving them, and has left a memory that the 'survivors' of the native population do not hesitate to define unique and unrepeatable.

Evaluation could well stop here, leaving to others the job of implementation analysis, and the updating of the planning processes to the new mixed economy system introduced in 1991. It is worth remembering, however, that the experience of this particular contact community, besides perfecting a planning language which is very different from those of the past, has defined a completely new horizon of discursive ethics.

The third evaluation design may be oriented towards the *plausibility of the programme*. The programme may be evaluated as a 'generator' of meanings (of meaningfulness, limits, or borders), but also for the efficiency and efficacy of the languages used to describe and motivate it; in other words, for its capacity to correct the original asymmetry caused by the logic of aid or the aforementioned idea of the 'inverted world'. These are the capacities that raise negotiations above the logic of mere exchange. This is what occurred concerning the safeguarding and conservation of cultural heritage in Ethiopia, a country with other, more important priorities.

Using the Master Plan the inventory of the capital's cultural assets was reconstructed (nearly 300 buildings and sites), seeking in the process to highlight the various stages of the evolution of the city and contextualize each object. Tecle Mahari sought in this light to update the meanings of two assets: the spa resort at Yirgalem in Sidamo (Tecle Mahari 1990) and a historical *ras* residence in Addis (Tecle Mahari 1993). The qualitative outcome of these two projects reaches well beyond simple renovation proposals. The first meant that environmental defence could be created by exploiting the symbolic value of the site, and the sacredness attributed to it by the Sidamo peoples. This is an action with far greater efficacy than any safeguarding legislation.

The second project sought to restore to the renovated building its context, reconstructing an urban centrality that the residence had lost, and a visibility that general urbanization had erased. This building was a royal concession in the foundations of the city, and is part of a network on which the plan and the urban structure of the city was developed, at the foot of the Entoto highlands.

The two projects together send a loud and extraordinary signal of continuity and pacification; they indicate how the history of the country must not be overwhelmed by the whims of politics: indeed, it is politics that delays the processes that are destined to occur. 'Suffering under God's environment', Wolde-Mariam (1991) would say. These projects have materialized in objects and reflect the strong ties that exist between tradition and modernism, thereby making less problematic the cohabitation of the two. They reproduce the benchmark to which the 'symbolic circuit' of the African civilization refers, so as not to die. Such a circuit represents the interaction of man and environment. A long time ago, E.A. Gutkind (1956) outlined three principal chains of transformation.

> The most general is the change in the interaction of man and environment from an 'I-Thou' to an 'I-It' relationship. The second chain of transformation is more explicit in character; it shows man's reaction to his environment in successive stages, ranging from fear and defence to confidence and aggressiveness and, finally to growing understanding and responsibility ... The third chain represents the widening scale and the changing experience of space, the latter intimately related to the notion of the universe'. 'When the first relationship became indirect and estranged, when it turned into an 'I-It', disintegrating the symbolic and magical bonds between man and environment and upsetting the oneness of man's functional and personal life, the growing abstractness and the ensuing disunity produced the amorphous character of all modern towns.

The fourth evaluation design takes into consideration the *outcomes*. The results concern the outcomes of contact in the specific context of the programme or in more general terms: assimilation, aloofness, or rejection. One undergraduate thesis (Cattaneo and Giordano 2003) proposed an examination of a minor urban centre, Melka Jebdu, and contested the official hypothesis that it is a mere offshoot of the regional capital, Dire Dawa. The study attempts to make a historic reconstruction and reveals how the urban effect (the supply of events otherwise absent) within rural contexts can be produced even by small residential areas, and are gravitational centres of commerce in areas crossed by multiple frontiers. This study has, in effect, recognized the settlement as a hub of opportunity and of multi-ethnic community value. Meetings held in the region were conducted simultaneously in Oromo, Kottu, Somali, Amharic, and, obviously, English and Italian. Failing to recognize place identity and its capacity for self governance means demolishing even the few resources that the place may activate independently, without external aid. Despite a sceptical interlocutor (the Municipality among all), this point was strongly pressed, and in the new municipal statute for Dire Dawa, Melka Jebdu has been recognized as *kebele* #1, with all the economic, social and organizational benefits that this entails. This experience helped in bringing the parties together.

Evaluation designs may also concern *cultural orientations* that guide the cooperation and condition the building of contact communities. Some studies offer interesting considerations. One of these (Piovesan 1995) evaluates a site and services project carried out in the area of Nefas Silk at Addis at the end of the 1980s: the building of 3,000 homes in an area of expansion for low-income families, funded by the World Bank. The project made a correct interpretation of the logic of the *key-projects* in the master plan; but the most interesting aspect is not so much conformity to the plan, as the fertility of the unforeseen shifting between objectives and realizations. What emerged, in fact, was that a drifting away from the original goals (a slight shift within the beneficiary target) enhanced traditional *mixitè* that is a co-existence of activities of various genres: artisan, trade and small-scale services, all of which produced positive effects on employment and integration of the neighbourhood into the city itself. The cross-financing model is also applied in developed countries within mixed-income communities and it seems to ease the programme sustainability. Today, the site and services approach is under scrutiny, if not blamed for a number of things: it is considered a top-down procedure, lacking participatory action; it asks for high costs of government procurement and contracting (that increase the overall transaction costs). Often, the design of housing typologies and public spaces does not meet population needs; projects are sometime implemented in critical locations or vacant lands lacking infrastructures and facilities. What's more, by facing difficulties in cost recovery, the project holders are willing to sell or rent to third parties, failing in their project target.

The project outcomes in Addis Ababa are less critical than expected. This experience seems to spell out how the struggle against poverty must be addressed by proper differentiation: below subsistence thresholds, it is to be fought with projects 'for the poor', but just above, mainly in urban contexts that are 'incubators of externalities', it is better off with interventions, right from the design phase, that aim at integration. The city as an incubator of externalities advises the contact communities.

A final design may be one oriented towards the building of a *general frame*, despite all the difficulties that this would involve. In this perspective, the evaluation design could respond to a variety of mandates all at once (criticism, building of contact communities, looking for plausibility, outcome and so forth). A design of this type, in the plural by construction, could aid wide-angle reflection on the sense of cooperation, and on the relationship between policies and contexts.

Conclusions

As a closing to this chapter, this section proposes to use a key notion from the classic book by William Easterly, *The Elusive Quest for Growth* (2001). 'In his iconoclastic book, which skilfully combined a history of economists' growth theories with a devastating empirical analysis of the failure of international efforts to spur third world development, Easterly discussed 'incentive matters' (Postrel 2006) like subsidies to help development or standard tariffs. In the 2006 book, *The White Man's Burden*,

> Easterly turns from incentives to the subtler problems of knowledge. He contrasts the traditional 'planner' approach of most aid projects with the 'searcher' approach that works so well in the markets and democracies of the West. Searcher treats problem-solving as an incremental discovery process, relying on competition and feed-back to figure out what works. 'A planner thinks he already knows the answers', Easterly writes. 'A searcher admits he doesn't know the answers in advance; he believes that poverty is a complicated tangle of political, social, historical, institutional and technological factors'. Planners trust outside experts. Searchers emphasize home-grown solutions.

If the planner as a searcher admits an aid project or programme cannot provide answers in advance, it means he is looking for a plausibility of what he's doing. The idea of a planner as a searcher might lead to several interpretations.

It acknowledges evaluation as a planning practice: i.e., the planner as a developer of conditions of reciprocity, as a promoter of different types of knowledge (either ordinary or expert), as a persuader and as a source of true or idle hopes. Within programmes and partnerships, looking at the intensity and the ways of such a searching exercise could help assess impacts, outcomes, or the process itself. But it is not simply a matter of programme effectiveness, nor of its 'text' (the attached documentation on analysis, projects and rules). Rather, and with a certain hesitation, we get closer to a type of Geddesian planner which 'thinks the plan, like a chess-board of the game, will be more effective the more is left unfinished with far-sightedness: the more qualified it is to mobilize further knowledge and action of citizens' (Ferraro 1998, 268).

Looking for plausibility the planner works in a condition of uncertainty: its design capacity and authority are both retrenched against increasing duties 'from listening [to] the voices of local communities, to negotiating through and with actors till a continuous updating of his own proposals' (Ferraro 1998, 269). Yet, looking for plausibility allows even for acknowledging planning as a pedagogical activity where the interaction between expert and ordinary knowledge does not warrant any outcome, but lightens the approach to the places.

Plausibility is a multidimensional perspective that is not only rarely valued, but often taken for granted for laziness or opportunism. This perspective asks from evaluation an effort that goes far beyond the programme theory. As indicated above, searching for looking plausibility puts the programme and its theory in the background. It may help to answer simple but crucial questions: Why and to what extent do both donors and recipients agree on that programme? What are they leaving out? Which contact devices do they use?

Mainstream evaluation literature is rooted in the idea of theory-driven design to the programme cycle, and is committed to development intervention hypotheses shaping the programme. As indicated by Scriven (1991, 286) a programme theory is 'a theory about the way that a programme brings about its effects (descriptive programme theories) or about ways in which it could bring about improved effects, or the same effects in an improved way (normative programme theories).[17] Moreover, 'criticism of evaluation theories is often based on a confusion between prescriptive and descriptive theories. For example, theories about how evaluation should be done in order to be valid cannot be criticized on the grounds that these practices are rarely followed or hard to implement. The only issue is whether there are some better alternatives. Otherwise, the original comment stands as correct, and the criticism is irrelevant – it is simply a way of saying that real-world evaluations will be invalid. Although they frequently make this mistake – by criticizing evaluation theories for not describing what evaluators do, even when the theory is intended to describe what they should do' (Scriven 1991, 56).[18]

---

17 The importance of normative programme theories has been stressed by Chen 1990.

18 For analyses of evaluation theories see Shadish (1991), Cook (1991), and Leviton (1991).

An aid programme is heavily affected by cooperation theory. That is a crucial issue for a 'developing world that has become more volatile and where violent conflict is common. Increasingly, resources for development assistance are mortgaged to meet the needs of peacemaking and peacekeeping' (Picciotto 1997, 207). The point is that aid programmes need to be designed and judged 'not only in cost-benefit terms as freestanding investments, but also as vehicles for policy reform and capacity building' (Picciotto 1997).

Unfortunately, the due attention to partnership often banishes the 'environment' that creates the conditions for its generation, an environment that has been called 'contact community'. Moreover, beyond the specific effects, an aid programme has a great influence on cultural relations, in the form of assimilation, aloofness or rejection.

A tentative evaluation design of an aid programme can be set up on the basis of the rationale indicated in the figure which provides some principal components for the analysis.

The evaluation designs briefly discussed in this chapter reveal how a programme can contribute to the building of contact communities with the aid of discursive logics, but also to what extent it may play a role in approaching or distancing cultures themselves – more than it could be recognized in the Country Programmes or in ex-post evaluation documents like the annual reports of the Operation Evaluation Department of the World Bank.

These are substantive questions which are all too often ignored, and that only rarely become an explicit object of evaluation. However, as the experience discussed suggests, the programmes are one thing, and the discourses that accompany them are another. It is a terrible shame that these discourses are lost in a din that deafens the timid appeal of cultures and their languages.

References

Amin, A. and Thrift, N. (2001), *Cities: Reimagining the Urban*, Cambridge: Polity Press.

Aymard,M. (2004), 'Empire and World Order', ed. S. Poli, 'The American Emperors', *La Repubblica* 23 April.

Bateson, G. (1972), *Steps to an Ecology of Mind*, New York, Ballantine Books.

Breheny, M. and Hooper, A. (eds) (1985), *Rationality in Planning*, London, Pion.

Bruschi, S. (1987), 'Il master plan di Addis Abeba', *Urbanistica Informazioni 1*, Rome, Istituto Nazionale di Urbanistica.

Cattaneo, A. and Giordano, M. (2003), *Melka Jebdu. Verso una nuova città*, Milan, Politecnico di Milano, Faculty of Architecture.

Ceccarelli, P. (1986), *Il piano regolatore di Addis Abeba. Un caso di collaborazione tecnica tra diversi livelli di sviluppo*, Bologna, Saie.

Chelimsky, E. and Shadish, W.R. (eds) (1997), *Evaluation for the 21st century. A Handbook*, Thousand Oaks, CA, Sage Publications.

Chen, H. (1990), *Theory-Driven Evaluations*, Newbury Park, CA, Sage Publications.

Chua, A. (2004), *L'età dell'odio. Esportare democrazia e libero mercato genera conflitti etnici?*, Rome, Carocci Editori.

Davidson, B. (1969), *The Africans. An Entry to Cultural History*, Harlow, Longmans.

Diamantini, C. and Patassini, D. (1993), *Addis Abeba. Villaggio e capitale di un continente*, Milan, Franco Angeli.

Easterly, W. (2001), *The Elusive Quest for Growth*, Cambridge, MA, MIT Press.

Easterly, W. (2006), *The White Man's Burden*, London, The Penguin Press.

Elster, J. (1986), 'The Market and the Forum: Three Varieties of Political Theory', in J. Elster and A. Hylland (eds) , *Foundations of Social Choice Theory*, Cambridge: Cambridge University Press.

Elster, J. and Hylland, A. (eds) (1986), *Foundations of Social Choice Theory*, Cambridge: Cambridge University Press.

Feinstein, O.N. and Picciotto, R. (eds) (2001), *Evaluation and Poverty Reduction*, New Brunswick, NJ, Transaction Publishers.

Ferraro, G. (1998), *Rieducazione alla speranza. Patrick Geddes Planner in India, 1914–1924*, Milan, Jaca Book.

Forester, J. (1985), 'Practical Rationality in Plan Making', in M. Breheny and A. Hooper (eds), *Rationality in Planning*, London, Pion.

Galimberti, U. (2004), 'Il gran mercato dell'odio', *La Repubblica*, 1 April.

Guba, E.G. and Lincoln, Y.S. (1989), *Fourth Generation Evaluation*, Newbury Park, CA, Sage Publications.

Gutkind, E.A. (1956), 'Our World from the Air: Conflict and Adaptation', in W.L. Thomas Jr (ed.), *Man's Role in Changing the Face of the Earth*, Chicago, University of Chicago Press.

Habermas, J. (1985), *Etica del discorso*, Bari, Laterza.

Hirschman, A.O. (1975), *I progetti di sviluppo. Un'analisi critica di progetti realizzati nel Meridione e in paesi del Terzo Mondo*, Milan, Franco Angeli.

House, E.R. (ed.) (1973), *School Evaluation the Politics and Process*, Berkeley, CA, McCutchan.

Ingram, G.K. (2004), 'Overview', in A. Liebenthal, O.N. Feinstein and G.K. Ingram (eds), *Evaluation and Development. The Partnership Dimension*, New Brunswick, NJ, Transaction Publishers.

Levine D.N. (1972), *Wax & Gold. Tradition and Innovation in Ethiopian Culture*, Chicago: The University of Chicago Press.

Liebenthal, A., Feinstein, O.N. and Ingram, G.K. (eds) (2004), *Evaluation and Development. The Partnership Dimension*, New Brunswick, NJ, Transaction Publishers.

Mahari, T. (1990), *Stazione termale a Yirgalem, Etiopia*, Venice, Iuav.

Mahari, T. (1990), *Restauro di una casa di ras*, Venice, Iuav.

Ministry of Capacity Building (2003), *National Capacity Building Plan 203/4–2007/8. Consolidated Strategic Public Sector Capacity Building Programme*, vol. I, Addis Ababa, Mcb.

MCB, (2002), *Project Proposal on Strengthening the Ministry of Capacity Building*, Addis Ababa, Mcb.

Moro, G. (2005), *La valutazione delle politiche pubbliche*, Rome, Carocci.

Moroni, S. (1997), *Etica e territorio. Prospettive di filosofia politica per la pianificazione territoriale*, Milan, FrancoAngeli.

Patassini, D. (2004), 'Cera e oro: prove di reciprocità dello Iuav in Etiopia', *Equilibri* 2, Bologna, Il Mulino, pp. 249–270.

Patton, M.Q. (1997), *Utilization-focused Evaluation*, London, Sage Publications.

Pawson, R. and Tilley, N. (1997), *Realistic Evaluation*, London, Sage Publications.

Picciotto, R. (1997), 'Evaluation in the World Bank: Antecedents, Instruments, and Concepts', in E. Chelimsky and W.R. Shadish (eds), *Evaluation for the 21st century. A Handbook*, Thousand Oaks, CA, Sage Publications.

Piovesan, V. (1995), *Politiche e progetti di low-cost housing in Etiopia: il caso di Addis Abeba*, Venice, Iuav.

Pitman, G.K. Feinstein, O.N. and Ingram, G.K. (eds) (2005), *Evaluating Development Effectiveness*, New Brunswick, NJ, Transaction Publishers.

Postrel, V. (2006), 'How to Solve the Foreign Aid Conundrum', *International Herald Tribune*, 18 March, p. 8.

Schwandt, T.A. (2004), 'The Centrality of Practice in Evaluation', keynote address presented at the inaugural meeting of the Swedish Evaluation Society, Stockholm, 22 April.

Scriven, M. (1973), 'Goal Free Evaluation', in E.R. House (ed.), *School Evaluation the Politics and Process*, Berkeley, CA, McCutchan.

Scriven, M. (1991), *Evaluation Thesaurus*, Newbury Park, CA, Sage Publications.

Sen, A. (1992), *Inequality Re-examined*, New York, Russell Sage Foundation.

Sen, A. (1999), *Development as Freedom*, New York, A Knopf.

Shadish, W.R., Cook, T.D. and Epstein, R. (1991), *Foundations of Program Evaluation: Theories of Practice*, Newbury Park, CA, Sage Publications.

Stame, N. (1998), *L'esperienza della valutazione*, Rome, Seam.

Stern, E. (2004) 'Evaluating Partnership', in A. Liebenthal, O.N. Feinstein and G.K. Ingram (eds), *Evaluation and Development. The Partnership Dimension*, New Brunswick, NJ, Transaction Publishers.

Thomas, W.L. Jr (ed.) (1956), *Man's Role in Changing the Face of the Earth*, Chicago, University of Chicago Press.

Unger, R.M. (1998), *Democracy Realized*, London, Verso.

Weiss, C.H. (1998), *Evaluation*, 2nd edn, Englewood Cliffs, NJ, Prentice Hall.

Wolde-Mariam, M. (1991), *Suffering under God's Environment. A Vertical Study of the Predicament of Peasants in North-Central Ethiopia*, Walsworth, Marceline.

Chapter 9

Evaluation in Area-Based Regeneration: Programme Evaluation Challenges

Angela Hull

Introduction and Problematic

There has been much soul searching in the social science academy concerning the impact of evaluation research on public policy. On the one hand this reflection has led to renewed attempts to strengthen the research methodology used (Pawson and Tilley 1997, Marshall and Rossman 1999; Armstrong et al. 2002) and, on the other hand, to question the relevance, and political nature, of evaluation research for policy makers (Blackman 1998; Davies et al. 2000; Flyvberg 2001; Sanderson 2002; Wilks-Heeg 2003). Recent methodological developments acknowledge the changing socio-political context within which public policy programmes and initiatives are inserted, including the 'decentralization' of public service delivery to public, private and third sector agencies and the simultaneous testing of new policy initiatives in the same spatial locality. This chapter is concerned with the evaluation of the cumulative impact of several spatially targeted policy initiatives in neighbourhoods that have been defined by the government as areas of multiple deprivation. The spatial dimension of area based initiatives raise several ontological and epistemological questions for evaluation including isolating the mechanisms, which produce the 'additional' anticipated impacts from the embedded interaction between goal-oriented actors and a complex web of sociopolitical structures existent in the area. The chapter also addresses a related theme that concerns the nature of 'evidence' for evidence-based policy-making.

Numerous interlocking initiatives inserted into areas of multiple social and economic deprivation bring added complexity to the evaluation process. Each initiative typically has multiple objectives in which both the clarity of purpose and priority are poorly codified and elaborated. The current trend in UK public policy is to typically pilot a package of loosely related initiatives in a spatially-defined area already populated with government structures and networks implementing 'mainstream' supportive services and strategies to change the behaviour of, or symptoms shown by, individuals and organizations. The task of evaluation research in area-based initiatives is to assess whether, and to what extent, the 'new' initiatives have indeed changed the behaviour of the specified agency or individuals. Isolating the mechanism by which the 'new' initiative(s) works and the outputs (even outcomes) derived solely from this initiative (or in combination with others) creates problems for programme evaluation. Sager (2001) has criticized planning theorists for being slow to reflect on the contextual changes in society and, in particular, the different and evolving public policy intervention modes which increasingly involve coordinated intervention across several policy sectors to address territorial issues.

The conundrums pertaining to structure and agency hold the key to understanding the questions of how policies and interventions work at different spatial scales on different social groups over time. Besides these ontological questions there are epistemological debates about what constitutes knowledge, how to interpret knowledge claims, and the validity of the evaluation research findings. The academic community may hold a common view on this but it is unlikely to be shared by the policy and practitioner community to which the 'evidence' will be addressed (Harrison 2000, 209). As Robson (2000, 2) noted evaluation is 'a very sensitive activity where there may be a risk or duty of revealing inadequacy or worse'. These issues impact on both the independence and credibility of the researcher and the opportunity to input effectively into policy making. This chapter is ultimately concerned with the issues of what constitutes evidence and research validity drawing on the author's experiences of leading a mid-point evaluation of a neighbourhood regeneration programme in six case study areas. The commissioning process, the technical problems of evaluation and sources of bias are all inextricably linked.

The chapter first sets the scene by categorizing the types of evaluation research carried out by social scientists and discusses the trend away from sole reliance on measures of cost effectiveness to evaluate programme success to more recent evaluation methodological approaches which involve the recipient as a participant. The second section raises issues about how evaluation is commissioned and typically used by the sponsoring government department. The challenge for the researcher is to stay detached and objective to ensure that the dialectic and critical validity of the data and analysis are upheld during the research process.[1] The chapter then goes on to identify the empirical challenges of undertaking a mid-point evaluation of a comprehensive neighbourhood regeneration programme, which had multiple project objectives, multiple beneficiaries, and interlocking initiatives with resident capacity building as a focus (DETR 2000; Hull 2006). In the final section, the author critically reflects on the validity of this evaluation research and examines the sources of bias inherent in the contract research model.

---

1 Waterman (1998) calls this 'dialectic validity' which recognizes the tensions within the field and the research process, and 'critical responsibility' which refers to the responsibilities of the researcher. Quoted in Cabinet Office (2003).

Typologies of planning evaluation

In his definition of planning evaluation, Khakee (1998, 359–60) used a twofold categorization based on the timing of the evaluation: an 'ex ante' (forward looking) appraisal of a policy, strategy, or scenario before implementation and the 'ex-post' (retrospective) evaluation after the decision is implemented. Faludi and Voogd (1985) classify the relevant methods as either monetary, overview, or multi-criteria methods (quoted in Khakee 1998, 261). A third classification might distinguish between the main focus of the evaluation research questions. For instance, the evaluation may be 'summative', i.e. whether and to what extent a policy is having its desired effect or impact on its intended target group, or 'formative', to determine why, how and under what conditions the policy may best be directed or implemented (Cabinet Office 2003). Taking this latter, functional, perspective a number of different genres of evaluation research can be identified:

- Studies that evaluate the impact of specific discrete policy measures within a defined area. The focus is on inputs and outputs, with evaluators typically collecting data on the number of new jobs and dwellings created, new infrastructure completed, and the leverage ratio between public and private sector investment. Outputs can then be presented in terms of the cost per job/ per dwelling created. Occasionally the analysis is extended with a survey of local company managers' perceptions of the impact the policy measures have had on the property market. In this genre, methods to evaluate the effect of interventions on a range of social and environmental factors are less well developed.
- Studies which evaluate the stated (strategic) intentions of a public policy programme. Robson et al. (1994) examined the evidence that spending follows need and the impact of policy on reducing deprivation by testing the statistical associations between additional government funding and the selected output measures of unemployment, job change, and firm creation in deprived areas. Few of these relationships were statistically significant. The approach has been recently developed in the UK through the performance management monitoring of local delivery agencies by a government agency (the Audit Commission) using economic methodologies of value-for-money and cost effectiveness.
- Studies primarily concerned with understanding and establishing the implementation processes of public sector policy programmes;[2] for example, to identify whether national policies for the social services have been implemented effectively and with what effects.
- Studies which take an action research approach simultaneous with the implementation of a range of policy initiatives to assess whether agreed

---

2   Hammersley (2000, 114) categorizes this work as 'policy clarification', which involves evaluating government policies in relation to their consistency with the ultimate values espoused by those promoting them, and also formulates more effective policies to achieve specific goals.

objectives have been achieved, to understand how the agendas of particular agencies can be influenced, and how, why and under what circumstances an initiative is working. Action research involving full collaboration between researchers, residents and service deliverers is often promoted to gather the tacit knowledge of residents and the practitioner knowledge of service delivery to enhance the design of programmes as they are implemented. In the UK, the New Deal for Community (NDC) is a multiple initiative (skill training, employment, social welfare services, environmental improvements, community capacity-building) programme which specifically uses this approach to enhance capacity building and service delivery to deprived neighbourhoods. In theory, this can empower residents provided that the evaluation research is underpinned by a theory of change that recognizes the power imbalance between professionals and residents.

Relevance and the Use of Research

There are disagreements amongst academic researchers about the value and relevance of each of these approaches. There is uncertainty also over what knowledge is relevant and whether researchers can produce the knowledge required by policy makers to generate reform. Academics themselves are mystified why so little evaluation research intelligence appears to have a direct impact on government policy-making. Sanderson (2002, 3–4) considers that the UK government is seeking two types of evidence:

- firstly, input and output data concerning the cost effectiveness (value for money) of government policy delivery agencies (principally local authorities) which holds these agencies accountable for the public money they spend; and
- secondly, evidence of how policy interventions work in different contexts to inform decisions on what policy actions to take.

Robson et al.'s review (1994, 4) of the ex-post evaluations of the UK government's urban policy programme over two decades concluded that the researchers found it difficult to say something sensible about how service delivery could be improved or how the effectiveness of services could be monitored that would be useful to decision makers. Evaluations therefore concentrated on analysing programme outputs using secondary data sources (aggregate GDP increases, employment increases) and primary sample survey data of beneficiaries. Researchers were, however, seeking methodological advances to address:

1) *deadweight*: the counterfactual question of what would have happened in the absence of public intervention;
2) *displacement*: whether jobs created in a targeted are removed from other areas;

3) *supply chain effects*: the extent to which expanding businesses buy significant amounts of their inputs from other local firms, thereby creating knock-on benefits elsewhere in the targeted area;
4) *multiplier effects*: the extent to which extra jobs (via the wages paid) lead to multiplier effects elsewhere in the region;
5) *leakage effects*: the extent to which in- and out-migration (on a daily or permanent basis), mean that the beneficiaries are not living or even working in the area.

Methodological advances were, and still are, constrained by problems of measurement and estimation. Robson et al.'s review found insufficient data availability to assess the two higher level objectives ('creation of employment opportunities' and 'cities that are more attractive places in which to live') and the ten principal programme objectives of the urban policy initiatives during this period. They were thus forced to measure the performance of combinations of the lower level objectives (Robson et al. 1994, 7). They found consistent data only on the following five indicators, none of which exactly measured any of the ten lower order or two higher order objectives:

i)   unemployment and long-term unemployment (1983–1991);
ii) net job changes (1981–1989);
iii) percentage change in the number of small businesses (1979–1990);
iv) house price changes (1983–1990);
v)   net change in the number/ proportion of 25–34-year-olds (1981–1990).

In many cases, aggregate datasets were not coterminous with the regeneration area. The same story appears with the next public policy programme (the Single Regeneration Budget) for which the independent programme evaluators were unable to assess the contribution that the programme actually made to the gross outputs they had identified in each programme area (ODPM 2002). They were not able to estimate the deadweight and displacement effects. Similarly, Armstrong et al. (2002) found insufficient data in the English regions to populate even a simplified regional econometric model and input-output model for his ex-post evaluation of four initiatives part-funded through EU structural funds between 1994–1999. Ex-post evaluations tend, therefore, to rely on generic indicators to track social and economic conditions within a larger administrative area, rather than those that more specifically measure programme objectives. Data quality and availability remain underdeveloped, despite concerted efforts by the government to produce disaggregated datasets from the 2001 national census for neighbourhoods which came available in 2005.

The UK government appears undeterred by these methodological and empirical 'failures'. Current mid-point and 'ex-post' programme evaluations are required to pay equal attention to quantitative measures of effectiveness, as well as outputs and outcomes and the implementation process. Current criteria to measure programme effectiveness are defined in two ways. Firstly in terms of cost effectiveness or value for money. The government Treasury department persists with analyses of cost-effectiveness and cost-benefit for public sector spending despite the technical challenges in deciding what to include in both costs and benefits, and then how to determine the amounts of money involved. Secondly, programme evaluation involves collecting evidence of which measures are effective via beneficiary surveys to improve service delivery. In the latter stages (2002–2005) of the Single Regeneration Budget programme effectiveness was partly assessed by surveys of residents' perceptions of the impact of policy measures as well as surveys of other beneficiaries such as service deliverers. New urban policy programmes are now subject to comprehensive evaluation to produce both summative and formative evaluations.

Political Interference in Evaluation Studies

The preoccupation with quantitative measurement and the turn to 'interactive' social learning evaluation has raised concerns regarding methodological reliability and validity and the political interference in what constitutes evidence. As Pawson and Tilley (1997) suggest evaluation is at a watershed.

Those who uphold the science of planning evaluation, such as Khakee, consider that the recent shift to analysing 'claims, concerns and issues' which 'consist[s] of discourse among all the stakeholders who are in some way affected by the evaluation of a specific policy' cannot be regarded as evaluation in the traditional sense since it 'takes the form of negotiations rather than "objective" effectivity measures' (1998, 361). This undermines the autonomy and independence of the evaluator and the assumption that policy implementation can be improved in a rational non-political technological way. Pawson and Tilley (1997), Khakee, (1998) and Sanderson (2002) are critical of the anecdotal evidence collected 'to spread experience and perspective from one stakeholder group to another' (Khakee 1998, 361) and designed to identify best buys for policy makers.

In the UK, there is a strong political sway to produce evidence-based policy which influences the form of evaluation and the interpretation of the impact or effectiveness of what are often complex interventions at the city, neighbourhood or the individual level. This has led for calls for clarification of what planning evaluation of pilot projects is meant to produce. Sanderson (2002, 13) asks whether they are intended to understand 'prototypes' and how they work or are they about disseminating good practice. If it is the former, he argues that the method of evaluation should be action research focusing on understanding the context, identifying the way in which the intervention works (mechanisms) and the outputs.

Qualitative methodologies, including action research, are being used to understand how and why programmes work for whom in what circumstances in an attempt to provide evidence to feed into improvements in policy and practice. Significantly, previous evaluation teams have concluded that the local context and a range of other factors have strongly influenced or modified the effects of the interventions they studied. The implementation process and the initial conditions (social, economic, institutional practices) are thus seen as explanatory factors to help understand how and why trajectories vary so much from place to place. Behavioural science and postmodern critiques point to the relevance of practitioner and resident 'cultures' that provide theories and templates of meaning that legitimize their actions. This recognizes that knowledge is subjective, relational, mediated by our perceptions and beliefs, and thus challenges the traditional distinctions between researcher and participants. Pawson and Tilley (1997) suggest that a realist perspective and a case study methodology is appropriate to synthesize the complexity of issues: the descriptive baseline socio-economic context, the theories in use (programme, practitioner, and residents' theories of change), the multiple programme objectives (policy mapping), the interlinking initiatives (the mechanisms for change), and the identification of emergent properties of systems (pattern recognition approaches). Realist methodologies draw on organizational and behavioural research to understand how effects have been produced and how the mechanisms work to produce tailored transferable theory. Part of the realist evaluation task is to surface the implicit or explicit theories underlying the programme, to identify the key assumptions and to test their validity.

This type of evaluation research often highlights issues that sponsors do not want to hear. For instance, the theory of change underlying the NDC suggests that the commitment to change and to action is inevitably a social-interactive process over time involving both the residents and service providers in the neighbourhood network and their interconnections to a wider network (Langley et al. 1995). Participation by multiple stakeholders is seen as central to the effectiveness of the regeneration through jointly constructing the programme for action through pooling their resources and knowledge and ensuring that agreed actions are achieved and monitored. Whilst the programme provides sufficient funding over a ten-year period to support this interaction and to make some improvements in service delivery, there is little funding for new infrastructure so expectations of change have to be dampened down. Delving deeper into the policy background, the rhetoric emphasizes the underused community resources in these neighbourhoods, deprived communities who can 'turn around' their own areas, and an employment policy to tackle worklessness. Which raises questions of what the residents will gain from the NDC programme?

The complexity of these area-based programmes with multiple objectives, interlocking initiatives targeted at multiple outcomes, multiple beneficiaries and multiple roles for the programme evaluation create significant challenges for the contract research team undertaking the evaluation. There are several obvious pitfalls. They may misinterpret the nature of the problem the programme is seeking to address since the 'agreed' objectives may not be the only significant ones for central or local government administrations. Urban policy initiatives in the UK are part of the public sector funding package to which local authorities can compete for top-up funding. To access this funding stream, they may overstate the problem to be addressed or the initiatives they will implement to address the perceived problem. Identifying cause and effect, through disentangling the separate initiatives and the mechanisms for change, and how, why and under what circumstances an initiative is working is problematic unless informed by, and testing, a specific theory of change. Even then the conclusions of the evaluation team tend to be tentative since it is rarely the case that all the decisions/actions flowing from an initiative can be isolated and tracked over the timeframe for evaluation.

Few of the programme evaluations using this social learning approach therefore provide, for government sponsors, the basis for a theoretically grounded understanding of transferable lessons about what works and why. In many cases there have been significant limitations on the ability of evaluation teams to understand the initial conditions (social, economic, institutional practices) into which the interventions were inserted. McCulloch (2000) raises the issues of data validity and researcher objectivity in his comparison of five different evaluations of the same community regeneration project in Newcastle in the UK that produced contradictory findings. One solution, preferred by the government, is to benchmark the net outcome changes during a specified time period against those of a comparator area not subject to the intervention. In practice, finding what looks like a similar control estate or area is difficult, as is accounting for how the trajectories vary from place to place within the evaluation timeframe.

To draw this discussion of the validity of evidence and the scientific probity of planning evaluation to a conclusion, it is necessary to draw attention to the constraints of undertaking research commissioned by government departments. Researchers and funders bring explicit theories on how to understand reality and change behaviour into their work. The exercise of power is involved in all human actions and activities. This is well-documented by Flyvbjerg (2001), Galster (1996), Hammersley (2000), Robson (2000), Sanderson (2002) and Wilks-Heeg (2003) and others. Political considerations steer the commissioning process constraining the length of time the evaluation can run and requiring the evaluators to share their preliminary findings with a steering group set up by the sponsors. It is often the case that the positive effects desired by government may be difficult to achieve during the course of a pilot project. The steering group will be composed of advocates of different evaluation approaches and government policies including the government department funding the project and the staff employed in project delivery. Given that organizations tendering to undertake evaluations depend on the income generated by such contracts, the dominant assumptions about the focus of the evaluation are rarely questioned in the commissioning process. If this is the case, it raises key questions about the extent to which an evaluator can claim to operate independently from the hierarchies present within the steering group.

The researcher can so easily become embedded in the resolution of competing values and political interests, that they incorporate specific interests, opinions, values and criteria into the evaluation and take on an advocacy role themselves:[3]

---

3   This corresponds to Nagel's (1961) category of 'appraising value judgments' which indicate the evaluators' feelings on the worthwhileness of action, i.e. approval or disapproval. He contrasts this with 'categorizing' value judgments which are factual judgments on whether some characteristic is present in some given instance.

Far from acting as detached and objective observers, evaluators may explicitly reinforce, subtly redefine, or, much more rarely, explicitly challenge the key ideological assumptions on which urban policy initiatives are founded. (Wilks-Heeg 2003, 205)

Case Study Introduction

The evaluation of the six Housing Action Trusts (HATs) in the late 1990s illustrates the empirical challenges raised by these debates on the validity of evidence and scientific probity. This section reflects on how these issues were resolved in the ex-post evaluation of one of the HATs, and the mid-point evaluations of the other five. The full report to the government sponsor (DETR 2000) and an analysis of the ontological issues raised by this area-based regeneration programme (Hull 2006) are in the public domain. The HAT programme was a politically charged initiative introduced by a Conservative government premised on a theory that the appropriate way to renew rundown local authority-owned housing estates was private ownership. Despite the government's offer of substantial resources and delegated powers and a direct appeal to the local authority tenants, only six HATs were established between 1991 and 1994. HATs were limited life agencies (eight to 13 years) with a governing board composed of three to four tenants/residents, one or two local elected councillors and five to six professional people (management consultants, public sector backgrounds in community development, architecture, etc.). The tenants were promised a rent freeze until the end of the renovation, the right to retain their tenure rights and to choose their future landlord following improvement work.

Each of the HATs evolved in different ways during the life of the programme due to the rights of engagement given to tenants, the supportive role played by a dedicated skilled team of practitioners, and the hands-off approach to performance management by the government sponsoring department. The HATs were required to publish annual performance indicators that record achievements directly attributed to their projects. A baseline analysis of the HATs was available, carried out by consultants in 1995, which mainly described the physical characteristics of each area, the public services on each estate, and the tenants' expressions of service satisfaction. Government policy at the time of the evaluation recommended that pilot projects should be assessed in terms of the socio-economic outcomes, most notably training and employment (Audit Commission 1999). It was not until 1997 that the HATs were advised of the 25 core (and non-core) performance indicators they would have to monitor. They were only ever assessed on their performance within their project area boundaries – although they had a much wider focus.

Monitoring by the sponsoring government department and the Government Offices in the Regions lacked a consistent and common approach. There was tacit acceptance of alternative performance indicators when used by the HATs. The core indicators do not show an adequate snapshot of what actually happened. In many cases, the core indicators (for example 'the number of community facilities', 'the number of sponsored training schemes' or 'the number of training weeks') are meaningless for comparison purposes since there was no agreed formula for measuring indicator outputs. The HATs, therefore, introduced their own training and employment indicators, which captured the type of job and the length of time in employment, for example. They expanded the core training indicator, which mentioned only construction, to include all types of training. The HATs invested heavily in their own monitoring employing market researchers to carry out annual or biennial household surveys of tenant satisfaction. One HAT chose to monitor residents going into training and employment through a computerized tracking system.

Sponsor Requirements

The role of evaluation in the contract model of research is to service the needs of a particular government sponsor department. The HAT research contract was won through a competitive bid on the basis of cost and quality. The tender brief invited bids for a six- months contract to assess the extent to which the six HATs were achieving their statutory programme objectives to:

* secure the repair or improvement of the housing held by the trust;
* secure the proper and effective management and use of that housing;
* encourage diversity in tenure and the identity of landlords;
* secure or facilitate improvement in living conditions in the area and the social conditions and general environment of the area.

The initially small departmental steering group for the evaluation was open to the evaluation team's approach to understanding what has worked and why using a range of qualitative methods with residents, HAT managers and service deliverers, as well as analysing quantitative indicators. They agreed also to a nine-month project provided the preliminary findings were available for internal government circulation after six months. Given that the six HATs were all at different stages in their programme, there was hardly sufficient time to gather and analyse the data and to ensure researcher clarity about the linkages and alignments between activities, and early, interim and long-term outcomes. The evaluation team also found it difficult to evaluate the statutory programme objectives because the emphasis moved away from dwelling refurbishment and diverse private sector landlords to effective estate and service delivery. This was due to the costs of dwelling refurbishment being too high and low interest from private landlords in managing the properties. HAT programme objectives also varied due to the involvement of residents in the design and implementation of the programme.

During the early contractual stages, the evaluation team were dependent on the departmental steering group for information, background programme documentation and support to gain site access. At this stage, the steering group did not question the evaluation team's type and style of evaluation chosen or the theory of change approach adopted, which aspired to undertake a systematic study of the links between activities, emerging outcomes and contexts of each

HAT. This would involve in-depth discussions in each of the HATs including with residents in focus groups, and interviews with 20 HAT officers, local authority staff, and service providers to articulate the HAT's theory of change and clarify how external and contextual factors have affected both activities and outcomes. The steering group did suggest people we should talk to and were eager to meet on a regular basis to hear our emerging findings.

Roughly half way through the contract, new requirements were introduced by our steering group, which had grown to include economists from the government Treasury department. The latter introduced a new problem focus: to benchmark the HAT achievements against other regeneration programmes in terms of cost of new job/training place, etc. This surfaced at the same time as lobbying by the HAT chief executives, who were negotiating with the sponsoring department for their next year's grant-in-aid from the Treasury. The steering group also requested a short report on the transferable lessons from the HATs for the Social Exclusion Unit to help inform the policy slant of the NDC programme being devised. The report on our initial thoughts on transferable lessons was duly dispatched, but the assessment of the cost effectiveness of HAT initiatives created some anguish amongst the evaluation team since it introduced a research approach that we had already rejected, but which satisfied the goals and values of the Treasury.

The request was to benchmark the case study areas against comparator estates and to benchmark the costs of policy outputs in relation to those achieved by other kinds of initiatives to provide some indication of value for money. The government Treasury department was very insistent that a value for money assessment of the numbers of new dwellings built and numbers improved, and numbers of people assisted into training and jobs, was carried out despite the problems. Benchmarking is seen as one solution to the counterfactual problem: the difficulty of taking account of what might have happened in the absence of public intervention. The methodology involves comparing the net outcome changes during a specified time period against those of a comparator area not subject to the intervention. The evaluation team and the steering group suggested several possible control estates for each of the HATs. After investigation, the evaluation team concluded that the control estates were not similar to the HAT estates because the initial conditions (social, economic, institutional practices, resident involvement) differed.

Empirical constraints

One of the key constraints of carrying out effective comparative evaluation is data availability. As already noted, performance measurement data was inconsistent because the monitoring framework developed incrementally. Table 9.1 shows the incomplete reporting by each HAT on the 25 core annual indicators, introduced by the sponsoring department half way through the HAT programme. It would have been helpful for the evaluation of the HATs if the basis on which performance targets and indicators, covering both impacts and value for money, were to be calculated had been agreed with the sponsoring department at the beginning of the programme.

Table 9.1   Annual reporting of the 25 core performance indicators by HAT

	Year of operation								
HAT	1	2	3	4	5	6	7	8	9
Hull	3	12	20	22	21	22	21	24	18
Waltham Forest	1	5	12	13	21	22	22		
Liverpool	10	9	11	11	25	25			
Castle Vale	2	11	8	19	19	19			
Tower Hamlets	1	1	7	15	25	21			
Stonebridge	1	4	19	22	21				

Source: DETR (2000, Table 5.1, 42).

Even for the 25 core indicators, where information was provided it was difficult to compare across the HATs. For instance, variation in the annual cost of repairs per dwelling can be partly explained either by the different approaches to defining these costs and the different circumstances of each HAT. For instance, the servicing costs of tower blocks, including caretakers and lift maintenance, were particularly high as they were allowed to empty whilst awaiting demolition. However, it was not clear whether the caretaking and warden costs should be included as a management cost or a service charge. In the early years of their programme, some of the HATs implemented short-term improvements to deal with the outstanding repairs to keep residents' homes safe, secure and watertight pending major development and improvement work. This covered items such as fire prevention measures, window replacements, emergency lighting, concrete and brickwork repairs, heating and lift improvements. These early 'wins' were seen as an important element in boosting tenants' confidence, demonstrating the HAT's commitment and contributing to altering the perceptions of the area in the long term.

It was possible to calculate average costs in relation to the number of dwellings contained within each HAT for comparison purposes but there are flaws in such an approach, especially when bearing in mind the different composition and needs of each HAT. In particular, the amount of rent collected as a percentage of the amount due, and the percentage (or number) of tenants with over 13 weeks rent due reflects the socio-economic characteristics of the HAT areas as well as the number of 'non-secure' tenants (for example, previously homeless) they have housed. One HAT was able to show that the average cost of their phase one new homes was 2 per cent higher than the benchmark costs of the Housing Corporation Total Cost Indicators. They explained their higher building costs on the high level of tenant involvement in design and the tenant choice of fittings and finishes, and the higher standards for thermal and sound insulation than required under Building Regulations prevailing at the time. Essentially it was difficult to disaggregate direct costs from global or broad category costs for any of the HATs because of the limitations of data, comparability, and the different time frames of the HATs.

It is because of these difficulties of interpreting the quantitative data available, that the evaluation team chose to focus on changed individual outcomes and perceptions of change as a more sensitive and accurate reflection of progress. The evaluation team used a theoretical analysis based on uncovering the theories of change held by government officials, the HAT board and officers, service deliverers and residents of the area. The evaluation team was particularly interested in their accounts of events, their responses to and interpretations of those events, and how they were involved in influencing and shaping neighbourhood change. A mixed method evaluation was used including indicator trend analysis, network analysis, documentary analysis, questionnaires, interviews and focus groups. In testing issues of reliability and validity we used multiple informants and multiple sources at each site to triangulate findings and obtained feedback on our findings from informants in a management focus group and at the national HAT residents' association conference.

On reflection, bias to some extent did creep into the work of the evaluation team. The team were reliant on HAT employees to gain access to the 100+ 'typical' residents we spoke to in the residents' focus groups we held. A small retail voucher in lieu of payment and a hot or cold buffet were important bribes. The team's approach to action research and the level of involvement we hoped for was not always forthcoming from the HAT employees as we penetrated the inner workings of each of the initiatives. The need to get clearances and permissions from government officials to use archived HAT reports held in trust by English Partnerships inevitably meant that not all the documented data were analysed in detail. We, therefore, as a team had to make sensible judgments of the relevance and validity of information we had access to in the time frame of the evaluation. The evaluation team worked hard to build up trust with all informants, explaining the objectives of the evaluation and the precise role of the evaluation team in any setting, what we hoped to learn and how the information would be used. We always sought consent to use information and protected participants' anonymity.

There were occasions when the evaluation team was expected to reciprocate. To the residents, the evaluation team was a source of advice on government policy, mediators with the HAT Board, conflict managers for rival groupings of residents, and presenters at the annual conference of HAT Residents and Tenants Associations. As a team we had to carefully think through our roles vis-à-vis the HAT Boards. There were clear tensions and expectations that we would play an advocacy role on their behalf. As an evaluation team we brought knowledge of other government regeneration projects and experiences of government officials to our empirical work. We were probably guilty of selecting what was to be counted as knowledge and valid claims although we tried to be independent observers. Our diverse presuppositions and cultural biases even led to a fall-out amongst certain members of the evaluation team.

The evaluation of the HATs was a requirement of the substantial public funding released by the government. Since the achievements of the HATs have not been reported widely by the government, the question arises of whether this was a pseudo evaluation (Suchman 1967) designed to bury the programme for whatever reasons, although John Prescott, the deputy prime minister, did hail the achievements of the HATs when he announced the NDC programme.

Conclusions

Programme evaluation is inevitably constrained by the availability of data. Traditionally, frameworks for evaluating regeneration programmes have relied on economic output measures, such as employment criteria and the (public-private funding) leverage ratio, to represent the key determinant of success or failure. The interlinking and multiple objectives of the HAT programme, the lack of agreement on how certain outputs should be measured, the difficulties of relating particular input costs to particular outputs, and the long time period that may elapse before the full impact of the programme can be identified, led to an emphasis on processes in the HAT mid-term evaluation. The HAT evaluation team acknowledged that the balance and the integration between the physical, social and economic regeneration were important. One without the other would have a limited impact. Residents in focus groups guided us as we explored how they felt they had influenced the agenda, their own criteria to evaluate programme success and how the programme has evolved in response to their views and the lessons learned. However, it was hard to disaggregate the physical outputs, the impact on residents and external perceptions of the area due to the multiple inputs of the intervention initiatives.

The activities of the HATs took place in very different contexts and therefore both costs and outputs varied considerably. The HATs developed their own alternative quantitative and qualitative measures of performance, which were more meaningful and closely related to local objectives. Several of the HATs developed long run measures to track performance. For example, these included tracking: how long in employment/movement to better jobs, school results, truancy levels, intervention impacts, and inter-generational shifts. Key features of this approach included measuring attainment against an average for the local authority area as a whole and attempting to introduce more qualitative assessments, for example, through community consultation, local evaluations and citizens' panels. This 'bottom-up' analysis is much more useful and robust to inform for local policy change. Advancements still to be made are how to sufficiently investigate the spatial distributional issues of public sector interventions, such as accessibility to customized services, recipients' assessment of the quality of service provision, or the cumulative effects of a regeneration initiative on disadvantaged communities.

The instrumental approach of government-sponsored evaluation typically underestimates the sufficiency of available resources (time, money, access to site or population of interest) required for undertaking an evaluation of multiple and interlocking initiatives. Universities in the UK have become more dependent upon government departments and government-financed research councils in the late twentieth century, which Hammersley (2000, 104–105) suggests is designed to make them serve national economic goals more effectively. Guided by their calculation of instrumental value, government funders control the choice of research topic, influence the scope of research and control the recommendations for action and the publication of findings (Robson 2000, 74). Whilst the research community has close contact with decision makers through contract research, this access comes with a price. Their autonomy in producing findings, the validity of their research and the pursuit of knowledge for its own sake has all been compromised.

References

Armstrong, H.W., Kehrer, B., Wells, P., and Wood, A.M. (2002), 'The Evaluation of Community Economic Development Initiatives', *Urban Studies* 39(3), 457–481.

Blackman, T. (1998), 'Towards Evidence-Based Local Government: Theory and Practice', *Local Government Studies* 24(2), 56–70.

Audit Commission (1999), *A Life's Work: local authorities, economic development, and economic regeneration*, London, Audit Commission.

Cabinet Office (2003), 'Assessing/Evaluating Pilots', report prepared by Roger Jowell, City University, London, Cabinet Office.

Davies, H.T.O., Nutley, S.M. and Smith, P.C. (eds) (2000), *What Works? Evidence-based policy and practice in the Public Services*, Bristol, The Policy Press.

Department of Environment, Transport and the Regions (2000), *Transferable Lessons in Regeneration from the Housing Action Trusts*, London, DETR.

Flyvberg, B. (2001), *Making Social Science Matter. Why Social Inquiry Fails and How It Can Succeed Again*, trans. Steven Sampson, Cambridge, Cambridge University Press.

Galster, G. (ed.) (1996), *Reality and Research. Social Science and US Urban Policy Since 1960*, Brookfield VT, Avebury.

Hammersley, M. (2000), *The Politics of Social Research*, 2nd reprint, London, Sage Publications.

Harrison, T. (2000), 'The History of Urban Policy Evaluation', in H.T.O. Davies, S.M. Nutley, and P.C. Smith (eds) (2000), *What Works? Evidence-based Policy and Practice in the Public Services*, Bristol, The Policy Press, pp. 207–28.

Hull, A.D. (2006), 'Facilitating Structures for Neighbourhood Regeneration: The Contribution of the Housing Action Trusts', *Urban Studies* 43(12), pp. 2317–50.

Khakee, A. (1998), 'Evaluation and Planning: Inseparable Concepts', *Town Planning Review* 69(4), pp. 359–74.

Langley, A., Mintzberg, H., Pitcher, P., Posada, E. and Saint-Macary, J. (1995), 'Opening up Decision Making: The View from the Black Stools', *Organization Science* 6(3), pp. 260–79.

Marshall, C. and Rossman, G.B. (1999), *Designing Qualitative Research*, 3rd edn, Thousand Oaks, CA, Sage Publications.

McCulloch, A. (2000), 'Evaluations of a Community Regeneration Project: Case Studies of Cruddas Park Development Trust, Newcastle upon Tyne', *Journal of Social Policy* 29, 397–419.

Office of the Deputy Prime Minister (2002), *Turning Areas Around. The Impact of the Single Regeneration Budget on Final Outcomes*, Urban Research Summary 4, London, ODPM.

Pawson, R and Tilley, N. (1997), *Realistic Evaluation*, 2001 reprint, London, Sage Publications.

Robson, B., Bradford, M., Deas, I., Hall, E., Harrison, E., Parkinson, M., Evans, R., Garside, P., Harding, A. and Robinson, F. (1994), *Assessing the Impact of Urban Policy*, London, HMSO.

Robson, C. (2000), *Real World Research*, 2nd edn, Oxford, Blackwell Publishing.

Sager, T. (2001), 'Planning Style and Agency Properties', *Environment and Planning A*, 33(3), pp. 509–37.

Sanderson, I. (2002), 'Evaluation, Policy Learning and Evidence-Based Policy Making', *Public Adminstration* 80(1), pp. 1–22.

Suchman, E.A. (1967), *Evaluative Research: Principles in Public Service and Action Program*, New York, Russell Sage.

Wilks-Heeg, S. (2003), 'Economy, Equity or Empowerment? New Labour, Communities and Urban Policy Evaluation', in R. Imrie (ed.), *Urban Renaissance; New Labour, Community and Urban Policy*, Bristol, Policy Press, pp. 205–21.

Chapter 10

Strategic Spatial Planning and Planning Evaluation: Developing an Entrepreneurial Urbanization Strategy in South-Holland

Johan Woltjer

Introduction

Regional planning currently is in a position of fundamental change and renewal. Conventionally, efforts by planners to establish some kind of regional plan have been predominantly oriented towards public agencies taking general rules or regulation as a standard for policy making. More recently, however, a more entrepreneurial and strategic planning style has materialized. This style emphasizes a socio-political view on planning and the need to 'frame mindsets', to 'organize attention', and to seek opportunities in a dynamic social context. Following Healey et al. (1999), strategic planning is understood to include a notion of providing regions with 'institutional capacity' and social, intellectual, and political capital. An entrepreneurial planning style refers to the intention of stimulating private investment and actively initiating development through planning. Entrepreneurial planning is related to an ability to identify opportunities in a public environment or existing market, and a willingness to take risks (see for instance Casson 2005).

A strategic, entrepreneurial style poses challenges and actions related to planning evaluation, which refers to the activity of comparing the impacts of alternative proposed plans, or the activity of assessing the likelihood and success of implementation. While much is known about how planning evaluation addresses economic, social and environmental impacts caused by planning intervention, comparatively few insights exist about how 'strategic thinking' in planning can take a part in evaluation practices. This chapter points out how strategic planning might be linked to planning evaluation.

The chapter aims to explore the experiences with efforts to establishing a strategic spatial plan in the South-Holland region, the Netherlands. The South Holland case shows evidence of a more strategic planning style, and of questions, given this strategic planning style, how to evaluate promising alternatives and implementation success. The chapter discusses options for the deployment of evaluation within strategic spatial planning in general, and within this specific case. It is based on a literature review, and on the empirical material collected during an explorative study concerning strategic spatial planning in the southern wing of the Dutch 'Randstad' (see Salet and Woltjer 2004).

Planning Evaluation in Regional Planning

Planning evaluation in regional spatial planning has been oriented on investigating how plans or policies might affect the spatial patterns of economic, social and environmental costs and benefits within an area, how they affect community interests (Lichfield 1996), issues of equity, participation and reflection (see for instance Dabinett and Richardson 1999), or issues defined by stakeholders (Guba and Lincoln 1989). An important change throughout the last decennia is the kind of indicators and criteria that are included in planning evaluation. The reliance on scientific rational principles for assessment has been considerable. Criteria related to economic evaluation (performance targets, capital, and so on) have had a strong position (Healey 2006). In the 1980s and 1990s, with the emergence of the notion of sustainable development and environmental assessment, criteria related to environmental and ecological quality have increasingly been used.

A conventional way to engage in regional planning would be to draw up strategic plans based on 'reliable' (demographic, economic, and so on) forecasts and environmental assessment, and to assume that they can be implemented and evaluated effectively, via legal structure and, predominantly, instrumental knowledge. It has now been acknowledged more and more that this kind of thinking does not operate well in rapidly changing environments that are increasingly hard to forecast. The problem is the assumption that the world would remain relatively unchanged during a planning process, that evaluation studies should be based on detached experts' observations and assessments, and, that strategies can be formalized and implemented in an instrumentally rational way (see Holloway 2004). The result of this assumption might be that intended plans become hard to implement (Wissink 2000).

The orientation now is focused increasingly on the strategic role that evaluation can play. It is not so much the substance and form of various evaluation methods (including cost-benefit analysis, planning balance sheet, goals achievement matrix, and multi-criteria analyses) that are key, but more their capability to influence dialogue and be persuasive (Voogd 1997). As Hildén et al. (2004) point out for strategic environmental impact assessment, effective evaluation exercises should contain efforts to bring in some more reflexive elements in the evaluation, thereby providing information on the social position of the actors involved (their power, their networks, their struggles, and so on). Successful assessments should then include criteria on these kind of socio-political attributes.

While evaluation in Dutch regional planning has been primarily concentrated on efforts of comparing the impacts of alternative proposed plans (see for instance VROM 2001), now the emphasis increasingly lays with efforts to evaluate the extent to which plans are likely to be implemented due to the strategic context of regional plans/planning. We are looking here at a separate kind of planning evaluation focused on including indicators of implementation success (Talen 1996) and the institutional realities of planning practice.

Strategic Spatial Planning and Planning Evaluation

Before we move into a discussion of possible evaluation criteria in strategic planning, this section first takes a general look into the link between strategic planning and evaluation. An essential interest in strategic planning by government agencies and non-profit organizations stems from the 1970s (see for instance Steiner 1979). Strategic planning, up till then, had been solely employed in private enterprise and by the military. Especially in the 1980s, many publications dealt with suggesting ways of using strategic planning in the public sector (see for instance Bryson et al. 1986; Kaufman and Jacobs 1987; Bloom 1986; Bryson 1988). Strategic planning was understood to be 'a disciplined effort to produce fundamental decisions and actions that shape and guide what an organization (or other entity) is, what it does, and why it does it' (Bryson 1988, 5–6). Strategic planning requires broad scale information gathering, an exploration of alternatives, and an emphasis on the future implications of present decisions.

More recently, new types of regional strategy making have occurred that include the emphasis on creating new institutional arenas, and on creating an integrative view on spatial strategies (see for instance Albrechts et al. 2003). Business literature generally emphasizes long-term visions for the future. In the public sector, a typical reason for designing new strategies includes the wish to convey a more coherent (integrative) strain of thought for some typical regional issues such as distributing infrastructure investments, protecting environmental values, and appointing new housing development. Another new purpose includes the notion in international planning practice of providing regions with institutional capacity and social, intellectual, and political capital (Healey et al. 1999).

Planning evaluation does not normally consider the strategic capacities present in a given institutional context. However, any planning evaluation, whether it is the comparison of alternative options to urban growth, the identification of promising places for business parks, or an assessment of the probabilities for carrying out major projects, inherently must address questions of institutional context. In practical terms, institutional capacity refers to the ability to successfully implement and complete a new plan or a project. When capacities such as public and political support are enhanced or more fully developed, the likelihood of successful implementation would increase (Harris and Ogbonna 2006).

Conventional regional planning practices generally have a permissive attitude, assessing the impacts of a proposed plan or project on typical sectors like finance, economy, and environment. Evaluation actions here typically address the relations between development activity and economic and environmental systems (Healey 2006). Determining appropriate plans or projects, however, not only depends on costs or impacts on the environment, but on identifying links between competing claims, assessing the opportunities given by political partnerships, judging contact networks between agencies, or considering levels of public trust. In short, strategic planning emphasizes the evaluation of the institutional context framing a plan or a project. In this sense, knowing about the prevailing arguments or perceptions on certain impacts is more important than the professional estimation of these impacts themselves.

Institutional capacities are specific to the context in which they emerge. Success is linked, accordingly, to the ability to 'read', or evaluate the opportunities and threats given by the context at hand. If, for example, a proposal for building a light-rail network in the south-wing of the Randstad is accompanied by an assessment of the richness of political networks supporting it, it would be easier to assess the chances for implementation.

A key thought in the emergent strategic planning literature is, in essence, an approach to strategy making whereby strategic planning decisions emerge from the specific institutional context at hand. The idea is that if planners link to this context, rather than the specific legal procedures or technical knowledge necessary for assessing plan-making quality, they would be better focused and have an enhanced likelihood of achieving policy objectives.

The strategic emphasis implies a move away from reliance on formalized planning evaluation towards planning evaluation built on criteria like, for example, 'strategic agility' (Holloway 2004). The success of a planning agency then is expressed through acknowledging the extent to which planning action is anticipating the institutional context and the worth of institutional capital. Clearly, strategic planning stands for a different approach to assessing and valuing proposals for land-use change. This comes to the fore in particular through the criteria used in planning evaluation for the assessment of alternatives.

Evaluation Criteria in Strategic Planning

The strategic planning approach brings new criteria for justification to the fore. While later sections will discuss some of these criteria for the South-Holland case, this section will explore in some detail the evaluation criteria suggested by literature on strategic planning. New evaluation criteria then include statements about, for example, capacity constraints, the richness of institutional webs available, the extent of social relations in an area, or dominant knowledge claims. The challenge for planning evaluation is to address these kinds of criteria explicitly.

Overall, I would like to refer to these kinds of new criteria as, first, the strength of regional development options to anticipate societal change, opportunities and market insights (for example, the principle of contextual anticipation), and, second, the extent to which strategic plans mobilize 'core capabilities', stakeholders and decision-making capacities within the region (for example, the principle of institutional capacity). The assumption in strategic planning is that attention to these kinds of criteria will make a plan or project more likely to succeed. In this sense, the chapter links to points by authors such as Talen (1996), who argued the need to establish a form of planning evaluation focused on evaluating the implementation success of plans.

Contextual Anticipation

Strategic planning has often included references to methods for identifying organizational strengths and weaknesses, and external opportunities and threats ('SWOT', see Miller and Holt-Jensen 1997). Thus strategic planning is concerned with an identification of and response to changes beyond the control of the organization, yet which eventually provides direction for that organization (So 1986). These include analyzing the current situation that an organization faces, and the internal resources it has to respond to this situation (Steiner 1979). Some of these resources include funds, and the quality of their employees. In addition, the organization should carry out a study of its environment and how this environment is changing. Obviously, for private companies, environmental factors include developments in the market, strategies by the competition, and governmental regulations. An analysis of these developments will make clear some opportunities and threats to the organization over the foreseeable future.

Since the 1990s, attention turned to strategic management issues (see for instance Hamel and Prahalad 1994), such as creating competitive advantages through insights into developments in the market. The term strategy refers to anticipating and assessing the possible behavior by other actors, including not only the direct usage of power resources (such as material resources, knowledge, legitimacy, formal position, personal relations or identity), but also improving one's point of departure or conditions (see for instance Mintzberg et al. 1998). Some examples of the latter approach include the ownership of land, efforts to add or change knowledge (research, marketing), coalition building or enlarging mutual interdependence. Following the strategic management approach, changes in the environment of planning agencies should not be viewed merely as threats, but as opportunities that can be employed to make possible reaching some long-term goals.

Obviously, we are dealing here with an entrepreneurial approach as well. An entrepreneurial planning style highlights an agency actively stimulating development across conventional borders and competency-levels (see for instance Brenner 2003). Entrepreneurial approaches are also oriented towards providing stimulating conditions for private investment. Entrepreneurial regional plans typically contain proposals for actively initiating development and encouraging project initiatives, even if these initiatives seem risky.

Contextual anticipation also refers to the ability for strategic planning to not be comprehensive in terms of addressing all the goals of an organization, but to be selective in focusing on the most essential ones (Gordon 1993; So 1986). Some arguments as to why such an approach is desirable include that it would be more practical and to keep the list with projects and activities limited, which would facilitate effectiveness. This is contrasted to the comprehensive planning ideal of the 1980s, which, in treating all possible goals, resulted in confusion (see for instance Olsen and Eadie 1982). In strategic business planning, selectiveness is typically explained in terms of developing 'core capabilities' that reflect the unique attributes of a certain company in their context (see for instance Porter 1996). The term 'core capabilities' refers to the collection of skills, assets, and practices in a company that provide its competitive possibilities and its comparative advantages in the market. By focusing only on selected critical issues and capabilities, it would be easier to identify important external developments, and to carefully getting necessary support from actors with the necessary means to implement strategic goals.

Strategic planning also involves the selection of new issue combinations. There is a long history of discussion about integration in planning (see for instance Innes 1996), especially on combining the social and economic policy fields within spatial planning. A new type of integration seems to be, as Friedmann (2004) observes, the amalgamation of territorial policy with environmental sustainability and with cultural identity. Accordingly, and in a summary of Albrechts et al. (2003), John Friedmann (2004, 52) argues that: 'Strategic planning is thus conceived as long-range planning for territorial development. It calls for new institutions of governance, and, in the long tradition of spatial planning, it calls for a new comprehensive, integrated approach.'

Strategic planning places emphasis on the 'longer-term perspective' and 'real vision'. Initially, business literature prescribed a shift from using a management time horizon of one or two years, to a view covering several decennia (Henderson 1983; Hickson et al. 1986). Strategies were to give a longer-term direction to shorter-term decisions, to incorporate a political vision for the future (Friedmann et al. 2004), but not to specify detailed steps to achieve the final destination. This is how strategic planning can be distinguished from implementation and control plans developed for individual projects. Real visions refer to a shared mental framework that helps to shape future decisions, and pictures what an organization or an area is intended to look like in due course.

If we look back at the points given above, one evaluation criterion that can be argued is evidently linked to strategic planning success is the degree by which it tries to anticipate societal change, opportunities, and identifies strength, weaknesses and external opportunities and threats. In this sense, strategic planning initiatives can only be implemented easily if the specific social context is receptive to them. A good indicator for this criterion would be whether a study of social environment change has been carried out.

Another criterion for contextual anticipation would be, for example, the extent to which a planning agency is oriented toward and understands the tactical and strategic capabilities of other public and private actors. In management literature, this kind of a criterion is usually given in terms of the orientation on competitors. The inability to grasp the social and political environment of other players, stakeholders, actors, and so on, is likely to affect planning success (Hamel and Prahalad 1989). Whether or not a strategic plan is grounded on market insights and an assessment of strategies by 'others' would be a good indicator for this criterion to contextual anticipation. Another distinctive evaluation criterion would be the question whether a strategic plan is a reflection of the unique attributes of a certain region in relation to its context.

Also the time horizon of plans or projects can be given as a criterion in evaluation. If alternative projects, plans or proposals have a time orientation that is focused on long-term horizons or longer-term issues, they may be more likely to get adopted eventually (Harris and Ogbonna 2006). A strategic plan, for example, would not generally specify detailed steps to achieve the final destination

Institutional Capacity

A fundamental effort in international planning practice has been to develop regional governance, and to equip regions with institutional capacity, including networks of interpersonal relations among economic actors. These networks could have a positive effect upon the innovative capacity of those economies possessing 'institutional thickness', for instance, regions whose governance structures are 'embedded in networks of interpersonal relations' (Amin and Thrift 1994). The central idea is that the region is the level at which institutional networks and institutional thickness are best developed. A region is small enough for regular face-to-face interactions, yet large enough to sustain a critical mass of interpersonal networks. Policy making, then, should be directed towards strengthening regional networks.

Accordingly, strategic spatial planning should be regarded as a social process. Its activities include using, influencing, and creating all kinds of institutional relations aimed at 'developing and implementing strategies, plans, policies and projects, and for regulating the location, timing and form of development' (Healey et al. 1999, 340–341). Therefore, strategic planning efforts are also exercises of will-shaping and persuasion. They include ideas, inspirations, invitations for dialogue, possible procedures for consensus building, and so on that should have some bearing on investment and regulatory activities. Generally, we refer to strategic planning as an activity that creates some 'soft' outputs such as social capital (Putnam 2000). Or, as Healey et al. put it: 'Strategic spatial planning as a social process for developing and maintaining territorial relationships (through resources of trust: social capital), infused with shared knowledge and cemented by common frames of reference (knowledge resources: intellectual capital), through which effective territorial mobilization is possible (political capital)' (Healey et al. 1999, 343).

An important element in institutional capital is the involvement of stakeholders in the strategic planning process. Stakeholder involvement can deliver such benefits as getting relevant knowledge and information, developing acceptance and support, and thus making implementation easier. Stakeholders are people and organizations who can affect in some way the performance of an organization (Bryson 1988). The emphasis by strategic spatial planning is on trying to advance the various levels of government within a region to 'work together and in partnerships with actors in diverse positions in the economy and civil society' (Albrechts et al. 2003, 114).

Hamel (1996) distinguishes planning from strategizing, emphasizing the need for more 'imagination applied to building a strategy'. A truly new perspective can only be found when a discussion of strategy is open to many people, especially those that are different from the 'guardians of the status quo'. It should not just be an 'intellectual exercise for the corporate elite'. A popular approach for involving more than just top management is called the future search, or use of future teams. The idea is that a large group of stakeholders assemble to create an action plan based on the vision that they discuss, and so to create and use personal networks that help build intellectual capital.

If we refer to our discussion on institutional capacity, a few clear criteria for evaluation come to the fore. In evaluation, the richness of network resources could be employed as a yardstick to evaluating the likeliness of planning success. According to the strategic and entrepreneurial approach, plans (and planners) that are rich in contacts would be more likely to be effective. The embeddedness of a strategic plan in influential (political, social) networks would help its likelihood to implementation.

A related principle is that of mobilizing decision-making capacities aimed at acceptability (the extent to which those affected are in accord with certain plan initiatives). Capacity indicators like social capital, intellectual capital and political capital may be presented as evaluation standards. The idea here is that strategy formulation can be understood as a social and political process. The implementation of strategic plans typically depends on interaction within social networks, discussions on ways of understanding problems and fitting solutions, and political manoeuvring and tactics. Some further criteria would include the extent to which strategic plans seek the involvement of stakeholders.

As a conclusion to this section, it is fair to say that the two key issues or characteristics of strategic planning discussed above (for instance, contextual anticipation, and institutional capacity) differentiate it from earlier regional planning, and so require changes in what have been daily planning evaluation practices. Table 10.1 summarizes the main principles for strategic planning as discussed above, and adds implications for planning evaluation, in terms of the criteria required. The practical basis for this table is illustrated by recent strategic, entrepreneurial spatial planning in South Holland.

Establishing a Regional Strategy in South Holland

The Dutch context provides abundant institutional and cultural resources for strategic planning efforts at the regional scale. The emphasis in this section is on the activities employed by the province of South-Holland to establish a regional urbanization strategy in the 'south wing' of the Randstad in the west of the Netherlands (see Figure 10.1). Regional planners here also do evaluation and are interested in using certain criteria to justify planning action. This section and the next will explore what these criteria might entail. This section will first provide a sketch of the approach by this province towards regional planning, explain recent changes, and an analysis of current regional planning efforts in this region.

In the Netherlands, regions are defined as the areas that are larger than municipalities yet smaller that the state, and in which there is a strong social coherence and a demand for policy coordination (see for instance ROB 2003). These regions constitute an intermediate domain that escapes a strict definition within the Dutch legal framework of the triad 'municipality-province-state', while often representing the (variable) spatial scale at which important socio-economic

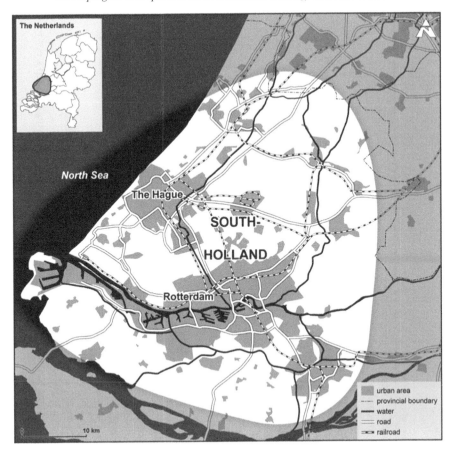

Figure 10.1 Indicative map of the Dutch 'South Wing', including The Hague and Rotterdam

dynamics occur. After the constitutional revision of 1848 and the subsequent Provincial Act (1850) and Municipal Act (1851), the Netherlands only has municipalities and provinces as legitimate types of general administration at the local and regional scales. Provinces, and especially municipalities, are still largely autonomous, while at the same time they participate in collaborations and a partly centralized conformity of policy and decision-making.

The provinces (12 in total) are responsible for a considerable amount of regional planning in the Netherlands. Especially since the 1920s, when there was a growing need for (supra municipal) plans for a region as a whole, with various policy interests involved (transport, water, recreation, nature, and so on), the province came into focus as the main regional planning agency. Only since the 1962 Spatial Planning Act has the 'regional plan' became the key task for provincial government. This regional plan was aimed at reaching full horizontal and vertical integration between plans at the various government levels (state-province-municipality), and across various policy fields (housing, transport, industry,

Table 10.1 Strategic planning and criteria for evaluation

Principle	Characteristics	Suggested evaluation criteria
Contextual anticipation	Evaluating societal change	The extent to which a spatial planning initiative: • proposes to anticipate changes beyond its control • monitors societal change, and identifies opportunities and threats The extent to which a planning agency: • is oriented toward and understands the tactical and strategic capabilities of other public and private actors • is identifying organizational strengths and weaknesses
	Evaluating market developments	The extent to which a planning agency: • seeks insights in market developments, and displays a willingness to take risks The extent to which a spatial planning initiative: • meets receptive institutions • is based on strategies by the competition, or behaviour by private actors • acknowledges international interdependencies and international quality standards
	Evaluating options for competitive, coherent visions	The extent to which a planning agency: • emphasizes its 'core capabilities' The extent to which a spatial planning initiative: • features a deliberate selection of the most essential goals/objectives/projects only • features new combinations of and coherences between, for example, economic potencies, sustainable development, water and urban identity
	Evaluating long-term policy horizons	The extent to which a planning agency: • assumes a long-term political vision and emphases a longer-term perspective The extent to which a spatial planning initiative: • provides a longer time horizon of plans or projects, adhering from specifying detailed steps to achieve final outcomes

Table 10.1 cont'd

Principle	Characteristics	Suggested evaluation criteria
Institutional capacity	Evaluating social networks	The extent to which a spatial planning initiative: • links to institutional webs available • is supported politically, intellectually, and financially by relevant actors The extent to which a planning agency: • has identified effective networks of interpersonal relations among key actors • contributes to the richness of social relations in an area, or to strengthening regional networks, in particular those networks connecting public (government) agendas, research (at universities and institutes) and business activity (at private companies)
	Evaluating social, intellectual and political capital	The extent to which a planning agency: • has identified decision-making capacities, and capacity constraints The extent to which a spatial planning initiative: • is aimed at creating 'soft' outputs such as trust and acceptability (social capital) • is based on shared knowledge and common frames of reference (intellectual capital) • is based on supporting coalitions and territorial mobilization (political capital)
	Evaluating stakeholder involvement and interaction	The extent to which a planning agency: • builds shared vision based on stakeholder involvement The extent to which a planning agency: • seeks citizen and interest group participation, and private partnerships • mobilizes stakeholders within the region

nature, agriculture, and so on). Another important characteristic stemming from the 1962 Act is the so-called 'touchstone' function for the regional plan: municipal plans all have to be tested on conformity against the provincial regional plan. The regional plan prescribes housing quota and zones, environmental protection zones, locations for industrial development, agricultural key areas, and so on. Municipal plan options have generally been evaluated here based on the extent to which they provide a contribution to, or fit within, these kinds of prescribed restrictions.

New Developments in Dutch Regional Planning

Only since the end of the 1990s has this spatial allocation of responsibilities come under heavy criticism. Overall, we can distinguish between four developments that have played a key role in these criticisms and on the rethinking about Dutch regional planning.

One development relates to the discrepancy between new geographical patterns and the institutional practices and arrangements that should respond to them. A key reason for this is that the territorial range of the labour market, the housing market and the mobility or commuter market has increased to a supra municipal level (see for instance Priemus 2002). Increasing mobility and telecommunication have led to notions about the network city or the urban network. Another reason is that insufficient space is available within cities for the construction of large numbers of dwellings, businesses, and related infrastructure.

A second development is that regional planning efforts in the Netherlands face a high administrative 'density'. As Louw et al. (2003) point out, the increase in scale of spatial processes has put pressure on the decentralized Dutch planning system, where municipalities traditionally have substantial executive powers. In an increasingly regionalized system, collaboration networks of private investors, foundations, societal groups, and regional authorities are looking for ways to get involved in the implementation of supra-municipal planning projects such as larger infrastructure, ecological networks, and regional business parks.

A third development entails the on-going debate on whether the three territorial levels of general government (state-province-municipality) should be retained. Related proposals including a possible fourth level of administration, and the discussion about subdividing the Netherlands into four state territories (see IPO 2002), have regularly failed to reach formalization. There is difficulty in matching democratic representation and interests representation in both legitimate and democratic terms at the regional territorial level (Gualini and Woltjer 2004). As a result, there is a discussion about the need for new institutional practices and capacities, often involving informal, loosely-coupled and 'experimental' settings. Since the second half of the 1990s, therefore, the Randstad region has been the scene for several initiatives in intergovernmental coordination, which have adopted a weakly institutionalized pattern as an alternative to constraints in statutory reform. Some examples include 'Randstad Region', 'Delta Metropolis', 'South Wing Platform' (see for instance Janssen-Jansen 2004).

The third development is related to the fourth: attention to an active, more governance-oriented role for the regional spatial plan. While in the past provincial regional planning was often primarily seen as a governmental activity aimed at the production and review of some kind of a spatial framework, now it implies regional coordination involving multiple local authorities and relevant stakeholders. A regional approach includes not only government but non-governmental actors too. Therefore, notions such as networks and governance have played an increasing role in the current position of regional planning in the Netherlands (see for instance WRR 1998). Governance efforts explicitly include private sector and individual stakeholders who together attempt to determine the progress of developments in a region.

Overall, the change in the character of provincial regional planning is due to a more fundamental change of orientation of Dutch planning into the direction of development planning (see also IPO 2002). In this notion, the active development of various project initiatives by means of a comprehensive plan should be strengthened at the provincial level (see for instance Koeman 1999). The provinces play a much more active, entrepreneurial role (IPO 2002), no longer limiting themselves to activities such as planning, testing, and monitoring, but also claiming a role in the actual realization of regional developments. Thus provinces become one among a multitude of actors involved in 'regional' governance, becoming involved in arenas in which the pattern of relationships is more and more defined in 'interconnected' rather than 'nested' terms (also see Gualini and Woltjer 2004).

Recent Policy Proposals

Recently, the Dutch province of South-Holland has launched several initiatives to engage in drawing up a comprehensive strategic regional plan for urbanization issues within the province. Some of the recent developments in the South-Holland urban area include combining work activities and housing, the intensive and multiple use of land, and interweaving water, nature and cultural activities. In addition, the province aspires to fulfil a strategic position in the Dutch Randstad and the European area. The Provincial Executives aim for a strategy with a horizon of 2020, which emphasizes the coherence between living, working, nature, and water (PZH 2002).

Recent regional planning policy in South-Holland focuses on strategy, a long-term perspective, and policy lines are set with a clear attention to context. The Spatial Image Memorandum (2002) mentions 'growing interdependence in a European environment', 'long term safety and well-being', and 'challenges related to strengthening economic potencies'. It also points out interrelations between issues of rising water levels, green space, and infrastructural investment. There also is a strong emphasis on selectiveness, particularly when decisiveness and implementation are involved: 'We are dealing with decision processes in which we would like to operate interactively, transparently, and verifiably. But we will also handle implementation and control in a vigorous way. The processes of choice will then also imply that we will have to say 'no' to principally useful projects' (PZH 2002, 3).

Key policy documents (PZH 2003 and 2004) state, as the key objectives for urbanization within the area, 'strengthening the coherence and identity of the South Wing urban network', and 'sustainable development of some specific economic clusters (knowledge centres, harbour of Rotterdam, glass horticulture and flower bulbs)'. There is a strong emphasis on fostering 'foci of innovation' related to knowledge and logistics. The means to this include developing high-grade business areas, good accessibility, a varied supply of houses, and a living environment with sufficient green and parks. Some projects include development for the harbour of Rotterdam, and two rail projects aimed at the integration of the area (Zuidvleugeloverleg 2003).

These new policy proposals feature interesting exercises in building strategic capacities and of using strategic criteria for evaluation. First, colleague government agencies are encouraged to subscribe politically and financially to the proposed projects. Second, regional actors within the South wing intend to coordinate efforts to deploy collective financial, administrative and personnel-related capacities. A third strategic element includes efforts by collaborating government agencies to seek citizen and interest group participation, and to seek partnerships with the private sector.

The focus of these regional efforts is towards 'getting things done'. Development projects are presented in terms of how they serve strategic ends: 'Working in projects makes it possible for us to realize integrated policy goals (housing, mobility, economy, green environment and water) in mutual co-operation with the region' (PZH 2004). The province makes clear it will contribute to the plan-making and development processes, but that it will also play a role in carrying out its policies. It will do so primarily via financial participation in spatial development companies and via risk-taking capital investment.

This kind of an emphasis on implementation is apparent, for example, in the so-called 'Green-blue string' project. This effort is aimed at developing a major regional park in close proximity to the cities of Rotterdam and The Hague. This project features a combination of unique attributes: green, water, recreation, mobility, and housing. These attributes are thought to be conditions for international quality and competitiveness. Typically, implementation here involves establishing a 'steering group' with local and regional government agencies, a 'reference group' including societal organizations, and five 'product groups' including private investors working on specific solutions. These solutions reflect practical proposals related to ecology, recreation, water, a highway connection, and shared financing.

Reflections by Key Stakeholders in the Region

Interviews with key private and public actors in the South-Holland arena provide evidence concerning what they consider to be major features of regional planning, including the necessary bases for evaluating promising options and implementation success. These considerations include references to notions of selectiveness, integration, networking, and capacity building (see Salet and Woltjer 2004).

A Selective, Integrated View

Interviewees underline the significance of new coherence in urban policy making, mentioning an integrative 'weighing of interests' related to issues such as the long-range determination of building locations, the combination of housing and healthcare, and the way that decisions about housing and commercial land-use are linked to decisions about establishing transport infrastructure. Respondents also emphasize the importance of selecting 'centres of gravity' and clusters, as opposed to simply providing an extended list with 'projects'. Strategic planning here is a matter of establishing a coherent spatial framework.

Our respondents point out the importance of developing a strategic profile for the South-Holland region, directed towards international quality standards such as safety (for example, 'Rotterdam: the safest harbour in the world'), environmental quality (for example, 'Dutch horticultural products: clean and sustainable'), consumer oriented public services, and 'quality of life'. It is assumed that using international quality norms as testing criteria for urban policy at the regional level would contribute to the standing and role of South-Holland worldwide.

Networking and Capacity Building

Respondents use various terms to typify capacity building related to the regional economy. One example is connecting research (at universities and institutes) and business activity (at companies). In this case the province would bring together private companies and universities, and assume direction in forming strong clusters between them.

Another striking finding is that key actors in the South Wing of the Randstad urge the province of South-Holland to stress its distinctive features in international networks and distinguish itself in the European arena. The province could evaluate, for example, the extent to which their spatial policies link to research and innovation in developing knowledge regions and international networks of private companies and universities. The province could utilize these networks as a breeding place for investigating new policy issues and subsequent policy initiatives, and to distinguish capabilities within the region compared to other international clusters.

If we follow the main lines of argumentation by our respondents, the strength of the South-Holland province, and possibly any regional administrative body, is not just determined by its regulative power, but by its ability to take assessment of relations within networks, and to monitor societal developments. The ability to appraise and facilitate establishing relations and networks, then, not the ability to accurately evaluate impacts of plans, is a large challenge for modern regional governance, even though these are not fulfilled very well today in Dutch regional planning practice (Salet and Molenaar 2004).

Some concluding Remarks

Generally, the study of regional planning in South Holland points to a clear need to giving more emphasis to a strategic and entrepreneurial approach in regional spatial planning. The attention to building and anticipating social relations, political networks, and market opportunity is crucial. A regional agency such as the Dutch province can have some essential tasks in assessing progress in relation to these strategic elements.

As a conclusion to this chapter, Table 10.1 recapitulates the key principles for strategic planning given by the literature review and the South-Holland case, and adds associated criteria for planning evaluation. The table shows how regional agencies could take the assessment of contextual change as a principle in evaluation. In strategic planning, plan options will not be evaluated formally on economic, environmental or social impacts, but, rather, on the strategic agility they can deliver. Regional planning would then sooner focus on explanations and interpretations of institutional capacities. Efforts within regional planning to evaluate policy proposals, then, would become more explicit in including criteria capturing the socio-political realities in planning practice.

Generally, dealing with new notions of regional strategic planning in planning evaluation implies rooting activities such as generating alternatives, assessing impacts, and comparing alternatives on the strength of regional development options to anticipate societal change and market insights, and to mobilize 'core capabilities', 'real vision' and institutional capacity within the region. Politically informed 'real visions' would be the basis for deriving alternatives options for action.

Also, the assessment of impacts would centre more around an explication of the conditions that make implementation more likely. In this sense, (evaluation) criteria to indicate success include assessment of societal change, market developments, options for coherent visions, long-term policy horizons, social networks, stakeholder involvement, and social, intellectual and political capital. Overall, contextual anticipation and institutional capacity would then become key principles in the assessment in regional-planning action.

The Dutch regional planning case shows an interest in assuming strategic ability, and in including contextual dynamics and institutional capacity building in their proposals for future urbanization in the south wing of the Dutch Randstad. Including elements of strategic action and entrepreneurship in planning evaluation for similar situations as in South Holland is not a straightforward task, however. If one of the key rationales of planning evaluation includes the provision of an explicit, replicable basis for public assessment (Miller and Patassini 2005), then we will have to look more closely at the options and feasibility of measuring and reporting the principles of contextual anticipation and institutional capacity (also see Khakee 2002). It would be a challenging task aimed at allowing new principles of strategic effectiveness in planning evaluation.

References

Albert, K.J. (ed.) (1983), *The Strategic Management Handbook*, St Louis, McGraw Hill.

Albrechts, L., Healey, P. Kunzmann, K. (2003), 'Strategic Spatial Planning and Regional Governance in Europe', *Journal of the American Planning Association* 69(2), pp. 113–29.

Amin, A. and Thrift, N. (eds) (1994), *Globalisation, Institutions and Regional Development*, Oxford, Oxford University Press.

Bloom, C. (1986), 'Strategic Planning in the Public Sector', *Journal of Planning Literature* 1(2), pp. 253–59.

Brenner, N. (2003), 'Metropolitan Institutional Reform and the Rescaling of State Space in Contemporary Western Europe', *European Urban and Regional Studies* 10(4), pp. 297–323.

Bryson, J.M. (1988), *Strategic Planning for Public and Nonprofit Organizations*, San Francisco, Jossey-Boss.

Bryson, J.M. and Einsweiler, R.C. (eds) (1988), *Strategic Planning: Threats and Opportunities for Planners*, Chicago, Planners Press.

Bryson, J.M. and Roering, W.D. (1988), 'Applying Private Sector Strategic Planning in the Public Sector', in J.M. Bryson and R.C. Einsweiler (eds), *Strategic Planning: Threats and Opportunities for Planners*, Chicago, Planners Press.

Bryson, J.M., Freeman, R.E. and Roering, W.D. (1986), 'Strategic Planning in the Public Sector: Approaches and Future Directions', in B. Checkoway (ed.), *Strategic Approaches to Planning Practice*, Lexington, MA, Lexington Books.

Casson, M. (2005), 'Entrepreneurship and the Theory of the Firm', *Journal of Economic Behavior and Organization* 58(2), pp. 327–48.

Checkoway, B. (ed.) (2001), *Strategic Approaches to Planning Practice*, Lexington, MA, Lexington Books.

Dabinett, G. and Richardson, T. (1999), 'The European Spatial Approach, the Role of Power and Knowledge in Strategic Planning and Policy Evaluation', *Evaluation* 5(2), pp. 220–36.

Friedmann, J. (2004), 'Strategic Spatial Planning and the Longer Range', *Planning Theory and Practice* 5(1), pp. 49–67.

Gordon, G.L. (1993), *Strategic Planning for Local Government*, Washington, DC, International City Management Association.

Gualini, E. and Woltjer, J. (2004), 'The Rescaling of Regional Planning and Governance in The Netherlands', paper presented at the AESOP Conference, Grenoble.

Guba, E.G. and Lincoln, Y.S. (1989), *Fourth Generation Evaluation*, Newbury Park, CA, Sage Publications.

Hamel, G. (1996), 'Strategy as Revolution', *Harvard Business Review* 74(4), pp. 69–83.

Hamel, G. and Prahalad, C.K. (1994), *Competing for the Future*, Boston, Harvard Business School Press.

Harris, L.C. and Ogbonna, E. (2006), 'Initiating Strategic Planning', *Journal of Business Research* 59(1), pp. 100–101.

Healey, P. (2006), *Collaborative Planning: Shaping Places in Fragmented Societies*, 2nd edn, Basingstoke, Palgrave Macmillan.

Healey, P., Khakee, A., Motte, A. and Needham, B. (1999), 'European Developments in Strategic Spatial Planning', *European Planning Studies* 7(3), pp. 339–56.

Henderson, B.D. (1983), 'The Concept of Strategy', in K.J. Albert (ed.), *The Strategic Management Handbook*, St Louis, McGraw Hill.

Hickson, D.J., Butler, R.J., Cray, D., Mallory, G. and Wilson, D.C. (1986), *Top Decisions: Strategic Decision Making in Organizations*, San Francisco, Jossey-Bass.

Hildén, M., Furman, E. and Kaljonen, M. (2004), 'Views on Planning and Expectations of SEA: the Case of Transport Planning', *Environmental Impact Assessment Review* 24(5), pp. 519–36.

Hoch, C.J., Dalton, L.C. and So, F.S. (eds), *The Practice of Local Government Planning*, Washington, DC, International City Management Association.

Holloway, D.A. (2004), 'Strategic Planning and Habermasian Informed Discourse: Reality or Rhetoric', *Critical Perspectives on Accounting* 15(4), pp. 469–84.

Innes, J.E. (1996), 'Planning through Consensus Building: A New View of the Comprehensive Planning Ideal', *Journal of the American Planning Association* 62(4), 460–72.

IPO (2002), *Op schaal gewogen, regionaal bestuur in Nederland in de 21e eeuw* [*21st Century Regional Governance in the Netherlands*], The Hague, Interprovinciaal Overleg.

Janssen-Jansen, B.L. (2004), 'Regio's uitgedaagd: "growth management" ter inspiratie voor nieuwe paden van proactieve ruimtelijke planning' ['Regions Challenged: "Growth Management" as an Inspiration to New Paths in Pro-active Spatial Planning'], Utrecht, PhD dissertation Utrecht University.

Kaufman, J. and Jacobs, H. (1987), 'A Public Planning Perspective on Strategic Planning', *Journal of the American Planning Association* 53(1), pp. 23–33.

Khakee, A. (2002), 'Assessing Institutional Capital Building in a Local Agenda 21 Process in Göteborg', *Planning Theory and Practice* 3(1), pp. 53–68.

Koeman, N.S.J. (1999), 'Verwachting: een Fundamentele Herziening van de Wet op de Ruimtelijke Ordening' ['Forthcoming: a Fundamental Revision of the Spatial Planning Act'], *Nederlands tijdschrift voor bestuursrecht* 4, p. 89.

Lichfield, N. (1996), *Community Impact Evaluation*, London, UCL Press.

Louw, E., Krabben, E. van der and Priemus, H. (2003), 'Spatial Development Policy: Changing Roles for Local and Regional Authorities in the Netherlands', *Land Use Policy* 20, pp. 357–66.

Miller, D. and Holt-Jensen, A. (1997), 'Bergen and Seattle: A Tale of Strategic Planning in Two Cities', *European Planning Studies* 5(12), pp. 195–214.

Miller, D. and Patassini, D. (2005), 'Introduction – Accounting for Non-market Values in Planning Evaluation', in D. Miller and D. Patassini (eds), *Beyond Benefit Cost Analysis*; Ashgate, Aldershot.

Mintzberg, H. (1983), *Structure in Fives: Designing Effective Organizations*, Englewood Cliffs, CA, Prentice-Hall.

Mintzberg, H., Ahlstrand, B. and Lampel, J. (1998), *Strategy Safari, a Guided Tour through the Wilds of Strategic Management*, Hemel Hempstead, Prentice Hall Europe.

Porter, M.E. (1996), 'What is Strategy?', *Harvard Business Review* 74(6), pp. 61–80.

Priemus, H. (2002), 'Combining Spatial Investments in Project Envelopes: Current Dutch Debates on Area Development', *Planning Practice and Research* 17(4), pp. 455–63.

Putnam, R.D. (2000), *Bowling Alone: The Collapse and Revival of American Community*, New York, Simon and Schuster.

PZH (2002), *Collegewerkprogramma 2003–2007* [*Council Working Programme 2003-2007*], The Hague, Provincial Government of South-Holland.

PZH (2003), *Nota Ruimtelijk Beeld 2015+* [*Spatial Image Memorandum 2015+*], The Hague, Provincial Government of South-Holland.

PZH (2004), *Provinciale Ruimtelijke Structuurvisie Zuid-Holland 2020* [*Provincial Spatial Structure Southern Holland 2020*], The Hague, Provincial Government of South-Holland.

ROB (2003), *Legio voor de regio, bestuurlijke antwoorden op regionale vraagstukken* [*Legion for the Region, Administrative Replies to Regional Issues*], The Hague, Council for Public Administration.

Salet, W. and Woltjer, J. (2004), *Verstedelijking Gewogen, standpunten verstedelijkings-strategie Zuid-Holland* [*Weighing Urbanisation, Views on a South-Holland Urbanisation Strategy*], Amsterdam, Amsterdam Institute for Metropolitan and International Development Studies.

Salet, W. and Woltjer, J. (2007), 'New Concepts of Strategic Spatial Planning, Dilemmas in the Dutch Randstad Region', *International Journal for Public Sector Management*.

Salet, W.G.M. and Molenaar, J. (2004), *Oefeningen in de Regio Amsterdam, netwerken van ruimte en bestuur* [*Exercises in the Amsterdam Region, Networks of Land and Governance*], Amsterdam, Amsterdam Study Centre for the Metropolitan Environment.

So, F.S. (1986), 'Planning Agency Management', in C.J. Hoch, L.C. Dalton, F.S. So (eds), *The Practice of Local Government Planning*, Washington, DC, International City Management Association.

Talen, E. (1996), 'Do Plans Get Implemented? A Review of Evaluation in Planning', *Journal of Planning Literature* 10(3), pp. 248–59.

Voogd, H. (1997), 'The Changing Role of Evaluation Methods in a Changing Planning Environment: Some Dutch Experiences', *European Planning Studies* 5(2), pp. 257–67.

VROM (2001), *Handreiking gebiedsgericht beleid* [*Manual for Area-oriented Policy*], The Hague, Ministry of Housing, Spatial Planning, and Environment.

Wissink B. (2000), *Ontworpen en ontstaan: een praktijktheoretische analyse van het debat over het provinciale omgevingsbeleid* [*A Practical-theoretical Analysis of the Debate on Provincial Comprehensive Planning*], The Hague, Sdu Uitgevers.

Zuidvleugeloverleg (2003), *De Zuidvleugel van de Randstad, netwerkstad van bestuur and recht, kennis en logistiek* [*The South Wing of the Randstad, Network City of Administration, Law, Knowledge and Logistics*], The Hague, South Wing Administrative Platform.

Chapter 11

Evaluating National Urban Planning: Is Dutch Planning a Success or Failure?

Willem K. Korthals Altes

Introduction

Planning on a national level performs a different function to planning on a local level. Uncertainty plays a larger role, as national planning is often more abstract, i.e., local government has its own responsibilities for the precise location of activities, and zoning of land use. National planning is a process in which public and private agencies each play a different role, and neither of these agents can control the outcome. This chapter is about the evaluation of one aspect of Dutch national planning: the concentration of urban development, in which there is a tradition of different government agencies each having a different role in planning (Faludi and van der Valk 1994). This chapter examines how to evaluate a complex national urban planning policy.

Measuring the success of planning starts with the question 'What is planning?' (Talen 1997). An answer given by Needham (2000) is that planning may be considered to be a 'design discipline', creating knowledge for the improvement of planning practice. This concept of planning as a design discipline or a 'design science' (van Aken 2004) has an impact on the evaluation, so this chapter will ask what a design science is, how design relates to processes, and how design may be evaluated. Based on these aspects there is, in relation to the development of technology, quite an extensive body of recent literature that has not previously been introduced in discussions on planning evaluation. This theoretical part of this chapter is concluded by a section on evaluating planning as a design discipline linked to sources in planning literature. In this section, a link will be made with the debate on planning literature between proponents of the criterion of performance (i.e. has the plan showed the way to better decision-making? (Faludi 2000)) and proponents of the criterion of conformance (i.e. is spatial development happening according to plan? (Talen 1997; Laurian et al. 2004)). The conclusions from the theoretical part of the chapter will be used to evaluate Dutch urbanization policies. Conclusions are drawn both on the implications this has for other evaluations of complex planning and on the results of Dutch national planning.

Design Discipline

Design sciences, such as engineering, medical sciences and planning can be compared to the formal sciences, such as mathematics and philosophy, and the explanatory sciences, such as the natural sciences, and major sections of the social sciences (van Aken 2004). The mission of formal sciences is to build systems of propositions that are internally logically consistent. Explanatory science is about the explaining and prediction of observable phenomena. Design sciences are not about explaining phenomena, but about changing the facts (Eekels and Roozenburg 1991).

> The mission of a design science is to develop knowledge for the design and realization of artefacts, i.e. to solve *construction problems*, or to be used in the improvement of the performance of existing entities, i.e. to solve *improvement problems*. (van Aken 2004, 224; see also van Aken 2005b)

In the literature on design sciences the term artefact (or artifact in American sources) is much broader than in plain English. 'An artifact is an intentionally modified tool whose modified properties were intended by the agent to be recognized by an agent at a later time as having been intentionally altered for that, or some other, use' (Dipert 1993, 29–30). Professionals, who must be able to translate general knowledge to 'the unique and specific case at hand', often use design knowledge (van Aken 2004, 226).

Design not only deals with observable phenomena, but it also involves the investigation into systems that do not yet exist: 'The main question thus becomes, "Will it work?" rather than, "Is it valid or true?"' (Romme 2003, 558; see also Simon 1996). Or, as Herbert A. Simon in one of his last papers states:

> Our task is not to predict the future; our task is to design a future ... and then to devote our efforts to bringing this future about. We are not observers of the future; we are actors who, whether we wish or not, by our action and our existence, will determine the future's shape. (2002, 601)

Based on Simon (1996) it can be stated that an artefact (the result of design process) has a dual nature. Artefacts are, on the one hand, physical objects that interact through causal connections, and which the laws of nature govern. On the other hand, an artefact has a function within a context of human action: it can be used as a means to an end (Kroes, 1998). Agents intentionally:

> ... represent the world and act intentionally in it, and whose behaviour is explained partly in terms of reasons (and not causes) . ... One aspect of this latter conceptualisation is that certain activities are interpreted in terms of realisations of goals and that functions are attributed to certain objects or activities. (Kroes 2002, 293–294)

Both function and physical characteristics together constitute the artefact or 'technological object'. 'The function cannot be isolated from the context of use of a technological object: it is defined within that context. Since that context is a context of human action, we will call the function a human (or social) construction' (Kroes 1998, 18). In a design process a bridge is made between a physical system and the function that will be performed by that system. The design not only gives a description of that system, but gives also an explanation of how the proposed system is able to perform the required function. This technological explanation is an integral part of the design and 'plays a crucial role in justifying a design' (1998, 21).

Engineers are, in this way, able to bridge the gap between structure, described in a non-intentional language, and function, described in intentional language. The operational principle is that it '… connects structure and function on the basis of causal relations and pragmatic rules of actions based on these causal relations' (Kroes 1998, 34). One physical object may be used for different functions and one function can be realized in different ways. Although functions of objects are observer-relative, and do not impact on the physical conditions of an object, '…the practice of engineering, and more generally the practice of everyday life, show that functional claims contain genuine knowledge about the world which is different from knowledge contained in structural claims' (Kroes 2001, 8). Normativity plays an important role in the evaluation of functionality.

By itself, the physical structure does not have a normative impact; the collapse of a bridge is not a bad thing, just as stellar collapse is not bad. 'A bridge that is prone to collapse is a bad bridge only because the physical structure plays a role in human practices, and has been designed to play this role by an engineer' (Houkes 2002, 262). In other words, an artefact must be placed in a context of human intentions, wishes and deliberations.

The research of both Kroes (2002) and of Vermaas and Houkes (2003) concludes that a theory that ascribes functions to an artefact on the basis only of the intentions of the agents that designed it is too limited, as many technological uses are based on the new functions of an existing instrument. Designers of an artefact may often have an idea of its proper use, and in this way are attempting to design human behaviour (Redström 2006; Kroes et al. 2006). Architects do not only design houses, but may attempt to design human activities. Urban designers have ideas on the proper use of cities, which have lead to debates with social science educated planners (Faludi 1996).

Consequently, this leaves us with the idea that the functions of an artefact must be part of an evaluation. We will later argue that evaluation of a plan (note that 'a plan' is an artefact in the definition used above), based on the intentions of a plan maker alone may be considered too limited as a plan may have new functions, outside the ones intended by the plan maker.

Designing as Process, Processes of Design

Design is a step in a process; it is a 'model of an entity to be realized' or 'an instruction for the next step in the creation process' (van Aken 2005a, 391). A design is an artefact (a design can be characterized using this term) to produce another artefact. van Aken indicates in this regard that designs may be over-specified. Or, the other way around, he proposes the 'principle of minimal specification' as 'a complete design should only specify what the makers of an artefact need to realize that artefact' (2005a, 391). So, design is a process of consecutive detailing, from rough sketches, via outlines to detailed designs. The principle of minimal specification does not only apply to the transition from design to realization, but also to the design process itself, and this principle of minimal specification is especially important for a design process. In a design process with high uncertainty '... one should put more effort in role definitions, leaving the actual scheduling of activities to the individuals and groups assigned to these roles themselves' (van Aken 2005a, 394). Design processes often must be sufficiently manageable, adaptable and robust to meet requirements in time, cost and quality.

The principle of minimum specification is, as stated above, especially important for the design of human action systems, as those are 'driven by the thoughts and feelings of the actors in question' (van Aken 2005a, 397). Human actors are not determined by the design of a human action system, although they can be influenced. 'A process-design for a design-process is realized through the *internalization* of the overall process-design by the designers in question and by a subsequent *redesign* by them of this overall design to a design of their own activities ...' (2005a, 397). A consequence is that process designers have much less control of the process than product designers have on a product. According to van Aken this has the advantage that it saves time in designing the process in detail, but it also opens the possibility of 'unmanaged deviations' from the process design, leading to coordination problems at a later date.

Planning takes place in such a human action system and most plans are made to influence these kinds of systems, posing well known planning dilemmas between flexibility and commitment, or between continuity and change.

Evaluation of Design, Artefacts and Planning

Design processes and artefacts can be evaluated. In a chapter on evaluation in his book on *Artifacts, Art Works, and Agency*, Dipert (1993) presents five types of dysfunctions in relation to artefacts:

1) the identification of an artefact is false;
2) although something is rightly identified as an artefact, there is no good conception of its purpose;
3) an artefact is not an effective instrument to its purpose;
4) the function the artefact may fulfil is currently not useful; or
5) this function is not useful at all.

An evaluation of artefacts – and thus of planning – may be designed in such a way that one or more of these dysfunctions are assessed.

Although there are many debates about whether something is a 'plan', the debate is not about whether or not something is an artefact (which an archaeologist may have with an odd-shaped stone), but about whether this artefact has the function attributed to plans. So an evaluation of planning usually starts with Dipert's second point, and goes all the way through to point five. This chapter deals particularly with evaluation in relation to points two and three. The debates on whether or not the containment of urban sprawl is desirable or not (in the present context, this fits within the debate on dysfunctions four and five, and are highly relevant) is not the focus of this chapter.

In relation to planning, the second point discusses: what is the function of plans? The question about the purpose of planning, both of plans as product and of planning as process, is however not always worked out in evaluation research. When we say that a certain planning instrument is not effective in pursuing a certain goal (failure type three in terms of Dipert), we are sometimes wrong, in that planning has been effective in addressing a goal which differs from the goal we assumed we were evaluating. So, using the Dipert classification, we have made a type two error in our evaluation.

Dipert gives another example from the performing arts.

> The particular conception of an assumed or intended audience was a component of the distinctive artifactual intention that characterizes its artifactuality. If the creator's conceptualization of audience and its qualifications was both reasonable and clearly conceived, then the failure lies with the later audience or context, not with the creator or artefact. The failure of Machaut's music to make students in Music History 101 think of God is hardly a problem with Machaut or his music. (1993, 143)

This kind of debate is also familiar in planning, for example when planning instruments are transplanted from one institutional setting to another. We lack in this instance a firm concept of the instrument's purpose in a certain context (de Jong 2004).

The dysfunctionality of type three can be judged by the standards of fulfilling a certain function (Dipert 1993, 145).

> The overall evaluation of whether an artifact of a certain sort does in fact fulfil its function (or how well it does so) is a relatively straightforward matter. We need only have a reasonably precise articulation of that function and some indication of how well the artifact succeeds in that function. A sufficient articulation of that function will be a description of the activity, or the contemplated change in the world, together with limitations or recommendations on the proper use of the tool for its purpose. This description must be cast in the terms of the description by which its maker understood the function. (Dipert 1993, 146)

Dipert's last point, that the description is in the artefact's maker's terms, is a matter of debate. Others argue (Preston 2003; but also Dipert 1993 himself, see also below) that the intentional use of an artefact may be relevant, for example, a screwdriver can be very functional to open cans of paint and a user may evaluate a screwdriver for performing this task alone. Dipert (1993), who wrote several papers on artwork, uses this argument because he wants to evaluate the quality of historic instruments in their own right. An eighteenth-century screwdriver must not be evaluated on the way it can be used on screws made of contemporary hardened steel. However, if you want to use a screwdriver, and look for a good one, you may prefer a contemporary screwdriver as a better instrument for contemporary screws.

Dipert (1993) identifies two possible goals for the interpretation of artefacts, instruments and tools. The first one is the understanding of its historical genesis and history; that involves finding out its 'deliberative history' (1993, 87), and is about its purpose in relation to its original purpose. The second one is about functionality; to find out the most useful function an object can have. This latter goal is not bound by considerations of its historical purpose. 'No agent or abstract loyalty to history ... binds us to think of objects in certain ways or to use them in certain activities – certainly not its maker only in virtue of this historical relationship to the object' (1993, 89). Dipert considers the pursuit of a function of an object as the chief aim, and pursuing the historical origins may help with this task. He makes a great effort to look at the relationship of how the inquiry into an object's history (and historical purpose) may sometimes be the best way to achieve an understanding of all its functions. Dipert considers looking at its original purpose as the best strategy to come towards present functionality. However, he also makes clear that in certain conditions and circumstances this may not be the case. Complicating this question, according to Dipert (1993), is the fact that not all of an artefact's intended functions can be observed by looking at the artefact itself. This is especially the case with the 'high-level intentions' that include the behaviour that an acting agent wishes to cause in other cognitive agents.

This question of intent in relation to evaluation is also relevant for the evaluation of plans. Do we have to evaluate it according to the intentions of the plan maker at the time the plan was made, or the intentions of the user of the plan at the time it is employed?

Is the original function of something its proper function? According to Preston (2003), using an artefact for its original function may not be superior to using it for new functions. It may not even be more creative, as a new use '... endows existing artefacts with new functions by using them in novel ways' (2003, 607). On the other hand; the design of new artefacts is not always very creative. Designs may be a variation of a standard type; many new cars are a variation of existing cars. Creativity cannot be the criterion that distinguishes design from use.

If we consider that the actual content of the agent's intentions define the artefact's 'proper function', it can consequently be stated that:

> ... if proper functions are derivable from the intentions of designers, it seems they must be derivable from the intentions of users as well. In other words, if the purpose of the designer establishes the proper function of the artifact designed, then the purpose of the user must equally establish the proper function of the artifact used. (Preston 2003, 608)

Based on her analyses, Preston (2003) argues for us to give up the idea that the proper function of a technical artefact should be attached to new artefacts. By doing this Preston stresses the importance of a social approach to the function of artefacts. It is not the designer of a new artefact that defines its proper function, but defining the proper function of an artefact is a social construction in which both the intention of the designer and the context of use may play a role.

Evaluating Planning as a Design Discipline

The insights and debates on design, its process and its evaluation presented above, leaves open the question of how to place planning in this context. From the above discussion, it is clear that plans and other products of planning can be seen as artefacts, an evaluation must therefore take place in the context of the functions of these artefacts: plans and other 'products of planning' (Faludi 2000). The functionality by which a plan may be evaluated may be both the plan maker's intended use, and the client's intended use. If we follow the reasoning by Dipert (1993; see also Faludi 1986), it may be useful in the context of discovery to reconstruct historically the intended use of a plan. However in terms of justification, actual use and present functions should play a decisive role in evaluating the functionality of a plan, or another product of planning.

Needham (2000) considers the following differences between planning and regular design. Planning is often about what measures to take in order to influence the decisions of others: 'Thus in most cases spatial planning is an *intervention in*, or an *influencing of*, the creation and use of the physical environment by others' (2000, 443). So planning is about policy design. The earlier discussion on design as a process shows that this aspect is not that different from other complex design processes, such as standardization of file formats or accountancy standards. The insights about minimal specification as a criterion for design in such cases (van Aken 2005a) may also play an important role in planning. The multiplicity of aims that a planning agency is pursuing adds to the complexity of planning as design. It is often territoriality (Vigar and Healey 1999) that makes it necessary to take different factors into account, as different sectors, such as transport, housing and the use of land in a certain territory.

A selection of the many functions of a document called a 'plan' (see also Needham 2000) may have to include the following:

- a description of the present state of affairs;
- a vision of a desirable future state of affairs;
- a legal rule, which has to be complied with by the agency itself;
- a legal rule, which has to be complied with by everyone;
- a legal rule that has to be taken into consideration by the agency itself;
- a legal rule that has to be taken into consideration by everyone;
- an investment in future decision-making by placing it in a wider field of choice;
- an overview of future subsidies;
- a statement about future decisions;
- a piece of propaganda aimed at the citizens;

- a piece of propaganda aimed at other governments, i.e. to acquire grants, new legal instruments;
- a vehicle for communicative action;
- an intention to act;
- a mixture of many of the functions described above and several other functions.

These functions are not always stated in the plan. Moreover, the actual function of a plan may differ from the plan maker's intended function. The above section has shown that the 'proper function' of a plan may differ from the function intended by its maker. Quite often plans have a variety of functions. Measures interact, and a plan may have a function as a package of mutually coordinated measures (Needham 2000). Planning agencies often work in a context with great uncertainty, in which a plan is often used as a flexible framework for subsequent action.

The evaluation of planning may take into account what function a plan has, both when it was made, and while it is being used. However a third kind of functionality may also come into the evaluation, the function that a plan should have according to the evaluator. This may not be the function of the plan according to the plan maker, nor the plan user. This is not necessarily an incorrect conclusion. It may be very fruitful to evaluate whether an existing plan may be used for a new function. It must be clear however that neither the plan nor its present use are being evaluated, but instead the contribution or potential contribution that a plan may have for a certain function defined by the evaluator.

The now popular practice of performance measurement may be an example of this. The objective is to focus government on improving results for citizens by measuring results in terms of outcomes the citizens care about (Osborne and Gaebler 1992; Helling 1998). These outcomes are not necessarily the same as the goals of policy makers. The idea is that an organization defines its products and services and develops indicators to measure its output in a planning and control cycle in order to improve the organization's performance (de Bruijn 2002). Performance measurement may have different forms, depending on the rationale behind measuring performance (Behn 2003), including for instance whether performance is measured in order to control subordinates or to celebrate successes. According to Behn (2003), there are eight different basic purposes of performance measurement, requiring different evaluation criteria.

Another example of an external criterion to evaluate planning is plan performance; does the plan contribute to an improvement of decision-making (Faludi and Korthals Altes 1997; Faludi 2000; Korthals Altes 2006)? The goal of planning is then considered to be a means of improving decisions by placing them in a wider field of choice (Friend and Jessop 1969; Faludi and Mastop 1982). The purpose of planning is not 'to draw pretty pictures of the future' (Faludi and Mastop 1982, 245), but to solve problems in the present. As time is limited at the time that decisions must be taken, it may often help to pre-empt this by placing the decisions in a wider picture, by making a plan. This conception of planning is a main theme in the work of Simon (for example 1997; see also Bratman

1987; Faludi 1973). Although makers and users of the plan may not intend to undertake a rational deliberation of alternatives and to choose the best one, it may, nevertheless, be worthwhile to evaluate whether a plan does contribute to a rational justification of decisions. This normative stand to improve present practice is also the principle Needham (2000) defends. Rationality is not an empirical concept, but a norm. Although the classical concept of rationality in planning has been rejected '… yet no one dares to argue for a methodology of planning in which people should act *irrationally*' (Needham 2000, 440–441). The idea is that the quality of a decision is based on its justification (see Faludi 1986), in which plan makers and plan users communicate to establish the meaning using different codes, partly because the decisions are made in an uncertain environment (Faludi and Korthals Altes 1997).

To return to the rather long, but incomplete, list of functions of planning noted earlier makes it clear that a holistic evaluation of planning, or the claim that a plan is fully evaluated, is not very fruitful. Moreover, evaluation may be specific about what functions of a plan are being assessed, and that the results of the evaluation are valid only for the functions that are explicitly taken into account in the evaluation.

As stated at the beginning of the chapter, two major measures of success in planning are the match between intention and implementation, and whether the plan resulted in better decisions. Conformance between plan and implementation, the first of these two criteria, assumes that the plan provides the preferred solution; and that the plan is there to make this solution real. The function of the plan is to be implemented literally, as the plan provides the best framework for action. Conformance may be measured in absolute or in relative terms, for example, 'how far the original intentions were carried through to the final result' (Knudsen 1988, 552) is often used as a measure of success in planning. The idea is that, with the conformance criteria, evaluators possess a 'tangible, objective measure of planning success' (Talen 1997, 577) in a context that plans function as blueprints (Laurian et al. 2004). The other approach to measuring success is the performance principle, the idea that planning is an investment to place decisions in a wider field of choice and must lead to an improved justification of decisions. The application of these two approaches for assessing the success of plans will be illustrated by the Dutch urbanization policies (see also Korthals Altes 2006).

The Case of Dutch Urbanization Policies

The Fourth Memorandum on Spatial Planning Extra (also known by its Dutch acronym VINEX, and a plan according to article 2a of the Dutch Law on Spatial Planning) is central to the traditional core issue in Dutch national urban planning: the management of urban growth (Korthals Altes 1992) by urban containment. The thought behind these concentration policies has been that urban regions would function as compact daily urban systems. Concentration policies were supposed to maintain the support for urban services, to limit mobility or commuting growth, to allocate housing, employment and facilities to optimize accessibility by bicycle and public transport, and to contain the further urbanization of rural areas. In addition to concentration policies to locate new housing and commercial areas within urban regions, there were also criteria for the choice of locations within concentration areas.

Based on the memorandum, contracts were drawn up directly with several urban regions (see also Korthals Altes 1994) and with the provincial governments for the other urban regions on the containment and development of urbanization between 1995 and 2005. The contracts covered funds for housing, public transport infrastructure and soil clean-up, providing that the city regions were to build 456,959 homes in the 26 urban regions between 1995 and 2005, in accordance with the wishes of central government (Needham and Faludi 1999; Korthals Altes 2006).

It was not obvious that the policy would be implemented, as it was based on a number of daring assumptions in relation to the cooperation of lower-tier governments, infrastructure provision and market demand (Needham and Faludi 1999). Notably the context on these three aspects was in flux as the Dutch welfare state was being restructured and undergoing decentralization. Between World War II and 1990, government policy was geared towards supply; producing homes, particularly social housing. Social rental housing in the Netherlands expanded from 12 per cent of the housing stock in 1945 to 44 per cent in the early 1990s (Boelhouwer 2002). Government subsidies for the building of social housing were abolished in the past ten years (Priemus 1995). The housing sector became more market oriented. The transformation process has been facilitated by a huge rise in house prices – 339 per cent – between 1982 and 2002 (NVM 2003).

Dutch municipalities have traditionally played an active role in land development; they engage in 'direct development' by acquiring land, providing infrastructure and selling the serviced plots to developers and housing associations (Needham 1997). All this is costly and takes time, and, in so doing, municipalities depend on cooperation and government grants, which were usually forthcoming (i.e., the implementation of spatial policies in developments seemed to be unproblematic). Consensus between governments combined with government grants was, in many cases, a guarantee for the implementation of spatial development plans.

Aspects such as neglect of the production of building land in the 1980s, changing market conditions (which made the housing development processes more viable) and the decentralization of government grants contributed to a change of the roles of municipal government and other players in urban development (Korthals Altes 2000; Louw et al. 2003; Verhage 2003). Growth in the proportion of market sector housing meant an average increase of sale prices for serviced land, which resulted in early acquisition of land by developers in certain locations (Needham 1997; Korthals Altes 2000) and private-sector parties abandoning their policies of not buying raw land (Korthals Altes 2000). Today, local government plays a crucial role, and is becoming a hybrid organization, as it not only acts as an independent government above market agents, but is also active on the market itself; servicing and selling land in competition with development companies.

New forms of public-private partnership are emerging and influencing strategic planning and land development practice (Louw et al. 2003).

Conformance analysis of concentration policies

The contracts for the concentration areas add up to a total growth of 456,959 housing units between 1995 and 2005. Between 1 January 1995 and 31 December 2004 the housing stock grew by 669,955 dwellings, of which 66.5 per cent is located in concentration areas (Table 11.1).

Table 11.1   Growth of housing stock in concentration areas and other areas

	Concentration areas		Other areas		Netherlands	
Area of land* in km²	7,219	21%	26,669	79%	33,889	100%
1995 housing stock	3,744,574	60.5%	2,447,348	39.5%	6,191,922	100%
2004 housing stock	4,189,995	61.1%	2,671,882	38.9%	6,861,877	100%
1995–2005 housing stock growth	445,421	66.5%	224,534	33.5%	669,955	100%
Correction for administrative changes	–33,178		33,178			
Corrected 1995–2005 housing growth	412,243	61.5%	257,712	38.5%	669,955	100%

*   This excludes the area of water (7,637 km²).

Source: CBS 2005.

The figures in Table 11.1 on the expansion of administrative borders are less impressive, but still show a net concentration towards the concentration areas. This containment of growth is reinforced by 'restrictive' policies. In other words, the approval of new building capacity outside the concentration areas has been more selective than in the past (Figure 11.1). This data indicates that conformance is high, although not complete, and that VINEX policies had a considerable impact on this. So we must evaluate this policy positively in relation to the conformance criteria (see also MVROM 2005, 6).

According to these figures, housing stock growth has made a slow start in the concentration areas, but has been on target since 2000. Housing stock growth elsewhere has been above targets. This is not because housing stock growth in other areas has risen, but because the reduction of house building in other areas has been less than planned for. However, in the Dutch debate, stagnation of house building is considered a major problem, because of the combination of exploding house prices and falling house-building output (Boelhouwer 2005). VINEX is often not considered to be a planning success story. This is not only because of the stagnation in house building, but there are also questions about the underlying

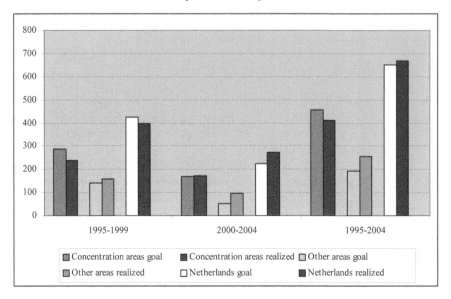

Figure 11.1 Conformance of corrected housing stock growth of VINEX areas in 1,000 dwellings in 1995–1999 and 2000–2004

Sources: CBS 2005; MVROM 2005.

targets of the policy, the quality of the locations, and the working of market forces on the locations (Korthals Altes 2006). Typically critics expect planning to meet other demands than conformance to pre-set targets. Is performance, as opposed to conformance, a better criterion to evaluate the functions of plans?

Performance Analysis of Concentration Policies

Performance analysis, that assesses whether a plan results in improved decision making, is less tangible than conformance analysis. A three-step research design devised by Faludi and Korthals Altes (1997; see also Korthals Altes 2006) is used to evaluate performance:

1) identification of the decisions that the plan should influence;
2) identification of the commitments that are being made in these decisions, identification of the arenas in which these decisions are justified;
3) assessment whether all or part of the plan helped in shaping the codes used in justifying subsequent decisions, and, if so, whether the quality of the justifications concerned improved in terms of taking account of the wider field of choice.

Identification of decisions The idea is that the plan is not only a framework for the implementation contracts, but also helps in setting subsequent policies to meet uncertain developments, such as the changing demands for housing. A strategic plan must also be able to adapt to changes in both spatial and sector policy views, as new governments take office. Although changes in demands and political priorities are likely, the direction of these changes remains uncertain.

Identification of commitments and arenas for justification   Between 1995 and 2005, there was an agreed commitment to conclude contracts with city regions and provinces on the housing programme. A new housing policy was then introduced, and for the period 2005–2010, local governments were asked to prepare plans based on high forecasts of need (Faludi and Korthals Altes 1996). The contracts for the 1995 to 2005 period did not cater for changes in demand. Moreover, because locations and housing size were specified, neither did they have the flexibility to meet known demands by preparing excess plan capacity when housing production on one location stagnated. One exception to this was the Amsterdam region, which requested flexibility, as they were unable to predict housing growth at a major location on a land reclamation site in a lake. The formal ratification of the contracts took place in parliament, which not only received copies of the contracts, but also debated the policy with the minister on numerous occasions. A parliamentary commission also put much effort into critically assessing the policy, including the idea of negotiating with local government to make contracts (see also Needham and Faludi 1999). The policy also attracted criticism among professionals and in articles in general newspapers and magazines. The government policy to overcome the stagnation in the production of housing has been the subject of debate in the same arenas.

Performance analysis of plan   There is a clear, strong relationship between the Fourth Memorandum Extra and the commitments made in the implementation contracts. However, it is questionable whether this strong commitment was necessary and whether it would have been more helpful if the plan involved a less prescriptive type of contract, allowing room for extra housing production in response to demand (Korthals Altes 1995, 301–302). The Dutch national government's ambition was to withdraw from urbanization policies for a decade by handing the responsibility to local governments, and paying them grants based on accountancy reports that established if performance targets were being met. However, new information is emerging. For example, housing demand has grown considerably in the 1990s, and political consideration of new knowledge on housing needs and new political aspirations are seen as necessary in the Dutch context. Moreover, financial analysis of green-field plans has shown that the government grants have been unnecessary due to the large increase in housing prices and the subsequent price of serviced land (Kolpron 2000). The Fourth Memorandum Extra plays hardly any role in the debate on housing stagnation, because housing growth complied with planning (Figure 11.1), and the stagnation relates to new political ambitions. A sign of this non-performance is that no information is available on the Ministry of Housing, Spatial Planning and the Environment's website.

Explanation of different outcomes in this case Working with ten-year contacts seems not to meet van Aken's (2005a) principle of minimal specification. The national government made binding agreements for a ten-year period and put all grant money into these covenants. By doing so, the government sacrificed flexibility for new policies. The VINEX plan does not contribute towards the discussion surrounding breaking the stagnation in house building; moreover, this stagnation was planned stagnation (Figure 11.1). This inflexibility was not only the case in relation to housing numbers, but also in relation to the grants. Over € 900 million has been allocated to the provinces and urban regions to fund the gap for the losses in developing the VINEX areas. These areas were not only on simple rural sites, but were sometimes much more expensive to develop, for example building on an airport at Ypenburg in The Hague, greenhouses, (e.g. Wateringse Veld in The Hague, Leidsche Rijn in Utrecht (Verhage, 2003) and Waalsprong near Nijmegen) and land reclamation at IJburg in Amsterdam. Due to a booming housing market in the 1990s, these grants were, generally speaking, not necessary as market developments made it possible to finance the costs of infrastructure provision, including cross subsidization of costs by providing plots for affordable housing (Kolpron 2000).

The background for the discrepancy between performance and conformance analysis is that we currently know a lot more about the present situation and present goals than we knew in 1990, when the draft of the Fourth Report Extra was made. Measuring conformance involves taking a plan based on old knowledge as a way of measuring success. To evaluate whether the plan has been a coercive force is an acceptable analysis and this analysis establishes to what extent agents in spatial development have obeyed the will of the planning agency as has been stated in the plan. The function of Dutch national planning documents is, however, different. These planning documents provide guidelines that, amongst other issues, must be considered when making decisions at a later date. The plan does not contain precise decisions on the location of facilities themselves. It is not a zoning ordinance.

The plan also has a role in organizing different agencies around a planning policy. Dutch national planning reports play an important role in sustaining Dutch spatial planning doctrine (Faludi and van der Valk 1994). This organization takes place before a planning report, and it is not necessary to freeze this process after publication of the plan.

Since the beginning of the 1990s many unforeseen and unforeseeable developments have taken place. Based on this new understanding the planned decrease in housing stock growth and the grant for the costs of infrastructure provision were no longer necessary. The new market demands have led to a different appreciation of the quality of the locations to be developed. The covenants of central government with local government did not cater for these developments, as they were not known when the negotiations took place.

Moreover, the difference in outcomes can be explained, as it seems that performance (rather than conformance) is in this case a criterion that fits better with the function of Dutch national planning. There were many developments in the context of the plan and there was a need to adapt to new information.

Conformance may be interpreted as a tangible criterion, a failure to adapt planning polices to a new understanding of uncertainty.

Conclusion

Conceptualizing planning as a design science opens the way to analyse planning based on a broader literature of theory on the work of design professionals in engineering, law, medicine, etc. in which technological rules (van Aken 2004) play an important role.

A plan can be considered to be a technical artefact, which not only has a physical appearance, but also a function within a certain context of human action (Kroes 2002). This context of human action is changing over time, so the function of a plan in the context of use, may be different from the intended context, as there is uncertainty about the development of this context. In other words; the function of a plan in use may differ from the original plan maker's purposes. The dual nature of such an artefact means that the function is not a property without value of the artefact itself, but is attributed to it. Based on different concepts of the function of a plan in a particular context of human action, different evaluations may be made. Therefore, evaluation is not a holistic exercise, but an analytical exercise, which does no more than establish whether plans meet the functions that are chosen for assessment. This evaluation may be based on the functions intended by the plan maker. It can also be based on the functions a plan has when in use. Or it can even be based on third party functions that the evaluator considers to be relevant.

At a national level, planning tends to be directed towards the longer term and many agencies have a role in subsequent decision-making after the plan has been validated. Using different valuation criteria may lead to entirely different outcomes. Dutch urbanization polices show that a rather high conformance between plan and later development can go together with a low performance of the plan in subsequent decision-making. It is therefore essential to discuss the functions of planning first before beginning any evaluation.

Based on the experiences of the Dutch case, it may be concluded that conformance is no sign of the possibility of being able to use new chances or to learn, but is an essential measure of the control of future developments by the plan, and the organization behind it.

This experience of the limitations of conformance has a parallel in design sciences in which there are opinions that over-specification of a design of a human action system is not very fruitful, as it hinders subsequent steps in the process. This does not involve not making commitments at all. For example, in the Dutch policies, there were large complex housing development sites close to the cities. Critics said that Dutch local government spent too long negotiating with central government on the urbanization covenants, instead of being active in developing these locations.

An alternative criterion, that looks at the way plans contribute to better decision-making later in the process, leaves more room for evaluating whether plans have the right degree of specification. It also meets the idea that the function of technical artefacts may go beyond the intentions of the producer of the artefact, because it considers whether a plan has contributed to a better justification for a decision. The case of Dutch urbanization planning shows that conformance and performance are independent criteria. A plan with a high conformance between the plan and later developments, may be evaluated negatively on the criteria of performance.

References

Aken, J.E. van (2004), 'Management Research Based on the Paradigm of the Design Sciences: The Quest for Field-Tested and Grounded Technological Rules', *Journal of Management Studies* 41(2), pp. 219–46.

Aken, J.E. van (2005a), 'Valid Knowledge for the Professional Design of Large and Complex Design Processes', *Design Studies* 26(4), pp. 379–404.

Aken, J.E. van (2005b), 'Management Research as a Design Science: Articulating the Research Products of Mode 2 Knowledge Production in Management', *British Journal of Management* 16(1), pp. 19–36.

Behn, R.D. (2003), 'Why Measure Performance? Different Purposes Require Different Measures', *Public Administration Review* 63(5), pp. 586–606.

Boelhouwer, P.J. (2002), 'Trends in Dutch Housing Policy and the Shifting Position of the Social Rented Sector', *Urban Studies* 39(2), pp. 219–35.

Boelhouwer, P.J. (2005), 'The Incomplete Privatization of the Dutch Housing Market: Exploding House Prices versus Falling House-building Output', *Journal of Housing and the Built Environment* 20(4), pp. 363–78.

Bratman, M.E. (1987), *Intention, Plans, and Practical Reason*, Cambridge, MA/London, Harvard University Press.

Bruijn, H. de (2002), 'Performance Measurement in the Public Sector: Strategies to Cope with the Risks of Performance Measurement', *International Journal of Public Sector Management* 15(6–7), pp. 578–94.

CBS (2005), *Regionale kerncijfers Nederland* [*regional key figures Netherlands*], CBS, Voorburg/Heerlen), <http://statline.cbs.nl>.

Dipert, R.R. (1993), *Artifacts, Art Works and Agency*, Philadelphia, Temple University Press.

Eekels, J. and Roozenburg, N.F.M. (1991), 'A Methodological Comparison of the Structures of Scientific Research and Engineering Design: Their Similarities and Differences', *Design Studies* 12(4), pp. 197–203.

Faludi, A. (1973), *Planning Theory*, Oxford, Pergamon.

Faludi, A. (1986), *Critical Rationalism and Planning Methodology*, London, Pion.

Faludi, A. (1996), 'Framing with Images', *Environment and Planning B: Planning and Design* 23(1), pp. 93–108

Faludi, A. (2000), 'The Performance of Spatial Planning', *Planning Practice and Research* 15(4), 299–318.

Faludi, A. and Korthals Altes, W.K. (1996), 'Marketing Planning and its Dangers: How the New Housing Crisis in The Netherlands Came About', *Town Planning Review* 67(2), pp. 183–202.

Faludi, A. and Korthals Altes, W. (1997), 'Evaluating Communicative Planning', in D. Borri, A. Khakee and C. Lacirignola (eds.), *Evaluating Theory-Practice and Urban-Rural Interplay in Planning*, Dordrecht, Kluwer Academic Publishers, pp. 3–22.

Faludi, A. and Mastop, J.M. (1982), 'The "IOR School": The Development of a Planning Methodology', *Environment and Planning B: Planning and Design* 9(1), pp. 241–56.

Faludi, A. and Valk, A.J. van der (1994), *Rule and Order: Dutch Planning Doctrine in the Twentieth Century*, Dordrecht, Kluwer Academic Publishers.

Friend, J.K. and Jessop, W.N. (1969), *Local Government and Strategic Choice*, London, Tavistock Publications.

Helling, A. (1998), 'Changing Intra-Metropolitan Accessibility in the US: Evidence from Atlanta', *Progress in Planning* 49(2), pp. 55–107.

Houkes, W. (2002), 'Normativity in Quine's Naturalism: the Technology of Truth-Seeking?', *Journal for General Philosophy of Science* 33(2), pp. 251–67.

Jong, M. de (2004), 'The Pitfalls of Family Resemblance: Why Transferring Planning Institutions between "Similar Countries" is Delicate Business', *European Planning Studies* 12(7), pp. 1055–68.

Knudsen, T. (1988), 'Success in Planning', *International Journal of Urban and Regional Research* 12(4), pp. 550–65.

Kolpron (2000), *Kostenverhaal in de grondexploitatie op VINEX-locaties* (Voorburg, Neprom), <http://www.neprom.nl/publicaties/publicaties.html>, accessed 22 March 2004.

Korthals Altes, W.K. (1992), 'How Do Planning Doctrines Function in a Changing Environment', *Planning Theory* 7–8, pp. 110–15.

Korthals Altes, W.K. (1994), 'Preparing for Performance: The Dutch Experience', *European Spatial Research and Policy* 1(1), pp. 77–85.

Korthals Altes, W.K. (2000), 'Economic Forces and Dutch Strategic Planning' in W. Salet and A. Faludi (eds) *The Revival of Strategic Spatial Planning*, Amsterdam, Royal Academy of Arts and Sciences, pp. 67–77.

Korthals Altes, W.K. (2006), 'Stagnation in Housing Production: Another Success in the Dutch "Planner's Paradise"?', *Environment and Planning B: Planning and Design* 33(1), pp. 97–114.

Korthals Altes, W.K. (2007), 'The Impact of Abolishing Social Housing Grants on the Compact City Policy of Dutch Municipalities', *Environment and Planning A* 39(6), pp. 1497–512.

Kroes, P. (1998), 'Technological Explanations: the Relation between Structure and Function of Technological Objects', *Society for Philosophy and Technology* 3(3), 18–34, <http://scholar.lib.vt.edu/ejournals/SPT/v3n3/pdf/KROES.PDF>, accessed 20 September 2006.

Kroes, P. (2001), 'Technical Functions as Dispositions: a Critical Assessment', *Techné: Journal of the Society for Philosophy and Technology* 5(3), 1–16, <http://scholar.lib.vt.edu/ejournals/SPT/v5n3/pdf/kroes.pdf>, accessed 20 September 2006.

Kroes, P. (2002), 'Design Methodology and the Nature of Technical Artefacts', *Design Studies* 23(3), pp. 287–302.

Kroes, P., Franssen, M., Poel, I. van de and Ottens, M. (2006), 'Treating Socio-technical Systems as Engineering Systems: Some Conceptual Problems', *Systems Research and Behavioral Science* 23(6), pp. 803–14.

Laurian L., Day, M., Berke, P., Ericksen, N., Backhurst, M., Crawford, J. and Dixon, J. (2004), 'Evaluating Plan Implementation: A Conformance-Based Methodology', *Journal of the American Planning Association* 70(4), pp. 471–80.

Louw, E., Krabben, E. van der and Priemus, H.(2003), 'Spatial Development Policy: Changing Roles for Local and Regional Authorities in the Netherlands', *Land Use Policy* 20(4), pp. 357–66.

Needham, B. (1997), 'Land Policy in the Netherlands', *Tijdschrift voor Economische en Sociale Geografie* 88(3), pp. 291–96.

Needham, B. (2000), 'Spatial Planning as a Design Discipline: A Paradigm for Western Europe?', *Environment and Planning B: Planning and Design* 27(3), pp. 437–53.

Needham, B. and Faludi, A. (1999), 'Dutch Growth Management in a Changing Market', *Planning Practice and Research* 14(4), pp. 481–91.

Osborne, D. and Gaebler, T. (1992), *Reinventing Government: How the Entrepreneurial Spirit is Transforming the Public Sector*, Reading, MA, Addison-Wesley.

Preston, B. (2003), 'Of Marigold Beer: A Reply to Vermaas and Houkes', *The British Journal for the Philosophy of Science* 54(4), pp. 601–12.

Priemus, H. (1995), 'How to Abolish Social Housing – The Dutch Case', *International Journal of Urban and Regional Research* 19(1), pp. 145–55.

Redström, J. (2006) 'Towards User Design? On the Shift from Object to User as the Subject of Design', *Design Studies* 27(2), pp. 123–39.

Romme, L.G.A. (2003), 'Making a Difference: Organization as Design', *Organization Science* 14(5), pp. 558–73.

Simon, H.A. (1996), *The Sciences of the Artificial*, 3rd edn, Cambridge, MA, MIT Press.

Simon, H.A. (2002), 'Forecasting the Future or Shaping It?', *Industrial and Corporate Change* 11(3), pp. 601–605.

Talen, E. (1997), 'Success, Failure, and Conformance: An Alternative Approach to Planning Evaluation', *Environment and Planning B: Planning and Design* 24(4), pp. 573–87.

Verhage, R. (2003), 'The Role of the Public Sector in Urban Development: Lessons from Leidsche Rijn Utrecht (The Netherlands)', *Planning Theory and Practice* 4(1), pp. 29–44.

Vermaas, P.E. and Houkes, W. (2003), 'Ascribing Functions to Technical Artefacts: A Challenge to Etiological Accounts of Functions', *The British Journal for the Philosophy of Science* 54(2), pp. 261–89.

Vigar, G. and Healey, P. (1999), 'Territorial Integration and "Plan-led" Planning', *Planning Practice and Research* 14(2), pp. 153–69.

Chapter 12

Evaluative Argumentation in Fragmented Discursive Planning: A New Theory-in-Practice of Environmental Assessment of Rational Distributive Policies

Angela Barbanente, Dino Borri and Valeria Monno

Introduction

It is increasingly clear how the shift in planning from strong to weak rationality in the 1960s and 1970s (Simon 1957; 1977) and to dialogic interaction in the 1980s and 1990s (Forester 1999) has innovated and in some sense demystified the meaning and role of evaluation. In rational planning evaluation is essential for choosing one action from among alternatives that are available for achieving a goal (ex-ante evaluation) or for controlling an ongoing action during the process of achievement of a goal (ongoing evaluation), or for understanding how much a goal has been achieved at the end of the related action (ex-post evaluation). In communicative, discursive planning evaluation is like any other component of a planning procedure that has become discourse: the discursive form of argumentation (Forester 1989; Healey 1997; Khakee et al. 1995; Sager 1994), a form whose rationality features remain highly controversial (Provis 2004). As rational planning and communicative planning coexist (Friedmann 1987) also the two above cited forms of evaluation coexist, without clear prevalence of one over the other.

As in any coexistence, these two forms of evaluation in many cases consist in hybrids, often without any clear awareness of this hybridization among those who are using a method thinking about its apparent specificity. This particularly holds for compulsory evaluations that are becoming increasingly more and more popular at institutional level for supporting the allocation of structural funds by EU. Institutional organizations in the governance era are becoming more and more complex and powerful and new decentralized powers coexist with techno-structures that are nested at the different levels of the organizations. Because of this variety of organizational agents, evaluation in planning has increasing become an unclear mixture of rational and discursive procedure, of technique and politics.

Many problems and questions arise concerning the changing nature of evaluation in planning, lying at the core of the theoretical curiosities of the present chapter. Is it possible to model the relationship between rational and discursive arguments in environment-oriented institutional evaluation? Which kind of rationality is appropriate for dealing with the complex and uncertain issues of environmental scenarios? Some case-studies of application under EU disposition of mandatory evaluation of regional allocation of structural funds are used in this chapter in order to try to give an answer to the above questions.

First, some theoretical considerations about the argumentative role of evaluation and about the logical structure per se of argumentation in environmental and/or ethical domains of knowledge-in-action are made. Secondly, shifting to case studies, the complex narratives evoked by the evaluation procedures are analysed for their 'logical' role in evaluation dialogue. Thirdly, the complex institutional governance structure of regional *vs.* EU evaluation procedures are criticized. Finally, some conclusions are drawn about the need for readdressing evaluation argumentation on the specific characteristics of the two knowledge-in-action domains that deal with environmental issues and with ethical issues.

Evaluation as Argumentation: In Between Formal and Informal Rationalities

In general terms, arguing seems to imply dealing with opinions and beliefs and hence with controversial issues and not with facts. In multi-agent contexts, arguing implies both individual and group actions that follow argumentation and involve 'public' consequences, which need to be managed in cooperation and/or with some preliminary consensus.

If evaluation in planning is the same as argumentation, and argumentation is an intriguing mixture of formal and informal rationality the same holds for evaluation. Since communicative planning, which is fundamentally based on dialogue and argumentation, has increasingly become mainstream planning, incorporating some remains of rational planning that continue to be useful and/ or that are difficult to replace, then evaluative argumentation (EA) in planning must use a contingent mixing of formal and informal rationales and must privilege methods that are easily interchangeable and that can easily activate interaction and mutual understanding among different evaluators-arguers and different EA users.

Are there principles and norms for evaluating arguments? In traditional rational plans should we think of evaluation as aimed at choosing the best of alternative actions or as evaluation of the ongoing or final consequences of implemented actions in the face of the given goals as if this evaluation is fundamentally argumentation? Today the influence of the communicative and discursive turns in planning would support this idea.

Mainstream literature on argumentation continues to assume some forms of distinction between argumentation via abstract rational logic and argumentation via what we could define as contextual or contractual rational logic, which maintains a distinction between true and false propositions (see Provis 2004).

From this perspective, there should be the possibility of distinguishing between logical-scientific and narrative argumentation. In the former, effectiveness would be reached through empirical proofs of truth; in the latter through resemblance to life evocations (Cala Carrillo and de la Mata Benitez 2004, who quote Bruner 1986). In the end, anyway, mainstream literature now accepts this also i.e. that argumentation processes are far from being compact as they contain complex alternations of 'rationales' and 'non-rationales'.

Advancing a little in our exploration of consistencies or inconsistencies of the equation, that is evaluation as argumentation, a further question arises. Should argumentation fundamentally deal with beliefs? Are beliefs possible objects of decisions on them or, in other words, in a sort of reflexive circle, are they objects of decisions of believing or not believing? If the answer is yes, then another distinction between evaluation as a domain of formal rationality and argumentation as a domain of informal rationality is denied.

According to Provis, argumentation is in line with formal rationality –in the sense that argumentation can provide room for judging true and false propositions– if it includes a concept of 'acceptance'. This means that an argument or part of an argument is true or false: which is completely different from believing that the argument or part of an argument is true or false (Provis 2004: 106). Current argumentation theories provide other useful insight on EA. Argumentation is strong and effective if it is based on logical relations *and* is in tune with the audience. The latter assertion is not uncontroversial. In fact, the ideal of rationality is the ideal of sound arguments: arguments whose conclusions are logical consequences of true premises that should be accepted by all rational people. Challenging this ideal rational vision, Jovicic argues that 'If persons in a group believe that the premises are true, and both know and accept the rules which determine the relation of logical consequence in the argument, and, moreover, accept soundness as the ideal of rationality, then they accept the argument as a good one' (Jovicic 2004: 2).

When a set of argumentations are neither precisely stated nor related to an explicit definition of rationality the acceptance of these reasons by the receivers of the argumentation is essential (Jovicic 2004: 3). Therefore, the EA process is also circular, as the audience evaluates what it receives and can accept or reject it so that any EA process is highly relationally fragmented. EA strategies aim at reducing the rejection of their reasons by the audience to reduce this fragmentation. Both senders and receivers of EAs are cognitively fragmented also because of their cognitive stories or gender differences. Fragmentation of cognitive positions within the new multi-agent governance organizations and institutional settings which constitute the premises of the contemporary efforts of EA evidently leads to further methodological (theoretical and practical) questions about the way in which rational resource allocation is conducted.

Following Jovicic's theses, we may say that aspects related to authority, relevance, and reliability, which here we call respectively influence power, pertinence power and demonstration power of the reasons that are brought by the senders to the audience, are essential for making sense of any EA. In granular EA, each reason conveys its amount of authority, relevance and reliability. If in the total set of reasons the number of acceptances is equal to the number of rejections then a need arises for differentiating such reasons on the base of their importance or intriguing abilities in managing dilemmas of reasons which might be both acceptable and unacceptable (Jovicic 2004).

Evaluation as Argumentation and Ethical Dilemmas

Evaluation as argumentation seems to present other interesting specificities for planning evaluators in the domain of the natural and artificial environment and of the impacts of actions on it. One question specifically arises: what are the different agents' ideas on nature and ethics in the complex and fragmented institutional settings of the mandatory environmental impact evaluations made for allocating EU structural funds in fragile areas and regions?

Fitzpatrick (2004), for instance, arguing about the instrumental or non-instrumental value of nature, of 'sentient' and in particular of 'non-sentient' parts of nature, underlines the theoretical difficulty in philosophy of giving value – an activity that in traditional rationality deals with justifying what we should do – to what is non instrumental and therefore should have no value. In fact, non-instrumental intrinsic values given to non-sentient, non-living things (like a lake, a mountain, the moon) would not even be arguable in these classical philosophical terms. Intrinsic values would have sense in the presence of the minds of evaluators who think about them and find an instrumental reason for considering them as such.

According to Fitzpatrick, a normative claim about valuing non-instrumental nature in non-intrinsic terms could be made by considering non-instrumental nature as having primitive relationships in terms of 'friendship' with human beings. This position could, in fact, avoid the aporias that come from realist theories of values (Korsgaard 1996, quoted by Fitzpatrick 2004). Another normative philosophical claim about the non-instrumental value of nature could be made by adopting appropriate semantics for the term 'non-instrumental value'. Among these expressivist semantics would avoid metaphysical commitment to reality of 'non-instrumental value, while cognitivist semantics would put the burden of commitment on the reality of non-instrumental values (Fitzpatrick, 2004).

Literature in cognitive science (see Stark 2004; Swan 2004; Damasio 1999; Veloso 1992) is increasingly showing interest in such an intriguing debate by focusing on emotional rationality, and we as planners and evaluators should assume this as highly interesting for us too. Do 'irrational' reasons such as beliefs, emotions etc. take part in the evaluators' (senders-and-receivers') ideas, apart from the rational-institutional reasons that originate the EA conduct of those agents who act both by mandate of someone and by self-reflection on their interests in the world?

Stark (2004: 359), for instance, argues that emotions are indispensable for correct moral judgment and for motivating agents to behave morally and stipulates that intrinsic values can be attributed only to *concrete particulars*. It is evident that, even if Stark's interest lies in human particulars, nothing prevents us from accepting her view as valid also for the non-human particulars of nature which are implied in planning EA on environmental impact procedures. Thus, in Stark's point of view, emotions are worthwhile considering also in our environmental planning EA perspectives, even if EA reduces emotions by seeing them as instrumental rather than primitive in representing and modifying the world by living agents. Yet, if emotions, as many contemporary philosophers argue, are fundamental constituents of an individual's good life, what about their roles in the good life of other individuals?

Exploring Evaluation as Argumentation

In the following sections, we try to follow this line of discussion in order to explore the complex and uncertain landscapes of rationalities which shape practices of evaluation. Our aim is to understand if and how rational and discursive evaluations are hybridized in multi-agent environments and the related implications in terms of value attribution and resources allocation when evaluations also concern integration of economic and environmental values. We focus on compulsory evaluations that are increasingly being used at institutional level for supporting the allocation of Structural Funds (SF) by EU in Southern Italian Regions. Specifically, of the various programmes financed by SFs in Southern Italy we have explored the evaluation process of the Territorial Integrated Programmes (TIPs).[1]

Our inquiry in evaluation practices starts by listening to the stories told by different policy analysts involved in TIPSs evaluation processes. In fact, as many scholars maintain, we can learn from storytelling since they bring to the fore 'an inside', nuanced and controversial environment of evaluation with its rationalities/ethics as they are played out in both closed decisional arenas as well as in every day practice. We refer to such stories as micro-narratives to differentiate them from macro-narratives, which operate at a tacit level on the logical structure of local evaluations. Before turning to micro-narratives, in the next section we present a short account of SFs evaluation process.

EU Structural Funds Programmes: The Institutionalization of Evaluation

The Structural Funds (SFs) are the main instruments designed to contribute economic and social cohesion at the European Union level. They are divided into programmes according to the main objectives that are addressed. Specifically, in Southern Italy (Objective 1) regions, the focus is on improving the position of low levels of GDP (and hence income) per head areas within the EU, as in some Mediterranean areas including the whole of Southern Italy.

---

1   TIPs are programmes aimed at sustaining/activating sustainable local development paths by guaranteeing that environmental concerns are integrated into the local development perspective.

The SF mechanism comprises several types of plans and programmes under different names: Programme Complement, Operational Programmes and Single Programming Documents. In the setting up and implementation strategies relevant actors from different policy fields should be involved, in a context that has been defined multi-level governance (Bache 1998; Hooghe 1995).

The new regulations for the funding period 2000–2006 has elevated sustainable development to a horizontal principle for all EU SFs programmes, which should crosscut and support all the programming activity as stated in the Environmental Policy Integration (EPI). EPI states that environmental objectives should be reflected in all policy areas, including those aimed primarily at economic sectors, in order to contribute to sustainable development. It should be pursued through a reasonable (dialogue) and rational (coordination) process.

The policy model for implementing the SFs contains a comprehensive approach to evaluation by defining a complete cycle of systematic evaluations for each programme. According to the well-established SFs rules, in order to gauge their effectiveness, all the programmes will be subject to ex-ante, mid-term and ex-post evaluation designed to appraise their impact with respect to the objectives of reducing disparities and to analyse their effects on specific structural problems.

Under the new Council Regulation,[2] the Member States must appoint a Managing Authority for each programme. Its tasks consist of implementation and management of the programme, including data collection, use of computerized monitoring and evaluation systems, preparation and transmission of annual and final reports to the Commission, organization of mid-term evaluation.

This examines the initial results of the operations, their consistency with the ex-ante evaluation, the relevance of the targets, as well as the soundness of the financial management and the quality of monitoring and implementation of the programme concerned. The evaluation of programmes should take place on a systematic, timely and rigorous basis. The evaluation process of SFs follows the well-known policy analysis model of evaluation and monitoring as shown in Figure 12.1.

The integration of sustainable development as a horizontal priority followed the first rounds of EU Structural Funding, whose common problem was that the environmental profile was seldom effectively incorporated into programmes and these were generally not framed in a manner suitable for evaluation of environmental factors, with negative consequences on later assessments (Clement 2000; Woodford 1991).

Thus SFs Regulation strengthened the requirements for the inclusion of the horizontal theme of environmental sustainability in the 2000–2006 programmes, making them more systematic and extensive. The integration of environmental issues is articulated around a comprehensive framework, with environmental considerations characterizing most of the main headings addressed by the Regulations: programme preparation, content, monitoring, evaluation and information.

---

2   (EC) No 1260/99.

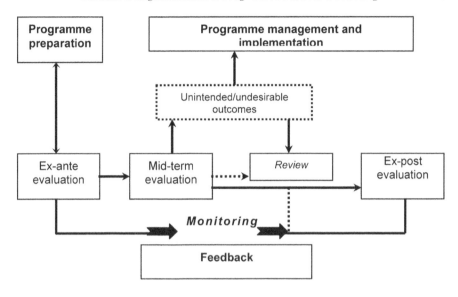

Figure 12.1 The general design of evaluation in SF programmes

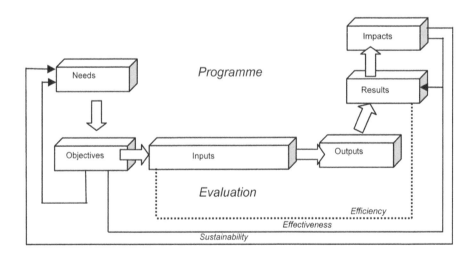

Figure 12.2 Key evaluation issues within a given programming framework

Source: EC 1999.

Plans and programmes are subject to an ex-ante evaluation of the 'expected impact' notably on the environmental situation. The ex-ante evaluation should include an appraisal of the environmental situation of the region concerned, which should address its main strengths and weaknesses to understand the opportunities for, and threats to, economic development in terms of the environmental assets and liabilities of the area (see Figure 12.2).

In order to comply with these requirements, Regional Environmental Authorities have been established in order to advocate environmental protection during all stages of programme management. These units have the role of both integrating the environmental dimension into overall policies funded by SF and guaranteeing coherence of these actions with the Community environmental policy and regulation.

Micro-narratives

In the following we report two among other semi-structured interviews since they represent the point of view of two analysts who had played different roles in the SFs evaluation. The first interviewee is the project manager of the Calabrian Management Authority. He is a planner and an evaluator who was charged with the responsibility of evaluating the TIPs. The second person interviewed is one of the members of the Apulian Task Force, a group of young experts who were employed in 2002 to support the weak Apulian Environmental Authority, which is responsible for evaluating the sustainability of SF programmes. We asked the evaluators to introduce themselves and to express their opinions on the role of evaluation in the multi-agent planning process under consideration. We also asked them if and how economic and environmental concerns had been evaluated during the TIPs implementation and what they thought could be changed in the evaluation process.

The Project Manager of the Calabria Regional Management Authority

The Calabrian project manager thought that the TIPs negotiation process was constructive, embedded in constructive argumentations and bargaining which led to broad consensus on all the programmes. In his opinion this was due to procedural and substantive norms ruling the governance system on which the policy making was based.

In fact, as in a sort of an Habermesian deliberative arena, before beginning the negotiation process the Regional Government established the rules of the game which all actors had to accept. These concerned both the people who could be considered decision makers and how to decide. Apart from the Regional Government itself, only the Mayors of the municipalities involved were recognized as decision makers while social and economic actors only had a consultative role. Furthermore, the Regional Government decided that the municipalities could only have one vote. This prevented power asymmetries and conflicts from arising between municipalities holding different territorial power. In any case, the

Regional Government had the final decisional power concerning the acceptability of programmes. Such rules proved useful for preventing interest and value conflicts from paralysing deliberation. Interest and values controversies were to be solved by local actors through informal agreements in informal deliberative arenas. As a result only argumentations which had been informally agreed on could be considered valid in the formal negotiations.

The second reason which made the negotiation process constructive, concerned the definition of a 'true premise' to evaluate any argumentation. To discern between acceptable and non acceptable arguments the Regional Government decided that each TIP was to converge towards a unique strategic goal: that of a new territorial development equilibrium aiming at reducing uneven territorial development. In a Southern Italian region, such a 'true premise', which acknowledges an action as valid when it is carried out in the public interest, was easily accepted since it sounds like an attempt to fight powerful local coalitions.

Although the project manager should have judged the acceptability of the programmes by evaluating the coherence and the legitimacy of arguments sustaining the proposed programmes by comparing them with the 'true premise', he felt that in a fragmented society neither policy makers nor himself as a planner and evaluator had the prerogative of knowing/telling the 'truth'. When can judgment be true or false if there are no shared values referring to it? To express his judgement he embraced a 'weak thinking' (Vattimo and Rovatti 1983). In his opinion, the virtues of a system of ethics are no more or no less than the virtues of the people practising them. He meant that whatever people find virtuous is morally good, which entails relativism. Given this conviction, he moved from a rational evaluation style to a more open and investigative one. Here the rational has to give way to reasonableness as intended by Rawls. An argument may be judged as meaningful if the various parties agree on it. Thus, the evaluation process of each TIP was not accompanied by any formal ex-ante intermediate or ex-post evaluation. It was rather an ongoing evaluation that ended in a shared final evaluation.

Following the myth of consensus and search for convergence, the evaluation process was permeated by arguments focusing on specific interests and based on an instrumental attribution of value. This was also the case of the environmental values which were always considered as implicit in the policy making process. In fact there cannot be sustainable local development without respecting the local environment given that the environment forms the endogenous base for the economic development of the territory. Thus, the attribution of value to the environment or any integration between economic and environmental ethics does not require any further investigation. In a local development perspective, what is really relevant is the way in which economy-environment integration is pursued and this depends on the specific context, local practices and the specific programme or plan and the governance system supporting it.

Consequently, no TIP was concerned with environmental protection but only with constructing infrastructures needed to support local development. In each TIP the environmental argumentations embedded a conception of the environment as a built-up landscape and proposed projects aimed at improving it by means of its physical reuse or rehabilitation of dismissed/underutilized areas and buildings. No environmental conflicts emerged, except in a few projects, such as that of a shipyard and a theme park. In these cases, in order to overcome conflicts the proposers of the specific TIP were asked to carry out an Environmental Impact Assessment based on technical rationality.

When from being an implicit dimension of the development the environment emerges as an ethical dimension, then integrating economic and environmental values becomes a difficult question. In the project manager's view 'it is simpler to talk about integration rather than to turn it into a reality'.

Put in another way, the values of the goods in conflict cannot be converted to a single measurement criterion, nor can any optimization be satisfying. If worlds generally cannot be compared, there is no way to determine whether, in terms of value, one is better or worse than another. To solve such a conflict, integration should be considered an eminently technical problem and solved by following a logic of optimization.

Yet, the manager thought that the required technical ex-ante evaluation of TIPs (which was prepared after the approval of TIPs, but before the programme agreement which makes any TIP operative), was a procedural obligation rather than a real support for programme formulation. His criticisms in fact concern the method adopted by the Environmental Authority which did not use strategic thinking:

> This approach mistakes the leaves for the forest. It was too narrowly focused on evaluating and eventually changing any single action proposed by each TIP. It tried to change single action by proposing some trivial solutions...like building or rehabilitating buildings in the 'local architectural style'. Such suggestions, besides being generic, seem actions intending to change only one single leaf in a forest of trees. In short this kind of evaluation doesn't take into account the strategic orientation required by integrated planning. Where is the strategy?

An Evaluator of the Apulia Environmental Authority Task Force

The young engineer started by saying that the Apulian Task Force had not succeeded in either influencing or orientating the Operative Regional Programme (ORP) implementation process towards a sustainable development as required by EU since it had not been recognized as a key actor during the implementation process of the ORP.

He and his colleagues are young and none of them is acknowledged as 'expert'. Moreover, the Task Force started working in an institutional context which had never before been proactively engaged in environmental protection programmes: environmental rehabilitation or protection had not been perceived or retained as 'public interest' by previous Apulian Regional Governments. The Task Force was set up when the Apulian ORP and TIPs policy making process had already started and nobody wanted to evaluate plans whose main goals and actions had already been negotiated and agreed upon.

Obviously, to become involved in deliberative arenas, the Task Force had to face a problem of authority, relevance, and reliability. Its members decided to influence the process through powerful expert-rational argumentations based on a well recognized accountable expert 'objective' evaluative method. The well-known pressure-state-response framework was their choice. After that, they began to give their suggestions in the TIPs negotiation process.

Their evaluations showed a profound fracture between the sustainability goals and the choices taken during the TIPs negotiation process. However, given the advanced phase of the TIPs implementation process, the Task Force was not allowed to correct, but only to give suggestions to make TIPs more suitable to EU requirements. So, whatever advice they gave was anyway useless. In fact, their advice was accepted only by local actors holding a weak technical and economic power in the decision concerning regional development futures. Reflecting on this experience the young engineer called for:

> deliberative arenas in which different rationalities should have the same dignity. Surely, privileging technical or discursive rationalities depends on the political will but deliberative arenas should be managed by equal people. Key actors must be equal in all decisional or negotiating arenas.

How can the young engineer's formal rationality be the same as non-formal rationality which dominates in a negotiation process? As the Task Force learnt during the ORP Strategic Environmental Assessment (SEA) process which involved many academic institutions, good argumentations can play a relevant role.

On the one hand, good arguments need to be scientifically true or legitimated. In fact, the SEA provided the Task Force with an undisputable multidisciplinary measurement criteria which gave it the desired scientific authority. Such authority is needed especially when we have to integrate:

> measurement criteria can simplify integration between economic and environmental concerns and interest thus making it more simple for the Task Force to convert political guidelines into integrative actions..

He continued arguing that good arguments should be able to make it clear to local actors (for example by drawing on European 'best practices') that environmental restrictions can generate many economic development opportunities:

> An environmental bond can be transformed into an opportunity for tourism development ... This can be used as a good argument to persuade local farmers to change their attitudes towards a sustainable development.

No expert advice can succeed without being mixed with a pinch of reasonableness. A maximizing logic, which does not imply giving value to different worlds but only seeking the best, should be the basis of a persuasive argument.

Micro-narratives as Landscapes of Institutional Misfits

One of the crucial aspects emerging from the stories we heard concerns the efficacy of evaluations in terms of the use[3] of their results for improving environmental integration in policy development. As we have seen in the previous section environmental evaluation only played a marginal role. This is due to different reasons.

The separation between the SFs Managing Authority and the Environmental Authorities both in Apulia and Calabria Regions was the origin of a number of problems. Firstly, there are few interactions between MAs and EAs. This prevents Environmental Authorities from being fully involved in the whole process, before and after the programming stage, which is managed above all by the Managing Authority. More important, this lessens possible development of a mutual dialogue and shared knowledge on the environmental integration in programme formation and implementation. Secondly, since MA is the authority accountable for the running of the SFs, its power, in terms both of the influence on political groups and of ability to control the whole process, is incomparable with that of the Environmental Authority. This power gap increases as the SFs programmes move from the setting up to the implementation phase.

Under the pressure of an increasing competitive environment introduced by the European style of policy-making, spending money as fast as possible has become the regional and local governments' main concern. As a result, the power of Managing Authorities becomes even greater than the Environmental Authorities, and it is tricky for EAs to ask for programmes to be changed to improve environmental integration.

Timing deficit and asynchrony between management and environmental evaluations is the most visible reason for the limited use of formal evaluation. Paradoxically, due to the misfit mentioned above, ex-ante environmental evaluation was carried out after the programme had been outlined. Due to delay, results in environmental evaluation were available when nobody needed them and probably some people were afraid of them. Once a strategy has been defined, it is difficult to re-define it in the light of environmental evaluation results.

As a result right from the conception of the programme no environmental values have been incorporated in the whole strategy. Later on, environmental consideration cannot be obtained: the risk of automatic loss of funds dictates the whole process. In this situation, formal evaluation risks becoming a symbolic, separate instrument, which does not influence the policy process.

---

3   The meaning of *use* here is not narrow. It refers to the idea of evaluation, and has a conceptual use and symbolic use. The former is used for enlightenment and the latter for persuasion: according to these notions evaluation can be used not only to direct action but also to form new views, perceptions and opinions in a pluralistic context, thus developing previous knowledge about a problem (Weiss 1998). In such a perspective, the evaluation process is to be considered as a forum for dialogue and mutual learning. From this standpoint it is worth investigating who has been involved, and how and when, in environmental assessment.

Micro-narratives as Landscapes of Rationalities and Ethical Dilemma

The two narratives seem to be brought together by a desire to reduce the complexity of evaluation as argumentation by dismissing substantive debates concerning the way in which we interpret and attribute value to humans, non human objects and relations.

On the one hand, the two analysts always avoid ethical and distributive dilemmas which emerge when the economy meets the environment by calling for procedures. They reduce the evaluation complexity by 'cleaning' (Bauman 1994) the interactive environment of personal beliefs and convictions or emotions and by opting for an impersonal procedural rationality. Environmental and social justice improvement is up to a specific governance system or the transformation of well established modes of governance. As Bauman would say, the two stories are 'ethically indifferent' (Bauman 1994). In them, we can find 'neither good nor bad; only correct or incorrect' (Bauman 1994). Once the evaluation scene has been cleaned, what is needed is to set up an agreed framework of rules for including/ excluding or accepting arguments. As a result, the evaluation processes are characterized by closure, and sometimes even reticence and power distortions.

On the other hand, the two analysts overcome environmental conflicts by invoking the dominant macro-narrative of local sustainable development, which allows them to avoid any ethical dilemmas concerning the attribution of value. In fact, such macro-narrative does not necessarily oblige us to challenge the social-ecological relations shaping our hybrid human and non-human life worlds. Thus, we cannot find in their stories any account about different argumentations carrying alternative interpretations of social ecological relations embodied in a specific development perspective. The value of nature is simply implicit in the sustainable orientation of SFs. Both analysts keep considering evaluative argumentation as a rational judgment to use to choose among alternative actions, rather than a process aimed at exploring possible consequences of actions in terms of distributive impacts.

Yet, as much as the two people interviewed attempt to evade ethical dilemmas, they always reappear on the scene. So, each time a distributional problem with its related dilemma of attribution of value arises, they resort to a technical rationality, which endows them with objective parameters of judgment and an undisputable formal logic of measure. Nevertheless, they are aware that in our fragmented environments neither procedural nor technical rationality if taken disjointedly can offer a secure anchorage for evaluating. Although the two analysts continuously mix formal and informal rationalities to evaluate transactions of values embedded in argumentations, they seem to be unable of rationally finding any solution to this puzzle. They do not know how to mix them, which of them to privilege and/or what methods to use to make such a mix easily interchangeable so that they can easily activate interaction and mutual understanding among different evaluators-arguers and different EA users.

Although the two analysts treat emotions as a source of bias and distortion and as barriers to rational deliberation, following Martha Nussbaum (2003), we can argue that, when they have no rational anchorages only personal experience and emotions seem to give them a source of judgment about complex choices. The Calabrian project manager following his disappointment with the ex-ante environmental evaluation opted for a strategic approach as a way to mix formal and informal rationality. In order to influence deliberative arenas the young Apulian engineer suggested that a hybrid evaluation needs good arguments.

In both TIPs evaluation processes, the mix between formal and informal rationalities should aim to persuade the audience rather than make sense of our actions in order to face environmental dilemmas better. But at the same time, emotions can play a relevant role in persuading since they embrace cognitive features that allow us to give meaning to our experience and actions (Hock 2006). Consequently, the two analysts do not hesitate to catalyse consensus in using instrumentally collective emotions by means of powerful images. In Calabria, powerful images like territorial equilibrium induced local actors to accept both the local Regional Government's development strategy and the governance system. The young Apulian engineer utilized winning images or images appealing to self-interest to make more attractive and reasonable arguments aimed at changing a development path into a sustainable one.

Here we can discern a difference between the two stories. In the project manager's point of view persuasion relies on the recognition of the strength of an argument in relation to its conformity to some pre-agreed interest, and not on the comparative strength of the values advanced by the arguments concerned in a specific context. Opening up the evaluative process to learning is seen as highly risky in terms of efficiency of funds allocation, especially when local practice and routines discourage and thwart any attempt at innovation. EAs should not involve ethical dilemmas, just interests. On the contrary, the young engineer tries to persuade his audience by introducing new powerful expert-rational EAs in order to induce local actors to reflect and change, even slightly, their 'economic centred view'. Evaluation can be also seen as a way to innovate by persuading.

Table 12.1 Evaluation as argumentation in the two analysts points of view

Evaluators	Role	Values	Rational	Discursive	Rational-discursive mix
EAs					
The project manager	Legitimating *factual* and *causal* arguments	Implicit	Measurement criteria	Conformity to a true premises Rhetorically effective	Strategic/ defensive
The Apulian engineer	Explaining/ affecting attitudes	Implicit	Measurement criteria	Good argument Rationally compelling	Reasonableness

Conclusions

The added value of policy development by means of open argumentation-evaluation rests on the possibility of opening up a dialogue and learning how to manage ethics or ideological contradictions that are already on the terrain. Evaluative argumentations, at least as used in SFs, seem to contradict the idea of evaluation as a constructive dialogue in which policy actors should accept the possibility that their own definitions of reality and policy theories will be challenged (Schön and Rein 1994).

Due to its strong affinities with the rational choice model and by emphasizing methodology and technique, the SFs evaluation process appears too limited to deal with the multiple forms of rationalities manifested in actual practices and ignores the dependence of programme preparation and management on the context. Furthermore, the economy and environment dichotomy embedded in the environmental integration principle – which is reflected in the separation between the Environmental and Management Authority – facilitates the adoption of persuasive and technocratic approaches. Consequently, SFs evaluation excludes both deep structures of values, assumptions and concepts underlying any environmental evaluation and impedes development of learning processes.

The multi-agent environments that we have explored do not seem to be able to experiment new forms of evaluation and face the challenges of an evaluation as argumentation as has been debated in first section of this chapter. As our research shows, the 'environmental failure' of SFs evaluations is also due to the absence of any theoretical and methodological references which can help evaluators to benefit from EAs as a way to deal with the complex and uncertain issues of environmental scenarios.

At the same time, the micro-narratives that we have reported suggest that in our hybrid, fragmented and powerful multi-agent environments we need to reconsider EAs as ethical and emotional landscapes which can enable us to understand how to mix informal and formal rationalities or which of them to privilege in order to integrate environmental and economic concerns. In this perspective evaluation would require articulating schemes of EAs offering:

> a kind of pragmatic purchase on how we construe the world, discuss our differences, act on some sense of what it is best to do, and evaluate the consequences of our action. (van der Knapp 2004, 28)

In turn, this would imply reinterpreting evaluation as 'learning and as a looking for better policies' which take into account the complex alternations of 'rationales' and 'non-rationales' shaping multi-agent interactions, as well as different ethics. As risky as this approach may appear in terms of policy efficacy and efficiency, it is a way to accept that:

> in a world where policy decisions have enormous impact on the planet and those who live on it, there is the constant need for doubt, dialogue and learning. (van der Knapp, 2004, 30)

References

Bache, I. (1998), *The Politics of European Union Regional Policy. Multi-Level Governance or Flexible Gatekeeping?*, Sheffield, Sheffield Academic Press.

Bauman, Z. (1994), *Alone Again – Ethics after Certainty*, London, Demos.

Bruner, J.S. (1986), *Actual Minds, Possible Worlds*, Cambridge, MA, Harvard University Press).

Cala Carrillo, M.J. and Mata Benitez, M.L. de la (2004) 'Educational Background, Modes of Discourse and Argumentation: Comparing Women and Men', *Argumentation* 18(4), pp. 403–26.

Clement, K. (2000), 'Environmental Gain and Sustainable Development in the Structural Funds', in T. Hilding-Rydevik (ed.), *Proceedings of the workshop on 'Regional Development Programmes and Integration of Environmental Issues: The Role of Strategic Environmental Assessment'*, Stockholm, Nordregio.

Damasio, A. (1999), *The Feeling of What Happens: Body, Emotion and the Making of Consciousness*, London: Heinemann.

Fischer, F. and Hajer M.A. (1999), *Living with Nature: Environmental Politics as Cultural Discourse*, Oxford, Oxford University Press.

Fitzpatrick, W.J. (2004) 'Valuing Nature Non-instrumentally', *The Journal of Value Inquiry* 38(3), pp. 315–32.

Forester, J. (1989), *Planning in the Face of Power*, Berkeley, CA, Berkeley University Press.

Forester, J. (1999), *The Deliberative Practitioner*, Cambridge, MA, MIT Press.

Friedmann, J. (1987), *Planning in the Public Domain. From Knowledge to Action*, Princeton, NJ, Princeton University Press.

Healey, P. (1997), *Collaborative Planning. Shaping Places in Fragmented Societies*, London, Macmillan.

Hilding-Rydevik, T. (ed.) (2000), *Proceedings of the workshop on 'Regional Development Programmes and Integration of Environmental Issues: The Role of Strategic Environmental Assessment'*, Stockholm, Nordregio.

Hoch, C. (2006), 'Emotions and Planning',< http://www.uic.edu/cuppa/upp/people/faculty/Hoch/Emotion&Planning%20Draft02.doc>.

Hooghe, L. (ed.) (1996), *Cohesion Policy and European Integration: Building Multi-Level Governance*, Oxford, Oxford University Press.

Jovicic, T. (2004), 'Authority-based Argumentative Strategies: A Model for their Evaluation', *Argumentation* 18(1), pp. 1–24.

Khakee, A., Elander, I. and Sunnesson, S. (1995), *Remaking the Welfare State*, Avebury, Aldershot.

Korsgaard, C. (1996), *The Sources of Normativity*, Cambridge, Cambridge University Press.

Nussbaum, M. (2003), *Upheavals of Thought: The Intelligence of Emotions*, New York, Cambridge University Press.

Provis, C. (2004), 'Negotiation, Persuasion and Argument', *Argumentation* 18(1), pp. 95–112.

Sager, T. (1994), *Communicative Planning Theory*, Avebury, Aldershot.

Schön, D.A. and Rein, M. (1994), *Frame Reflection. Towards the Resolution of Intractable Policy Controversies*, New York, Basic Books.

Simon, H.A. (1957), *Models of Man*, New York, Wiley.

Simon, H.A. (1977), *The New Science of Management Decision*, Englewood Cliffs, NJ, Prentice-Hall.

Stark, S. (2004), 'Emotions and the Ontology of Moral Value', *Journal of Value Inquiry* 38(3), pp. 355–74.

Swan, K. (2004), 'Moral Judgement and Emotions', *The Journal of Value Inquiry* 38(3), pp. 375–81.

Van der Knaap, P. (2004), 'Theory-based Evaluation and Learning: Possibilities and Challenges', *Evaluation* 10(1), pp. 16–34.

Vattimo, G. and Rovatti P.A. (1983), *Il pensiero debole*, Milano, Feltrinelli.

Veloso, M.M. (1992), 'Learning by Analogical Reasoning in General Problem Solving', PhD thesis.

Weiss, C. H. (1998), 'Have We Learned Anything New about the Use of Evaluation?', *American Journal of Evaluation* 19(1), pp. 21–33.

Woodford, J. (1991), 'Conflict or Convergence? Environmental Priorities and the Structural Funds', *EPRC Environmental Policy Discussion Paper No. 1*, Glasgow, University of Strathclyde.

PART 3
Interactiveness/Communication Principles

Chapter 13

Democratic Concerns and Governance in Planning Evaluation

Luigi Fusco Girard

Introduction

Are we still able to produce images of the future for our cities and will we be able to stimulate hope for change through effective actions? It is these kinds of questions that are central to the link between issues of democratic governance and planning evaluation throughout this chapter. It is not a straightforward question if we consider that the city can be an ambivalent place. It is the place of hope, new ideas, of creativity, but it is also the place of inequality and desperation. The city is contradictory because it mirrors human contradictions.

With the shift from urban government to urban governance, there are more and more spaces opened to 'participation' for new and old actors and people. City success is strictly linked to the capacity to project its future (the vision) and to build a large consensus on it. Goals and values are more and more multiple and often in conflict.

Strategic urban/metropolitan plans have introduced several kinds of (deliberative) forums to manage conflicts between private and public interests and objectives/values to build a city vision. Generally, strategic plans rest strongly on established participation by multiple actors for its implementation. The idea is that new forums or arenas could stimulate communication and coordination among different subjects: aggregation, coalition, agreements, pacts and partnerships (Dublin City Council 2005; Grand Lyon Communauté 2003).

Sometimes in plan-making processes for these strategic plans, the 'conflict' between local and national concerns becomes more and more evident: when national interests have precedence over local interest (localization of a nuclear power plant, highway infrastructure versus urban park etc.), participation can easily turn into an involvement 'against' a project. This might happen in some cases, for example, in nuclear plant localization, or a nuclear waste management site, or in the practice of incinerator localization with regard to a public green park or tourist use of a territory. Nevertheless, this kind of participation can stimulate sometimes new creative actions /proposals, new alternatives, which in turn reduce conflicts (Fusco Girard and Nijkamp 2004).

But in general there remains a deficit in participation, which reflects a deficit in evaluation. It occurs on distinguishing between, on the one hand, strategic priorities about objectives/values and interests, and, on the other hand, about different actions/projects and their comprehensive impacts.

This chapter proposes that the notion of 'participatory evaluation' would make it possible to better understand the comprehensive characteristics of a project. It allows the identification of the most satisfying solutions in relation to numerous, heterogeneous and conflicting interests/objectives/values (environmental preservation and expanding industry; pollution reduction and growth of jobs, etc.) (Nijkamp 1980).

A 'dynamic' impact matrix is the core element for comparing different actions, projects and plans. Typically, this matrix is considered fixed at a certain time and restricted to direct impacts, ignoring indirect, induced impacts and intangible, and primarily long term impacts. New indicators are absolutely necessary to spread this knowledge about qualitative (perception of quality of life, etc.) and long term impacts (Nijkamp et al. 1990).

A second kind of shortcoming central to this chapter is related to the point that evaluation cannot be reduced only to a technical problem of approach and indicators, but that it also reflects institutional and cultural matters. Frequently, public institutions do not make full use of evaluation processes, which might be prescribed by laws or rules, but undergo them or reduce them to 'technical' moments involving only 'expert knowledge' and not citizens (Fusco Girard and Nijkamp 2004).

A key problem here is that the general public might not be educated adequately to make critical comparisons of different/opposite impacts of proposals, or, more generally, engage in critical thinking. Critical knowledge is a precondition for evaluation.

One of the instruments to perhaps deal with this problem includes the establishment of 'deliberative forums'. Deliberative forums can be introduced in particular in strategic plan making, when public places become platforms for critical reflection. It should then be possible to create capacities for listening, organization, competence and persuasion. Deliberate forums could be public places for a free comparison of ideas, and where citizenship and common knowledge flourish (Popdziba 2006).

But are we 'really' able to move away from a kind of passive reception of information from mass media, to balanced 'vertical' and 'horizontal' (interpersonal) communication? Through interactive processes, these forums could help to structure data and information in knowledge, and could be open to stimulate a different rationality of various actors, and not be restricted to instrumental/functional rationality of 'here and now' (the economic rationality). This would be an hermeneutic/interpretative and a relational and emotional rationality. They are important to build a *strong* vision.

A move towards participatory evaluation and deliberate forums will remain very important if our cultural context is dominated by a 'short period approach', by the absence of past, and the incapability of building a perspective of future in medium/long terms. Other reasons to perhaps make clear the need for participatory evaluation and deliberate forums include, possibly, current indifference in society with regard to going beyond efficiency and private utility and the lack of attention and space for general concerns and common good. Clearly, evaluation should go beyond the calculation of costs and revenues.

Our thesis is that it is possible to change if we are able to face this kind of cultural knot. If we accept the arguments given above, the realization of a project of sustainable *human* development then will make it necessary to:

- make people reason and discuss in a reflective, critical way (starting from students in classrooms and lecture-halls of universities);
- promote evaluation in an 'enlarged public space' and not only in specialized fields;
- introduce a values approach to go beyond the comparison between acquired economic benefits and lost economic benefits through certain choices of thinking through values, which helps 'horizontal' communications (Keeney 1992);
- support the establishment of networks in civil society, aimed at the attention to general concerns.

The change this chapter suggests is even more urgent if we recognize that the new urban 'economic base' is represented by knowledge, know-how, education, and the creative capacity of people. It looks like the new twenty-first-century urban 'industry' might be founded on the cultural, immaterial, symbolic and spiritual capital of inhabitants (Hall and Pfeiffer 2000; Fusco Girard et al. 2003b). We think that it is possible, through participation, to link physical space with its inhabitants. We could refer to this relationship as the 'soul' of the city.

Participation and Evaluation

The process of evaluation generally makes possible the identification of different economic and meta-economic values. This process usually reflects an effort to display some configuration of the general interest or common good. In this way, evaluation expresses specific 'visions' of a local community, for example, composed of many differences.

Through participation it is also possible to evaluate shared priorities: priorities between the satisfaction of economic and meta-economic values; between the satisfaction of use and market values; between economic and environmental values; between the satisfaction of needs of this generation and of future generations; between investment in natural capital and investment in other forms of capital (man-made, human and social) (Serageldin 1996); or between the satisfaction of needs of people already inserted in economic activities and excluded people. By participation it is also possible to better understand values and interests at stake, to multiply information and knowledge, and to promote and produce new alternatives. But, above all, if a participatory process runs well, its result can be represented by the construction of a more general, less individualistic point of view: that is of citizenship. Citizenship refers here to the perception of a public spirit and of rights and duties. It is an essential ingredient to face the dilemma of increasing social disparity, of ecological crisis and of an increasingly intense competition. Citizenship is also a crucial element in the design of new governance practices.

The resulting point for this chapter, of course, is that it is evaluation that makes a shift possible from participation 'against' to participation 'for', hereby reducing conflict. Participation against a project can create new ideas/alternatives that are checked for feasibility through assessment.

There are various examples one can think of here. Some are represented by the localization of power stations, wind energy parks, incinerators of urban solid waste, or plants for energy production. High environmental impacts can be reduced only by the use of particular technologies and they can also be a risk factor for the environment and public health. They are often rejected, therefore by the general public, and they are commonly accompanied by rising emotion and passion, more so than a rational approach would suggest. But we need to realize such plants now and then, and to designate space for them in order to safeguard services for a larger series of communities, facing a problem of distribution between costs and benefits (Bobbio and Zeppetella 1999; Bobbio 2002).

In addition to this problem of the unequal distribution of benefit and cost, there is also often an 'irreversible uncertainty' linked to the lack of adequate information to evaluate alternative proposals. If we look at medium-long term impacts, only qualitative information prevails.

We are talking here about examples of evaluation where different elements are combined: the element of risk might be included, while economic elements mix with values, emotions, and even hope. These are also cases where 'facts' appear paradoxically 'uncertain'. Besides, the conflict between economic interests and non-economic values may be very high. But, in spite of this uncertainty, it is more and more urgent to make integrated choices (Funtowicz and Ravetz 1993). The following challenge is how to build priorities in the light of a combination between rational, relational and emotional intelligence.

Technical support alone will not be sufficient to legitimate a choice. From the perspective we have provided above, evaluation practices need to open decisions to participatory and deliberative processes aimed to compare private and public net benefits and their redistribution. In this way, a greater distributive justice can be introduced and positive (win-win) strategies identified.

Participation can allow a better combination of the distribution of costs and benefits starting from the trade-off between ecology and economy; that is, between ecological and economic values. Participatory spatial planning aims to guarantee a better balance between these different orders of values, which can not be achieved only through the economic calculation, because public goods and extra-economic values are at stake.

Within processes of deliberative democracy, people not only participate, but also decide on evaluation criteria, deduce their range of priorities, and choose among alternatives. Evaluation techniques can be useful to deduce a shared priority among different objectives/criteria (Saaty 1990).

It is evident that the relation between participation and evaluation is linked to the level of planning elaboration. In the strategic phase, for instance, participation typically concerns institutional/political authorities, the business world, trade unions, and civil society movements. During the implementation/management phase, participation normally concerns technicians, real estate agencies and investment businesses, other public institutions, and direct users of the project.

But the potentially most important contributions of participation are represented by the promotion of critical knowledge and of social-civil capital (Fusco Girard et al. 2003a). The community could identify an immaterial form of energy that is able to somehow reflect a 'collective spirit', which then promotes social cohesion and relational values. These relational values should be able to increase the resilience capacity of cities. It promotes auto-sustainability and the establishment of further governance practices, thereby opening new possibilities for collective action.

New Tools for Participation

As a continuation to the considerations given above, we can now say that we should widen the amount of options to establish 'public spaces' for participation and debate in our cities, and work on building new individual and collective responsibilities in improved citizenship. This process is all about creating open, vital and interactive spaces of participation, connected by networks in cooperation, so that people get used 'to reason together', in the light of a pluralist approach, considering different 'reasons'. The line of reasoning for this chapter would be that the notions of participation, evaluation and responsibility are strictly linked. We have to create new institutions (rules, processes, approaches,) to stress this interdependence. For this reason we will now ask the question in what way and by what rules, tools and approaches we can stress this relationship.

One way is that the awareness of all impacts of an action, and not only about economic ones, allows for more responsibility because it is easier to compose alternatives. Economic impact knowledge is very important for communication to everyone. But, they tend to underestimate all costs and benefits in the medium to long term, such as environmental and cultural impacts (Nijkamp 1977; Nijkamp and Voogd 1989).

Another angle is that, through participation we can reduce urban poverty. Participative processes, after all, could give voice to interests that otherwise would be excluded. New networks of associations of marginalized people should be involved in participation process. The point here is that participation and justice are interdependent: the former recalls the latter.

In addition to these linkages between participation, evaluation and responsibility, we should also point out that cities that are open to participation processes are more careful, generally, with distributive dimensions.

The openness of decision-making processes to traditionally 'absent' subjects allows building choices that appeal to notions of truth are could well be more coherent with the notion of sustainable human development. The promotion of a stronger equality in accessibility to decision-making resources is linked to the good functioning of democratic institutions (Young 1990; Bohman and Rehg

1997). Democracy is made by subjects proposing, discussing critically, arguing about good reasons of one option versus another, and it is also about setting priorities that are more desirable compared to others.

Societies promoting forms of participatory and deliberative democracy, through which they stimulate citizens to discuss in public about alternative proposals, tend to be more 'attracted' to objectives of equality/equity/justice, compared to those where 'privatized' decision making processes prevail, where partial interests easily prevail (which could reproduce and multiply disparities) (Cristiano 1997).

Justice is promoted through participation via the construction general interests, which are discussed in public arenas, forums or public juries. It is all about the establishment of 'good reasons' about what is preferable and what is not, what is good and less good in a collective sense, and what is of common interest or only of individual interest.

An element fundamental to producing 'good reasons' is embedded in the provision and systematic comparison of all of the impacts of each alternative option. The general framework proposed by CIE (Community Impact Evaluation) (Lichfield 1996) is a good way to order multiple, multidimensional and conflicting impacts in a clear and communicative perspective.

Critical knowledge of best practices, through ex-post evaluation is very useful because people would be able to make comparisons more easily (Fusco Girard and You 2006). Through this shared knowledge it is possible to convince and to promote consensus on one project over another. An active citizenship where members of the public take on responsibilities is essential. If this can be established, then projects of maintenance or of the management of public spaces, such as green areas, squares, streets, and so on should be more and more open to participatory processes. In its turn, they stimulate new civic capital.

Participatory Budget

The implementation of strategic plans is linked to the capacity to make people and community grow culturally. This capacity depends on institutions (North 1990). We feel that it is necessary to re-design in a creative way our institutions, in order to promote civil society, intermediate social formations, and civil energy.

'Participatory Budget' (Allegretti 2004; Allegretti and Herzberg 2004) and 'Ecobudget' (ICLEI 2004), could be fertile tools to promote 'a rich way of reasoning' in processes of participative/deliberative democracy, which integrate institutions of representative democracy. They allow the economic culture in urban government to be overcome, introducing new meta-economic dimensions in thinking and in choices (Fusco Girard and You 2006).

At the base of the 'Participatory Budget', there are several processes of evaluation. A first evaluation refers to the selection of four or five 'thematic areas' from 12 to 15 identified on the whole, which are assessed on their comprehensive priority compared to others (that are then excluded in the successive analysis). It is an evaluation based on the perception of the intensity of urban needs by inhabitants, in light of their income, culture, social composition, and so on.

A second step is a multi-criteria evaluation, which is commonly characterized by a rigorously technical approach and is founded on the prevision of impacts of different projects in the different selected 'thematic areas'. It would combine a check of technical/functional/financial feasibility with sustainability. Coherence then refers to the whole urban value. It is based on a comparison of impacts in relation to each evaluation criterion. It should not be characterized by a synthetic aggregation in order to avoid the introduction of weighted coefficients.

A third evaluation is that of relativizing the definition of priorities of projects in each region/neighbourhood of the city. This would occur on the basis of a comparison of impacts by different actions on objectives. It expresses itself as a voting exercise on behalf of each assembly/forum.

The last evaluation refers to the construction of a 'comprehensive priority' at the urban level and it is directed at deducing the distribution of disposable resources among different regions/neighbourhoods of the city, in the light of a multi-criteria approach. For instance, criteria are represented by the lack of public services in the area; the present population compared to the whole; the demand from poorest groups; the priorities recognized by each region/neighbourhood, and so on. Such a comprehensive priority is in essence a social/cultural construction, founded on economic/technical arguments.

If specific indicators are available through which it is possible to compare different alternatives, then it would be possible to understand whether quality of life can be improved by a certain action or project. 'Harder' indicators aimed at comparing one city to others, are generally useful for investors and not for people. They are not congruent with a humanistic approach to development. Indicators about perceived qualities of life are more useful to facilitate effective participation by people. Overall, some notable efforts should be undertaken to improve existing indicator sets.

Ecobudget

'Ecobudget' is another tool which many cities are adopting in order to reduce negative environmental impacts and to improve governance (ICLEI 2004). It is aimed at internalizing environmental considerations in choices regarding economy and land use. This tool introduces an 'environmental coverage' of city development programmes to avoid that the carrying capacity of the territory is overloaded (Coizet et al. 2000). It also helps to manage natural resources, water and energy in a more effective way, ensuring that urban development does not produce ecological poverty.

In addition to the point given above, ecobudgets are oriented to making more explicit the 'hidden costs' (including indirect and induced costs). More in general, they promote a way of reasoning going 'beyond' the economic/financial dimension, and capturing the whole set of values of the territory: use value (and not only market values), non-use values and intrinsic values. These values cannot be expressed in monetary terms, but are the preconditions of use value. This tool proposes a set of quantitative indicators to assess the impact of city development on climate stability, air quality, and beauty of landscape that can not be expressed in monetary terms. 'Ecobudget' stimulates new creative ideas and actions in the field of energy technologies to reduce the contribution of the city to climate change (considered the most important objective of the strategic plan).

Most importantly, perhaps, is that such an approach would 'open' a long-term and multidimensional way of thinking that is based on fluxes of material and energy and not on stocks, linking together economic wealth and ecology (Fusco Girard and You 2006). While being more clear than a financial balance sheet, it can stimulate participation and communication by people, creating awareness on the relationship which links the scarcity of financial resources and the scarcity of environmental resources and on the values produced by the territory against the values lost (of use and non-use).

The problem with financial budgets is that they do not reflect the voice of qualitative aspects, or ecological and environmental indirect/hidden costs. It would be fair to say that a financial budget is a useful instrument at the accounting level, not for promoting cooperation or dialogue, responsibility or citizenship.

We have pointed out here that there are different experiences displaying a move towards the integration between economic/financial and ecological/environmental dimensions. A trajectory seems to be opening, showing a shift from economic/ financial budgets, to environmental budgets, and to a social one. The state of the environment, the state of the economic system and the conditions of the welfare system are a promising starting perspective to elaborating an 'integrated environmental budget', as an effective tool for implementing strategic urban plans.

'Local Agenda 21 for Culture'

A 'Local Agenda 21 for Culture' (Fusco Girard and You 2006) can be another fertile tool to promote the cultural awareness of people and growing citizenship, both of which are fundamental elements of city wealth. A 'Local Agenda 21 for Culture' has the characteristics to become the new 'public space' of dialogue and evaluations (see art. 19, 25, 49, and so on). It would be one that improves the cultural dimension of sustainable development and urban governance.

Through 'Local Agenda 21 for Culture' it is possible, potentially, to build better answers to the city's three great interdependent challenges of our time: economic global competition; social exclusion (in a world of increasing social disparities with a need for social cohesion); and climate change.

In 'Local Agenda 21 for Culture', evaluation and participation can be strictly connected to the promotion of economic development in the city, based on knowledge, know-how, education, and the creative economy. The economic regeneration of a periphery neighbourhood or of an ancient city centre leans more and more towards an investment in the cultural sector, because of its direct, indirect and induced impacts. This sector can deliver a series of positive multiplier effects. These effects could be generated within a process started and sustained by 'Local Agenda 21 for Culture'.

An urban strategic plan for culture would be an appropriate kind of output. Urban strategic plans for culture put culture at the core of urban policies (land use, environmental, economic, work policies), and consider culture as an engine for economic development, through creative capacity, innovation and artistic creativity (as in Bilbao, Glasgow, Barcelona, and so on) (Fusco Girard and You 2006).

There has not been a real interest for the promotion of 'Local Agenda 21 for Culture' by public or private institutions. Civil society would be a fitting actor to promote 'Local Agenda 21 for Culture' in the city, together with some specific cultural associations and institutions (universities and schools, first of all). All kinds of universities, cultural actors (museums, galleries, concert halls, libraries etc.), associations, NGOs and so on have a role here to be connected in a new vital network, to stimulate synergies in impacts through coordination of actions and cooperation. They should sustain in the long term their vision of strategic plan.

The Role of Cultural Heritage in 'Local Agenda 21 for Culture'

Recently, cities more often rely on proposing new strategic urban/metropolitan plans as a general tool to increase economic development (economic regeneration), to re-balance urban assets in an integrated way, connecting different areas, and to reduce negative environmental impacts. The basic starting point here is to give identity (a soul) to each urban area, in the light of its specific characteristics (Fusco Girard and You 2006).

Within this context, the culturally built heritage can become a 'new pole of development' for each area: a 'new catalyst' of sustainable development. It could give 'centrality' to an area, improving its attractiveness, and serve as an incubator of new activities (in traditional and new sectors).

Beauty is a general characteristic of cultural heritage. It is directly 'useful' because it produces economic wealth through the tourist sector (Fusco Girard 1987). We can speak of a 'beauty economy'. But beauty also helps to rebuild a relationship between physical space and inhabitants: to rebuild identity and the social capital of the city.

In 'Local Agenda 21 for Culture', if possible, we should link the 'beauty economy' with the 'social/civil economy' (Zamagni 1998), or with the 'creative economy' (Florida 2002). The third sector can have an increasing role in cultural heritage conservation/management, because of some of its characterizing factors: long time thinking; multidimensional thinking; high consideration of intrinsic values. In this sense, new development strategies should be conceived to link the globalized creative economy (that is innovation, research, and so on) with the localized economy, that is social/civic economic (linked to identity, to specific spaces and 'places') and the 'beauty economy' (tourism economy, artistic production, and so on).

Recognizing different urban values involved in a place is the first step of a 'Local Agenda 21 for Culture'. The production of new values (not those already 'given') would be set most effectively through processes of participatory evaluation. The subsequent steps involve an economic approach and a social approach, as discussed below.

An economic approach can identify some aspects of value: user value, option value, bequest value, existence value. It considers WTP (willingness to pay) deduced values, but it also tends to underestimate the flows of values in the long run: the value over a whole series of generations (Fusco Girard 1987; Fusco Girard and Nijkamp 1997). This kind of approach uses a language more easily understood by all the participants, however, since the monetary dimension (although being a reducer of heterogeneity) succeeds in communicating better than other indicators.

Also additional values should be introduced and assessed (also in qualitative terms) within the public debate. Social values, non-use values, intrinsic values, spiritual/symbolic values, religious values, and so on should be recognized in their interdependences with user value (the chain of values). Social value is linked to social groups living in cultural heritage sites, and therefore to the tissue of society, its social glue, social identity and cohesion. The symbolic value of cultural heritage is linked to the capacity of fixing the collective memory and producing consciousness and memory of the territory.

Overall, the set of these combined values express the capacity of attraction of a place, which could be improved through new cultural activities, services, actions. It communicates the value as the whole, and would be able in its turn to also stimulate economic value for investors.

Other Conditions for Success in a 'Local Agenda 21 for Culture'

As discussed, this chapter takes on a line of reasoning emphasizing that 'Local Agenda 21 for Culture' should identify some key projects/programs/actions, including a rehabilitation of the cultural heritage, new architecture, new cultural services, and so on. These projects would produce multidimensional impacts: some values are stimulated, but they also give up other values, reflecting existing interdependencies and conflicts. It is key, therefore, to investigate the conditions for an integrated project/action/programme identifying all of the values that are produced and lost, and to balance and manage them in the best possible way.

Recognizing All of the Impacts

The restoration of historical/artistic heritage and of cultural landscape through projects with a high symbolic value and through new cultural activities/services can become the entrance point to new city development strategies. In the globalized economy, beauty, arts, and quality of landscape are not only goods of tourist fruition, but can also become export activities to produce wealth: the promotion of an 'economy of places' integrating economic flows (Castells 1996; Greffe 2005).

As pointed out earlier, cultural economics has underlined the role of cultural activities and of cultural heritage for local economic development. The emphasis is on the identification of direct, indirect and induced impacts, depending on (Greffe 2002):

- the nature of the activities (continuous or discontinuous);
- the involvement of people (and not only of tourists);
- the concentration of other cultural activities and people in the same site;
- the level of activities specialization in the site and their relationships;
- the degree of interdependence among different impacts.

These multiple complex impacts should be compared for each proposal through specific indicators. New indicators would be required for capturing these direct, indirect and intangible *long-term* impacts.

Clearly, nowadays – and ever more in the future – the factory is no more the catalyst of economic development. New catalysts include the investment in cultural facilities, activities and services. We are referring to investing in culture creation, culture transmission and culture innovation. As already underlined, the approach used in CIE (Community Impact Evaluation) helps to organize in a systemic and comprehensive way the different impacts and indicators (Lichfield 1996), thereby stimulating communication.

'Local Agenda 21 for Culture' should stress the attention and the assessment of *long-term* impacts. Thinking in the *long term* seems to be the new priority in a time of 'here and now'. However, we need time to communicate, to produce sense, and to think critically. Key ingredients of a 'Local Agenda 21 for Culture' would be to listen to the others, to think together, to be sympathetic with people, to reintegrate/connect the economic rhythms with the ecological ones and to be open to a distant future way of thinking.

A New Challenge: to Connect Beauty Economy, Creative Economy and Social Economy

Conventionally, the relationship between beauty and utility has relied only on tourism, through which beauty is transformed in economic wealth, employment, and income. Nowadays the beauty of an urban site, landscape or territory not only attracts, but also allows the export of goods/services abroad, because it reflects a fundamental 'strength factor' of a territory. It is important to recognize that the aesthetic dimension has increasingly become the entrance point for economic development. Differences in shape, sign, colour, for example, seem to better satisfy needs and become central in the identification of new market niches (Greffe 2005).

In this context, architecture and restoration assume particular relevance. These fields inherently have the ability of increasing the values of places, their identity, and their diversity, giving sense and role to different areas and stimulating innovations and economic growth.

A key challenge, then, would be a 'Local Agenda 21 for Culture' that sustains, with a bottom up approach, new development strategies by the promotion of a 'creativity district'. A 'creativity district' refers to the organization of an integrated network, starting from specific elements of excellence (museums, theatres, art laboratories, research departments, hubs, spin-off incubators, and so on) with an integrated model. Similarly, a strategic plan for culture should be characterized by a systemic approach and by capacities to stimulate synergies among the poles of the network, linking culture development, culture communication and culture production with a network use of ICT innovations.

But this is not enough. Beauty and creativity certainly open new perspectives as well. It is also necessary to engage in efforts to reproduce social capital, trust, hope, stimulating processes of social exchange, such as 'banks of time' (Amorevole et al. 1996), cooperative enterprises, LETS (Local Exchange Trading Systems) and SEL (Systeme d'Exchange Local) initiatives.

Creativity and Innovations in the Technological Field

We have underlined that a City Strategic Plan should be founded on an 'Urban Strategic Plan for Culture' (as an output of 'Local Agenda 21 for Culture' process). The main reason for this is that culture can sustain and improve material and immaterial well-being. But any strategic plan should be also founded on technological issues as well. One of these issues is energy.

A key concern nowadays has become the question how we can conserve climate stability as a fundamental goal. This has become a new priority at any level of planning/project. A key condition to improving urban sustainable development is represented by the capacity to integrate urban strategic plans with the 'solar city strategy' (Stadt Friburg in Breisgau 2004). The idea that is central to this notion is that if we do not change the position of conventional energy in our development plans or projects, we could realize economic growth in the short term, but may end in a completely unsustainable development over the longer term.

These remarks imply that we will have to use renewable energies. They are a key element of sustainable development, while a large amount of environmental damages are the result of fossil energy production and consumption. Integrated cultural heritage conservation, new architecture, and urban planning, therefore, should be strictly linked to innovations in the energy field: biomass, solar energy, wind energy, geothermal energy, micro-hydroelectric energy, and be linked to the use of water (Fusco Girard and Nijkamp 2004) and the re-use of waste.

Good practices show that innovation in energy, water and waste recycling field is able to produce strong economic impacts, spin-off effects, and new direct/indirect jobs. Energy, water and community participation are linked (Fusco Girard and Nijkamp 2004). A final point here, therefore, is that the management of energy, water, and green spaces should include participation of associations and civil society to be really effective. This challenge is in line with our points on integrating in a participatory fashion cultural matters in urban strategies.

Conclusion

This chapter has provided some selected lines of argumentation on how participation and evaluation are strictly interdependent. Participation allows us to improve the quality of an action/project (increasing different values of a place by the project), to build new partnerships, pacts among different actors, and to control results for better future actions. Through participatory evaluations it is also possible to re-shape the project or programme and to guarantee the conditions for implementation, thereby reducing conflicts.

This chapter has also called the attention to the way in which notions such as 'Participatory Budget', 'Ecobudget' and 'Local Agenda 21 for Culture' facilitate potentially effective processes to stimulate new approaches of evaluation and to produce social capital. They have some elements to stimulate the assessment of economic use and non-use values, and of 'intrinsic' values of a place. However, certain new indicators would be required to include the perceived level of the achievement of specific values (including cognitive evaluations and so on). Another point made is that the entrance point to enhancing participation and governance could be represented by taking a culturally strategic perspective in urban and environmental heritage conservation management.

In general, evaluation helps to build a new way of thinking, with new priorities. It also helps to combine interests and values, rights and duties in new ways, thereby developing a public spirit, a critical sense and overcoming 'citizens unpolitics'. A 'Local Agenda 21 for Culture' offers an interesting perspective because it underlines the centrality of the cultural dimension in the elaboration of strategies of city development. In addition, it emphasizes the importance of processes of participatory democracy, as well as the fundamental role of evaluation in its relation to cultural processes.

Obviously, it is necessary to make these democratic concerns in planning evaluation not an exercise of 'empty values'. It should be a process in which values are recognized, discussed, added, produced and re-produced. In this way, participatory evaluation can help to rebuild a hope to our cities, be it founded on creativity and positive-sum strategies.

References

Allegretti, G. (2004), *L'insegnamento di Porto Alegre. Autoprogettualità come paradigma urbano* [*Porto Alegre Teaching. Self-planning as Urban Paradigm*], Firenze, Alinea.

Allegretti, G. and Herzberg, C. (2004), *El Retorno de las Carabelas* [*The Return of the Caravels*], Madrid, Foundation de Investigationes Marxistas.

Amorevole, R., Colombo, G. and Grisendi, A. (1996), *La Banca del Tempo* [*The Time Bank*], Milano, Ed. Franco Angeli.

Aznar, G., Caillé, A., Laville, J.L., Robin, J. and Sue, R. (1996), *Vers une economie plurielle* [*Towards a Plural Economy*], Paris, Syros.

Bobbio, L. and Zeppetella, A. (1999), *Perché proprio qui? Grandi opere ed opposizioni locali* [*Why Here? Big Projects and Local Oppositions*], Milan, Angeli.

Bobbio, L. (2002), 'Le arene deliberative', *Rivista italiana di politiche pubbliche* 3(2), pp. 5–29.

Bohman, J.F. and Rehg, W. (1997), *Deliberative Democracy. Essays on Reason and Politics*, Cambridge, MA, MIT Press.

Biudè, J. (2004), *The Future of Values*, New York, UNESCO Publishing and Berghohn Books.

Castells, M. (1996), *The Rise of the Network Society*, Oxford, Blackwell Publishers.

Cristiano, T. (1997), 'The Significance of Public Deliberation', in J.F. Bohman and W. Rehg (eds), *Deliberative Democracy. Essays on Reason and Politics*, Cambridge, MA, MIT Press.

Coizet, G. (2003), *Il Metodo CLEAR* [*The CLEAR Method*], Milano, Ed. Ambiente.

Coizet, R, Giovannelli, F. and Di Bella, I. (2000), *La natura nel conto: Contabilità ambientale* [*Environmental Accounting*], Milan, Ed. Ambiente.

Cotturri, G. (1998), *La cittadinanza attiva. Democrazia e riforma della politica* [*Active Citizenship. Democracy and Political Reform*], Fondazione Italiana per il Volontariato, Roma.

Dublin City Council (2005), *Dublin City Development Plan 2005–2011*, Dublin, Nicholson and Bass.

Funtowicz, S.O. and Ravetz, J.R. (1993), 'Science for the Post-Normal Age', *Futures* 25(3), pp. 568–582.

Fusco Girard, L. (1986), *Risorse architettoniche e culturali. Valutazioni e strategie di conservazione.* [*Cultural and Architectonic Heritage. Evaluation and Conservation Strategies*], Milano, Ed. Franco Angeli.

Fusco Girard, L. and Nijkamp, P. (1997), *Le valutazioni per lo sviluppo sostenibile della città e del territorio.* [*Evaluation for Sustainable Development of City and Territory*], Milan, Ed. Franco Angeli.

Fusco Girard, L., Forte, B., Cerreta, M., De Toro, P. and Forte, F. (eds) (2003a), *L'uomo e la città* [*The Man and the City*], Milan, Ed. Franco Angeli.

Fusco Girard, L., Forte B., Cerreta M., De Toro P. and Forte F. (eds) (2003b), *The Human Sustainable City: Challenges and Perspectives for Habitat Agenda*, Ashgate, Aldershot.

Fusco Girard, L. and Nijkamp, P. (2004), *Energia, bellezza, partecipazione: La sfida della sostenibilità,* [*Energy, Beauty, Participation: The Challenge of Sustainability*], Milan, Ed. Franco Angeli.

Fusco Girard, L. and You, N. (2006), *Città attrattori di speranza. Dalle buone pratiche alle buone politiche* [*City Magnets of Hope. From Best Practices to Best Politics*], Milan, Ed. Franco Angeli.

Florida, R., 2002), *L'economia della creatività* [*The Creative Economy*], Milan, Mondadori.

Grand Lyon Communauté Urbaine (2003), *Charte de la Partecipation*, Lyon.

Greffe, X. (2003), *La Valorisation Economique du Patrimoine* [*Economic Valorisation of Heritage*], Paris, La Documentation Française.

Greffe, X. (2005), *Culture and Local Development*, Paris, OECD.

Hall, P. and Pfeiffer, U. (2000), *Urban 21: A Global Agenda for 21st Century Cities*, London, Spon.

ICLEI (2004), *The Ecobudget Guide*, Vaxjo, ICLEI.

Keeney, R.L. (1992), *Value Focused Thinking*, Cambridge, MA, Harvard University Press.

Lichfield, N. (1996), *Economics in Urban Conservation*, Cambridge, Cambridge University Press.

Nijkamp, P. (1977), *Theory and Applications of Environmental Economics*, Amsterdam, North Holland.

Nijkamp, P. and Voogd, H. (1989), *Conservazione e Sviluppo.* [*Conservation and development*], Milan, Angeli.

Nijkamp, P., Voogd, H. and Rietveld, P. (eds) (1990), *Multicriteria Evaluation in Physical Planning*, Amsterdam, North Holland.

North, D. (1990), *Institutions, Institutional Change and Economic Performance*, Cambridge, Cambridge University Press.

Moro, G. (1998), *Manuale di cittadinanza attiva* [*Handbook of Active Citizenship*], Rome, Carocci.

OECD (2005), *Cultural and Local Development*, Paris, OECD Publications.

Podziba, S. (2006), *Chelsea Story*, Milan, Mondadori.

Saaty, T.L. (1990), *The Analytic Hierarchy Process for Decision in a Complex World*, Pittsburg, RWS.

Sen, A. (1999), *Development as Freedom*, New York, Alfred Knopf.

Serageldin, I. (1996), *Sustainability and Wealth of Nations*, Washington, The World Bank.

Stadt Friburg in Breisgau (2004), *Solarfuhre Region Friburg* [*Solar Cart-load Region in Friburg*], Friburg, Vollmer and End.

Young, I.M. (1990), *Justice and the Politics of Difference*, Princeton, NJ, Princeton University Press.

Zamagni, S. (1998), *Non profit come economia civile* [*Non-profit as Civil Economy*], Bologna, Il Mulino.

Zeleny, M. (2005), *Human System Management: Integrating Knowledge, Management and Systems*, World Scientific, London.

Zolo, D. (ed.) (1994), *La cittadinanza* [*Citizenship*], Bari, Laterza.

Chapter 14

Using a Communication Audit to Evaluate Organizational Communication in Planning

Elaine Hogard and Roger Ellis

Introduction

It might seem self-evident that the evaluation of social programmes should include some attention to communication. After all, the nature of most interventions is that some form of communication takes place between the providers and recipients. Furthermore, the nature of social provision is that it usually depends on a network of communication between the various providers. Surprisingly, however, the literature explicitly linking evaluation and communication is sparse and within that there is no mention of communication audits.

Planning is a highly complex social activity requiring effective communication at a number of levels both within planning processes themselves and between planners and various stakeholders and organizations in the community. Any evaluation of planning should, we would argue, identify, describe and analyse these communication activities.

The purpose of this chapter is to describe the method of enquiry known as communication audit and to argue for its place in evaluation studies of both planning and delivery. This argument in principle is reinforced by a case study of communication audits used in the evaluation of the planning and delivery of a complex social and educational programme.

In researching this chapter it became evident that relatively little evaluation of the processes and outcomes of communication has been reported explicitly as part of evaluation studies. While there is a substantial literature concerned with communication within organizations, this is not reflected in the evaluation literature. Searches of evaluation journals demonstrated that within the burgeoning literature of evaluation activity, communication principles and purposes receive little attention. Focussing on programme evaluation in particular increasingly highlighted this area of apparent inactivity.

There have, however, been a number of studies that address communication in organizations. They could have been described as the evaluation of communication but are not. Such studies would be of direct relevance to an evaluation of the organization and its work but tend not to be described in that way. Typically, such studies highlight the importance of good communication and its consequent effect on organizational structure and management but do not specifically advocate the study of communication as a key element in programme evaluation.

There is an equally marked dearth of studies linking evaluation, communication and planning. The published reports of evaluations of planning, and the remaining chapters in this book, exemplify an indifference to the direct evaluation of communication processes in planning. It is as though the predominant paradigms for analysis and the generation of research questions are blind to the study of the communication activity that must permeate planning as all social activities.

There are therefore two parallel and non-interacting bodies of literature: that addressing communication in organizations and that concerned with programme evaluation including the evaluation of planning. This article aims to bridge these literatures through identifying research and evaluation of communication conducted so far, the value of the evaluation of communication and how the Communication Audit a tool developed by Hargie and Tourish (2000) can be imported into evaluation research.

Evaluation and Communication

We started with the intention to contextualize our interest in communication audits as an evaluation tool in the evaluation literature. An extensive literature search was conducted using the following electronic databases for the social sciences and business communication: Applied Social Sciences Indexes and Abstracts (ASSIA); the Bath Information and Data Services for the Social Sciences (BIDS); the Ingenta database; and Blackwell Publishing's Synergy database.

Nothing was found linking the two concepts theoretically or methodologically thus justifying the theme of this article. In part, this may be explained by the fact that those who study communication in organizations tend not to describe these studies as evaluation although they could and perhaps should have. For example, Quinn and Hargie (2004) studied communication in the Royal Ulster Constabulary and their study could have been described as an evaluation of the effectiveness of that organization with a particular emphasis on communication processes.

Surprised by this apparent lacuna in the evaluation literature we then decided to review a selection of basic texts in evaluation to find what they had to say about communication. Significantly the term communication receives scant attention in standard texts on evaluation. For example, Ovretveit (1998) in a substantial overview volume has no reference to communication in a comprehensive index nor do Pawson and Tilley (1997) in their more idiosyncratic but theoretically challenging and influential monograph.

Fearful that we might have missed something we then extended this search to the majority of available texts including both relative classics such as Scriven's Evaluation Thesaurus (1991) and Rossi and Freeman (1989), and recently published encyclopaedic compendia such as Stern (2005) and Mathison (2005). In each case we looked for communication in the index, chapter headings and chapter subheadings. The results are shown in Table 14.1. Since we were beginning

Table 14.1 Communication in evaluation: search for 'communication' and 'process' in basic texts

Text	Communication reference	Process reference
Stern, Eliot. (ed.) (2005) *Evaluation Research Methods*. 4 Volumes. London: Sage.	None (In 75 titles or abstracts)	Minor
Mathison, Sandra. (ed.) (2005) *Encyclopaedia of Evaluation*. London: Sage.	Entry stresses communication to gather data and communication to disseminate findings	Brief one paragraph entry
Chen, H.-T. (2005) *Practical Program Evaluation*. London: Sage.	None	3 pages
Kazi, M. (2003) *Realist Evaluation in Practice*. London: Sage.	None	1 page
Patton, M.Q. (2002) *Qualitative Research and Evaluation Methods* (3rd edn). London: Sage.	None	9 pages
Clarke, A. (1999) *Evaluation Research*. London: Sage.	None	2 pages
Ovretveit, J. (1998) *Evaluating Health Interventions*. Buckingham: Open University Press.	None	5 references
Pawson, R and Tilley, N. (1997) *Realistic Evaluation*. London: Sage.	None	None
Phillips, C., Palfory, C. and Thomas, P. (1994) *Evaluating Health and Social Care*. London: Mcmillan.	None	3 pages
Scriven, M. (1991) *Evaluation Thesaurus* (4th edn). London: Sage.	None	2 pages
Rossi, P and Freeman, H. (1989) *Evaluation and A Systematic Approach*. (4th edn). London: Sage.	None	3 pages
Rutman, L. (1977) *Evaluation Research Methods: A Basic Guide*. London: Sage.	None	3 pages

to suspect that the lack of attention to communication might be associated with the relatively light emphasis on process (as opposed to outcomes) in the literature we also looked for references to process in the texts.

As will be apparent from Table 14.1, our view that communication is a neglected area in the evaluation literature is substantiated.

It may be that this neglect reflects the relative emphasis in evaluation studies on outcomes rather than the process, identified as significant by Ellis and Hogard (2003). Such an emphasis would tend to ignore communication unless it was an explicit intended outcome of an intervention or programme. The review of texts tends to support this with brief references to process in the context of volumes as a whole.

However, returning to the literature as a whole, the term communication does appear in reports of evaluations albeit not in the explicit theoretical manner sought. A rough typology of communication in evaluation points to three uses. First are evaluations of initiatives which are primarily about communication. Second are evaluations of initiatives where communication is a significant element. Finally there are a small number of evaluations where communication is identified as a significant factor in the success or failure of the programme.

As an example of the first type Altamara et al. (2004) evaluate a two-way communication checklist as a method of heightening patient and doctor awareness of communication processes. The checklist improved patient-reported quality of communication. As an example of the second type Kavanagh McBride (2003) in evaluating a Dublin hospital's accident and emergency services highlighted, ab initio, communication channels as an important feature of the services and found that there was reciprocal influence between forms of service and communication. As an example of the third type Allaz et al. (2003) in evaluating the appropriateness of hospitalization as perceived by patients and health care professionals identify communication between patients and health care providers about decisions regarding hospitalization as an important factor in determining patient satisfaction.

Organizational Communication

Simply defined, communication is the transmission of information from one entity to another. Entities may include persons, organizations and various mechanical and electrical devices. The content of communication may be verbal, non-verbal or symbolic. Elements in the process of communication include the notions of transmission, reception, storage and transformation. Communication may be studied with regard to dyads, small groups and organizations. So there is no shortage of concepts and models that might be applied to communication in the programme that is being evaluated.

An organization is conventionally defined (Tubbs and Moss 2003) as a group of individuals who seek to achieve a predetermined goal. Whilst this would allow for a range of group sizes there is usually a further expectation that the organization will be organized hierarchically and with a division of labour and roles. With regard to communication it is unlikely if not impossible that all members of an organization could achieve face-to-face communication. Organizational communication is a major subdivision of communication studies with its own distinctive set of theoretical, practical and methodological issues (Tourish and Hargie 2004). Programme evaluation inevitably involves the evaluation of at least one organization, that which is delivering the programme hence the significance of organizational communication and its evaluation. We would argue that every programme evaluation should, at least in part, focus on organizational communication.

In social programmes communication is important between individuals and groups in the context of the organization as a whole. In such programmes vast quantities of information flow between individuals, groups and organizational units and appropriate content, transmission and reception are essential to the operation of the programme. Various media are involved in the information transmission including emails, letters, telephone calls, documents, presentations and videos.

There is a substantial literature regarding organizational communication and its significance. It is beyond the scope of this chapter to review this literature but the following examples should be illustrative.

Communication is considered to be the cornerstone to effective organizational activity in both commercial and public service settings (Odell 1996). Recognizing the importance of communication, much research has been conducted into specific ways it can affect organizations, particularly in the health and social sectors and with regard to inter-agency working.

Effective communication between all levels of personnel and between professionals in organizations may lead to faster provision of care, successful co-ordination and higher quality of care (Fakhoury and Wright 2000) and is a prerequisite for effective leadership (Flauto 1999). Used ineffectively it can damage professional relationships and affect the quality of care (Smith and Preston 1996). Brownlee et al. (1996) emphasize the benefits of conducting such research into communication issues, believing that by identifying areas of difficulty solutions could be found which in turn would improve understanding, communication, relationships and ultimately the quality of care.

Effective communication is, then, essential in the organizations that provide programmes or interventions that are being evaluated. Research into organizational communication shows where particular problems may arise and how strategies can be developed to deal with them. The methods used in such research are crucial to this article which focuses on the use of communication audits.

In his research on the communication between nurses and their manager, Bolden (1996) used MBTI profiles to label the 'drivers' of the participants in an attempt to locate the issues and to work out ways to combat them. For example, where the manager was assessed as 'J' (having a strong sense of structure), this will not agree with the nurses' trait of 'P' (preferring to be flexible). This is a much more behavioural method of identifying communication issues as opposed to analysing large amounts of qualitative data about people's own perceptions. It explores further the reasons why problems may arise, attempting to demonstrate 'that many of the simplest communication problems which arise can be explained and, more importantly, strategies developed to improve them at the individual and team level' (Bolden 1996, 21).

Brownlee et al. (1996), in their research into staff perceptions of the relationship between midwives and doctors working on a labour suite, explored various aspects of inter-professional communication as perceived by the individuals involved. This was largely to allow solutions to be developed to impact on the quality of care, recognizing the impact it can have. Structured interviews were conducted and analysed by grade with the main focus being on verbatim comments.

Wiemann and Backlund (1980) identify two categories of research strategies for evaluating communication competence. They believe the first, third-person observation, has a major disadvantage – the observer is removed from the context of the communication episode. Without this knowledge the observer cannot be aware of the interpersonal relationships that form an essential part of all human communication. The second category they define as participant observation. With this method they highlight that observers are limited in the number of communication situations in which they may participate. The scope of such observations limits their ability to compare communication competence from one episode to the next.

Fakhoury and Wright (2000) sampled 200 psychiatric teams across the country sending questionnaires to community psychiatric nurses attached to the teams to investigate their communication and information needs. Of the respondents, 84% reported communication problems as a barrier to their work, reinforcing the importance of such research. The questionnaire was heavily focused on communication with other professional including the working relationship they had with GPs, psychiatrists, approved social workers, registered mental nurses and counsellors. This mirrors the increasing need within the public health sector for multi-professional collaboration and the consequent issues for communication. This study revealed that increased communication between professional 'is not only essential to the success of the co-ordination of various health and social services for these patients, but also to the harmonious integration of these services in ways that serve the patients' needs' (Fakhoury and Wright 2000, 878).

In response to the question of how communication and leadership are linked, Flauto (1999) conducted research into employee's perceptions of their supervisor's leadership effectiveness. He used a three-dimensional model looking at transactional leadership i.e. realizing the independent goals of them and their followers; transformational leadership in which they change the goals of their follower, usually to a higher level, and the quality of exchange in leader-member dyads. Self-report instruments were for employees to rate their leaders' leadership behaviour across the three dimensions and their leader's communication competence across two dimensions.

The study revealed that transformational factors, such as charismatic leadership, individual consideration and intellectual stimulation increase perceived unit effectiveness and subordinate satisfaction. The implication is, therefore, that communication competence is a prerequisite for effective leadership. Communication, whilst important at all levels, is crucial in the management of organizations. Perception of employees into the effectiveness and qualities of a leader will have a large impact on the quality of service and their communication with colleagues.

Across the various studies of organizational communication the communication audit approach stands out for its demonstrated validity, reliability and, given motivated respondents, feasibility (Hargie and Tourish 2000; Tourish and Hargie 2004).

Communication Audit

As part of an evaluation, communication can be approached in three ways using the trident approach developed by Ellis and Hogard (2006). This approach advocates attention to outcomes, process, and multiple stakeholder perspectives. Improving some aspect of communication may be an explicit objective for the programme or intervention and will thus be an outcome to be measured. Second the focus might be on capturing the particular processes occurring or established to facilitate communication. Third it may be important to identify the views on communication of all the participants. Whichever approach is adopted, and it will probably be a combination of all three, a tool will be needed to measure communication in some way. Methods are needed that allow us to analyse communication in-depth, validly, reliably and feasibly. This leads to the idea of a communication audit which can identify the main strengths and weaknesses of organizational communication.

Tourish and Hargie (2004) review the justification for and nature of communication audit. They identify a number of reasons for audit, grounded in the communication research literature and highlight three main methods of data gathering: survey questionnaires, interviews and focus groups. The particular method described bellow is the use of a specially constructed questionnaire instrument to audit organizational communication.

A communication audit has been defined as: 'a comprehensive and thorough study of communication philosophy, concepts, structure, flow and practice within an organisation' (Emmanuel 1985, 50). It is essentially a process for acquiring data regarding communication for analysis (Downs 1988). It can assist managers by providing them with knowledge of what is actually happening at the communication level, rather than what they thought, or were told was happening (Hurst 1991).

A survey of available instruments for communication audit highlighted the work of the International Communication Association during the 1970s (Goldhaber and Krivinos 1977) and the scholarly interest this had attracted (e.g. Greenbaum and White 1976). Goldhaber and Rogers (1979) identified key objectives to be achieved by a communication audit. These included the description of the major topics, sources and channels of communication, consideration of the quality and utility of information, and the identification of the positive and negative experiences of users with a view to effecting improvement.

Communication audit is, then, a broad concept that might include a number of approaches. Goldhaber and Krivinos (1977) in describing the International Communication Association's (ICA) Communication Audit identify its five measurement tools. These are: questionnaire survey, interviews, network analysis, communication experiences, and communication diary.

The questionnaire survey section of the audit has been widely used in the United States (Goldhaber and Rogers 1979) in the study of organizational communication particularly in business settings However, we were unable to find any examples of its use in the context of programme evaluation. The ICA questionnaire is a comprehensive instrument in thirteen sections. Completion depends heavily on the recollection and reporting of critical incidents in communication. Hargie and Tourish (2000) produced a simplified version which focussed on one critical incident rather than requiring a critical incident for each section. Further work is under way to produce a short form of this simplified version (Hargie 2004).

Goldhaber (1990) describes the instrument as covering nine topics these being:

1) amount of information received and needed from others on selected topics;
2) amount of information sent and needed to be sent to others on selected topics;
3) amount of follow-up or action taken and needed on information sent to others;
4) amount of information received and needed from selected sources;
5) timelines of information received from key sources;
6) amount of information received and needed from selected channels;
7) quality of communication relationships;
8) satisfaction with the major organizational outcomes;
9) demographic information.

Goldhaber reports the reliability of the scales on the 134-item set as ranging from a low of 0.73 to a high of 0.92. The validity of these scales was based upon their self-evident relationship to organizational communication; their ability to predict organizational outcomes, and their consistency with previously validated measures of organizational communication (1990, 355).

The questionnaire tool is split into thirteen sections and employs several forms of question. A major part of the instrument consists of seven main sections and their subsets which aim to identify satisfaction with various forms and topics of communication. Satisfaction is gauged by a comparison between what is happening and what the respondent would like to happen. These seven questions are designed to find out how much information is sent or received on various topics and using different methods, and how much would be required to be effective. Quantities of information received and desired are judged by respondents on two scales of one to five ranging from 'Very Little', through 'Some', to 'Very Great' and the difference between the two scores is used to obtain the raw data. The closer the difference is to zero the less of an issue that method or topic is. For example if the amount of information received on a particular topic is scored one for very little but the amount desired is scored five for very great there will be a difference of four. Clearly a difference of zero implies satisfaction with amount of information received. Any difference scores from one to five, plus or minus, indicate some degree of dissatisfaction. Goldhaber and Krivinos (1977) refer to confidential data banks which in effect offer some standardized comparisons for difference scores and profiles. In our experience difference scores of two or larger highlight areas that merit further consideration.

Two further sections, based on a five point Likert scale from 'Very Little' to 'Always' ask whether information received from different sources is usually timely and how much the respondent trusts various people within the organization. The latter is based on the belief of Hargie and Tourish (2000) that a good level of trust provides a starting point to enhance communication.

Another section focuses on a critical incident in which a certain communication has been effective or ineffective. As well as providing specific information about the person involved and the consequences of this incident it allows insight into whether staff feel positive or negative about the organization in general. The remaining three sections are open questions that identify specific strengths and weaknesses in communication, knowledge of the challenges and priorities of the organization and the individual, and suggestions for improving communication. This produces a mass of information that reveals recurring trends and comments on the functioning of the organization. So a completed questionnaire offers a profile of satisfaction for an individual and, taken in aggregate with other questionnaires, for an organization or organizational unit.

Both the original ICA questionnaire and the Hargie and Tourish (2000) adaptation need specific topics to make the instrument applicable to actual situations such as those which will be encountered in particular evaluations. The major development work for the instrument used in the studies described below consisted therefore of devising specific references to topics relevant to the areas being evaluated. The customized instruments are described in greater detail below.

Using a Communication Audit in an Evaluation Study

We now give an example where the authors have used a communication audit as part of an evaluation of two Sure Start programmes. Sure Start is a major United Kingdom government initiative, analogous to Head Start in the USA, aimed at providing a better start in life for children under four years of age and their families through working towards targets based on improving health, education, social and emotional development and strengthening communities. One key objective for these schemes is to improve inter professional collaboration in order to deliver more focussed and coherent support to parents and children. Clearly, communication is central to this collaboration and we decided to use an audit to assess its present effectiveness and to provide a baseline against which improvements might be detected. At the same time communication was a vital area within the schemes involving professional staff, Sure Start unqualified staff and volunteers. Much of the communication within Sure Start and between

Sure Start and external agencies is concerned with planning. Each programme is encouraged to innovate and find new ways of reaching and meeting the needs of parents, carers and children. Planning these innovations requires intensive discussion within the Sure Start scheme itself and between the scheme and external agencies with whom the scheme has to collaborate.

The audit thus looked at communication within a multi-professional team who are working in a fairly innovative environment. Each programme has the autonomy to form their organization based on the needs of the deprived area they serve and manage it according the resources available such as being spilt between different sites or having limited information technology systems.

The scheme managers were particularly interested in staff development and the effectiveness of their staff appraisal and support in the context of a complex multi-layered organization involving professionally qualified staff and volunteers. The topics selected for the audit tool were therefore chosen in part to reflect issues related to supervision, support, appraisal and development. Other topics were concerned with communication between the Sure Start team and external agencies. Sure Start staff completed the entire tool whereas external agencies completed only the sections covering inter-agency communication.

The instrument used was based structurally on that devised by Hargie and Tourish (2000) but with specific references to topics, methods and issues within the Sure Start schemes and between the scheme and external agencies. The instrument was tailored in close consultation with the manager of the schemes to ensure its relevance.

Highest levels of dissatisfaction were expressed with communication between the staff of the Sure Start scheme and specific outside agencies, particularly the social services departments. Sure Start staff were more dissatisfied than agency staff which suggests an asymmetrical level of expectation.

Within the Sure Start organizations staff were generally dissatisfied with communication from managers on issues concerning their performance. Insufficient information was being provided on positive and negative aspects of performance and too much reliance was placed on informal channels. In particular a much-vaunted e-mail system was perceived as under-used and inaccessible. Supervision was seen as arbitrary and inadequate. Group sessions were seen as providing insufficient guidance and feedback on individual tasks.

The audit allowed the evaluators to form specific recommendations which each programme has begun to act upon to improve communication. For example, one programme had a specific issue with the consistency of information being passed to each sub-team by the managers, which in turn was creating a certain level of distrust. Another identified a specific need for more face-to-face communication with the programme manager and equality for all team members who are often employed by other organizations and receive line management from varying sources. Both evaluations highlighted problems with computer systems; one not having enough technology available, the other often using it too much thus replacing the valuable personal contact. Each audit demonstrated areas for improvement that would benefit the organization if implemented.

Communication with outside agencies posed less tractable problems. For example several outside agency that were highlighted for poor communication declined to fill in the questionnaire so their views of the process remain obscure.

We now describe the communication audit questionnaire survey tool in more detail. The broad aim of the communication audit in the study was to identify effective and less effective features of organizational communication with a review to recommending improvements. The audit tool had two major foci; one to evaluate communication within the organization and the other to focus on communication between the organization and other related organizations.

The instrument drew on the demonstrated validity and reliability of both the original international communication audit questionnaire (Goldhaber and Rogers 1979; Downs 1988) and Hargie and Tourish's adaptation (Hargie and Tourish 2000; Quinn and Hargie 2004). The customizing of the instrument for this purpose enhanced its face validity and utility.

The modified questionnaire was divided into 13 sections, each one dealing with a different aspect of communication together with a covering sheet for demographic information on participants. In seven of the sections respondents were asked to rate the amount of communication received or sent and communication needed on a broad topic and its subsets according to the following values: Very Little (VL), Little, (L) Some (S), Great (G) and Very Great (VG). The broad topics for these seven sections were:

1) amount of information received on eight topics grouped under major themes relevant to the internal working and external relations of Sure Start;
2) amount of information received from 11 sources including individuals, organizational units and external agencies;
3) amount of information received through 11 channels including face-to-face communication, documents and e-mails;
4) amount of information sent on seven aspects of Sure Start operation;
5) amount of information received on six important issues;
6) amount of information sent on six important issues;
7) action taken on sent communication by seven groups.

Each section was further subdivided into specific topics relevant to Sure Start operation. In all the seven sections included 56 specific topics. Levels of satisfaction with communication for each section and its subtopics were determined by calculating the difference between the two raw scores for each question-the ideal score being zero which would indicate no difference between what was received and what was wanted. In order to identify potential problems areas, it was first necessary to compare the main question areas overall and then in relation to the different specific topics.

Two further questions referred to how quickly information was received (timeliness) and to working relationships and trust. Within each question seven specific groups or individuals were included. Respondents were asked to rate their answers in terms of a five-point Likert scale. For timeliness communication could be rated as (N) never on time, (R) rarely on time, (SOT) sometimes on time,

(MOT) mostly on time and (AOT) always on time. For working relations the seven participant groups were rated for trust using markers of trust VL (very little), L (little), S (sometimes), O (often) and A (always).

One section focused on a critical incident in which a certain communication has been effective or ineffective. As well as providing specific information about the person involved and the consequences of this incident it provided insight into whether staff feel positive or negative about the organization in general. The remaining three sections were open questions that identified specific strengths and weaknesses in communication, knowledge of the challenges and priorities of the organization and the individual, and suggestions for improving communication. This produces a mass of information that reveals recurring trends and comments on the functioning of the organization.

Four sections, of a more open-ended nature, asked for strengths and weaknesses in communication regarding Sure start operation: for a description of a positive or negative communication experience (the critical incident); for comments on challenges ahead including the greatest challenge, the biggest priority in the workplace, and manager expectations; and suggestions for making communication better.

Typically each instrument took 30 to 40 minutes to complete and represented a significant investment in time and effort. Briefing sessions were organized for Sure Start staff to explain the significance of the work and its potential impact. Participants volunteered to participate after a briefing. Forms were completed anonymously.

On the comparative items, differences were computed for all items overall, for each section and for each item in each section. A difference of zero indicated satisfaction with minus or plus scores (up to a maximum possible of four) indicating dissatisfaction with either too little information or too much. A difference of two or more was taken as indicative of dissatisfaction. Overall most groups were dissatisfied and believed insufficient information was being received or sent on a range of topics. The overall average difference for all sections and items was 2.1 indicating a significant shortfall between what was received or sent and what was wanted. The survey highlighted that different groups would like to see improvements in communication from different areas. For example the Sure Start staff felt the need for greater communication from outside social services agencies. Despite this general level of dissatisfaction, more positive critical incidents were reported than negative, and there was generally a positive level of trust within the Sure start organization but not consistently, with outside agencies.

All respondents felt they were not receiving enough information from any source. In an analysis of sources of information it was felt that all did not provide sufficient information with the exception of the so-called 'grapevine' – informal communication channels – where many staff felt they received an excess of information.

In the context of the evaluation the use of the audit tool provided detailed information on what had initially been identified broadly as dissatisfaction with the accomplishment of a particular objective, that was the improvement of communication within Sure Start and between health and social service organizations. The results highlighted specific areas of dissatisfaction where it was possible to take specific remedial steps. Finally it provided a baseline against which anticipated improvements could be measured.

Conclusion

We have identified a surprising absence of explicit reference to communication in evaluation studies, theories and methods. We have attributed this in part to the relative neglect of process in evaluation studies.

We have argued for the central importance of organizational communication both within the organization and between the organization and external partners and stakeholders in programme planning and delivery and hence the need to evaluate communication as part of any evaluation. As a means to do this we have advocated the communication audit and in particular the use of a survey questionnaire tool based on that developed by the International Communication Association. This communication questionnaire can be modified to suit the needs of particular organizations being evaluated. It highlights areas and issues that are relevant to each individual organization and indicates ways in which these might be addressed. On the assumption that all programmes rely on effective communication within and between organizations this should be assessed as of any programme evaluation. The audit instrument provides a way to do this.

References

Allaz, A.F., Luthy, C., Perneger, T.V. and Rentsch, D. (2003), 'Hospitalization Process Seen by Patients and Health Care Professionals', *Social Science and Medicine* 57(3), pp. 571–76.

Altamara, A.C., Bobes, J., Gerlach, J., Hellewell, J.S.E., Kasper, S., Naber, D., Robert, P. and Van Os, J. (2004), 'Evaluation of the Two-way Communication Checklist as a Clinical Intervention: Results of a Multi-national, Randomised Controlled Trial', *British Journal of Psychiatry* 184 (January), pp. 78–83.

Bolden, K. (1996), 'Communication: Theory and Practice', *Practice Nursing* 17(7), pp. 19–21.

Brownlee, M., McIntosh, C., Wallace, E., Johnstone, F. and Murphy-Black, T. (1996), 'A Survey of Inter-professional Communication in a Labour Suite', *British Journal of Midwifery* 4(9), pp. 492–95.

Chen, H.-T. (2005), *Practical Program Evaluation*, London, Sage.

Clarke, A. (1999), *Evaluation Research*, London, Sage.

Downs, C. (1988), *Communication Audits*, London, HarperCollins.

Ellis, R. and Hogard, E. (2006), 'The Trident: A Three-pronged Method for Evaluating Clinical, Social and Educational Innovations', *Evaluation* 12(2), pp. 372–83.

Ellis, R. and Hogard, E. (2003), 'Two Deficits and a Solution? Explicating and Evaluating Clinical Facilitation using Consultative Methods and Multiple Stakeholder Perspectives', *Learning in Health and Social Care* 2(1), pp. 18–27.

Emmanuel, M. (1985), 'Auditing Communication Practices', in C. Reuss and R. Desilvas (eds), *Inside Organisational Communication*, New York, Longman.

Fakhoury, W.K.H. and Wright, D. (2000), 'Communication and Information Needs of a Random Sample of Community Psychiatric Nurses in the United Kingdom', *Journal of Advanced Nursing* 32(4), pp. 871–80.

Flauto, F.J. (1999), 'Walking the Talk: The Relationship between Leadership and Communication Competence', *Journal of Leadership Studies* (Winter–Spring), pp. 86–96.

Goldhaber, G. (1990), *Organizational Communication*, 5th edn, Dubuque, IA, Wm. C. Brown Publications.

Goldhaber, G. and Krivinos, P. (1977), 'The ICA Communication Audit: Process, Status and Critique', *Journal of Business Communications* 15, pp. 41–64.

Goldhaber, G. and Rogers, D.P. (1979)n *Auditing Organizational Communication Systems: The ICA Communication Audit*, Dubuque, IA, Kendall/Hunt.

Greenbaum, H. and White, N. (1976), 'Biofeedback at the Organisational Level: The Communication Audit', *Journal of Business Communication* 13, pp. 3–15.

Hargie, O. (2004), personal communication, 10 December.

Hargie, O. and Tourish, D. (eds) (2000), *Handbook of Communication Audits for Organisations*, London, Routledge.

Hogard, E., Ellis, R., Ellis, J. and Barker, C. (2005), 'Using a Communication Audit to Improve Communication on Clinical Placement in Pre-registration Nursing', *Nurse Education Today* 25(2), pp. 119–25.

Hurst, B. (1991), *The Handbook of Communication Skills*, London, Kogan Page.

Kavanagh McBride, L. (2003), 'Making the Link: An Impact Evaluation of One of Dublin Hospital's Accident and Emergency Department's Liaison Nurse Service', *Accident and Emergency Nurse* 11(1), pp. 39–48.

Kazi, M. (2003), *Realist Evaluation in Practice*, London, Sage.

Mathison, S. (ed.) (2005), *Encyclopaedia of Evaluation*, London, Sage.

Odell, A. (1996), 'Communication Theory and the Shift Handover Report', *British Journal of Nursing* 5(21), pp. 1323–26.

Ovretveit, J. 1998. *Evaluating Health Interventions*. Buckingham: Open University Press.

Patton, M.Q. 2002. *Qualitative Research and Evaluation Methods.* (third ed.). London: Sage.

Pawson, R. and Tilley, N. (1997), *Realistic Evaluation.*, London, Sage.

Phillips, C., Palfory, C. and Thomas, P. (1994), *Evaluating Health and Social Care*, London, Macmillan.

Quinn, D. and Hargie, O. (2004), 'Internal Communication Audits: A Case Study', *Corporate Communications: An International Journal* 9(2), pp. 146–58.

Rossi, P and Freeman, H. (1989), *Evaluation and A Systematic Approach*, 4th edn, London, Sage.

Rutman, L. (1977), *Evaluation Research Methods: A Basic Guide*, London, Sage.

Scriven, M. (1991), *Evaluation Thesaurus*, 4th edn, London, Sage.

Smith, A.J. and Preston, D. (1996), 'Communication between Professional Groups in an NHS Trust Hospital', *Journal of Management in Medicine* 10(2), pp. 31–9.

Stern, E. (ed.) (2005), *Evaluation Research Methods*, 4 vols, London, Sage.

Tourish, D. and Hargie, O. (2004), *Key Issues in Organizational Communication*, London, Routledge.

Tubbs, S. and Moss, S. (2003), *Human Communication: Principles and Contexts*, 9th edn, New York, McGraw-Hill.

Wiemann, J.M. and Backlund, P. (1980), 'Current Theory and Research in Communication Competence', *Review of Educational Research* 50, pp. 185–99.

Chapter 15

Policy Network Theory: An Ex-post Planning Evaluation Tool?

Shinji Tsubohara and Henk Voogd

Introduction

According to Sandercock (1998) the study of planning history is much more than the recorded progress of planning as a discipline and a profession. Planning history usually celebrates government and its traditions of city building and regional development. In this chapter it will be investigated if research into the planning history of a city will benefit from using a modern research paradigm of systematic ex-post evaluation.

Academic planning literature shows in recent decades a growing awareness of the complexity of decision-making processes (de Roo 2003). This complexity is often said to resemble the metaphor of a network (for instance, see Albrechts and Mandelbaum 2005). Bureaucrats, politicians, and interest group representatives usually discuss public problems and devise means for their solution. They have common interests that lead them to co-operate, voluntarily or forced. Over time these interactions and exchanges form networks of interrelationships. The network concept has often been promoted by policy scientists to better understand this complexity (e.g. Jordan 1990; Kickert et al. 1997; Thatcher 1997; Klijn and Koppejan 2000; Teisman 2000). The central focus of *Policy Network Theory* (PNT) is on decision-making. It provides a framework for analysing policy-making processes in relation to the role of actors and their interactions. How fruitful is this theory for use as an ex-post evaluation tool for research of urban planning history?

This chapter aims to investigate the usefulness of policy-network theory for an evaluation of the planning history of the city of Groningen in the northern Netherlands. Groningen has been in the Netherlands in the past a frontrunner in many urban planning innovations, such as the introduction of a traffic circulation plan, urban bicycle routes, etc. (Tsubohara and Voogd, 2004). It will be examined in this chapter if PNT is a proper framework for evaluating such planning processes.

The structure of this chapter is as follows. First, attention is paid to the main characteristics of PNT. In addition, an empirical illustration is given of this theory applied to inner city planning processes in Groningen. This chapter is finished with a critical examination of PNT for its use in planning evaluation.

Policy Network Theory

Principal Features

A network exists if because of dependencies actors can not escape from each other, can obtain a benefit or have another interest in entering into a coalition with other actors. There are different conceptions of policy networks (e.g. see Borzel 1998; John 2004). For instance, the distinction between qualitative, interpretative networks versus formal networks is important, whereby interrelations are quantified. The latter offer possibilities for 'hard', i.e. numerical analysis (e.g. see Stokman and Zeggelink 1996). This chapter will not focus on formal networks, but on qualitative networks, since quantification of urban policy networks is very difficult due to, among others, the lack of adequate data and the stochastic nature of such data.

The main characteristics of a policy network are:

1) It deals with a configuration of actors who primarily because of mutual dependencies and secondarily because of an (assumed) advantage or an interest or any other concern are relating to each other.
2) This is expressed in decision-making processes that are built upon numerous 'smaller' decision-making processes with a basic feature that they can not be known in advance.
3) Actors have means that are unequally distributed among them; these means can be very varied, i.e. tangible means like money or ownership of ground or real estate, but also intangible means like knowledge and experience, competencies and tasks.
4) Next to the means of actors, decision-making is affected by external influencing rules (e.g. higher government policy) and network rules (e.g. habits defined by past behaviour).
5) Actors take different positions during decision-making and this also affects the decisions.
6) If actors have sustainable relationships, specific network rules may exist that provide a network with certain stability, but this is not a necessity.

Previous characteristics are interrelated and result in complex decision-making including a series of decisions taken by various actors involved.

The Analytical Framework

What might the analytical framework of PNT look like? Decision-making in a complex network is influenced by a number of conditions as outlined in Figure 15.1.

- **Social framework**
  Influencing factors in this framework are:
  - Interest organizations (viz. their goals, demands, initiatives);
  - Public support (media);

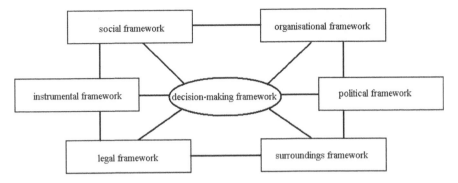

Figure 15.1 Conditions influencing a complex policy network (cf. Kräwinkel, 1997)

- Availability of means.
- **Organizational framework**
  Influencing factors in this framework are:
  - Actors with a direct, own or professional concern (e.g. housing corporations, architects);
  - Goals and objectives of these actors;
  - Distribution of means over these actors.
- **Instrumental framework**
  Influencing factors in this framework are:
  - Formal policy instruments (e.g. tasks and competences);
  - Distribution of means, including tasks, competences and finances.
- **Legal framework**
  Influencing factors in this framework are:
  - Court of law;
  - Legal rules;
  - Legal protection.
- **Surroundings framework**
  Influencing factors in this framework are:
  - Economic situation;
  - Social-cultural situation;
  - Physical situation;
  - Spatial situation;
  - Housing situation;
  - Demographic situation;
  - Technological situation.
- **Political framework**
  Influencing factors in this framework are:
  - Representatives of the people;
  - Political parties;
  - Goals and priorities
  - Availability of necessary means.

- **Decision-making framework**
  Influencing factors in this framework are:
  - Participating actors
  - Position and roles of actors;
  - Interactions between actors.

About Decision-making

Teisman (2000) distinguishes three conceptual models of decision-making: the phase model, the stream model and the rounds model. Each model is based on specific assumptions about what decision making is and how it should be analysed. The phase model focuses on successive and distinctive stages in a process, i.e. defining a problem, searching for, choosing and implementing solutions. The stream model emphasizes concurrent streams of participants, problems and solutions, defining decision making as the connection between these streams. The rounds model combines elements of the other two models, assuming that several actors introduce combinations of problems and solutions, and create progress through interaction. Each model generates specific insights. The phase model concentrates on decisions taken by a focal actor; the stream model focuses on the coincidental links between problems, solutions and actors; and the rounds model on the interaction between actors.

About Actors

PNT is about actors and their interrelationships. Urban development processes are characterized by the presence of many actors. By an 'actor' is meant a person, group or organization with common interests and/or objectives. An actor becomes a stakeholder if he/she/it may gain or lose from decisions taken in such processes. According to Teisman (1992, 55) actors may take different positions in a policy network: viz. interaction, incentive and intervention positions (see also Ike et al. 2004):

- An *interaction position* is taken if actors try to realize their own objectives by co-operating with other actors who have powers to influence a process. An example is the co-operation between project developers and landowners.
- An *incentive position* is taken by actors who are not directly involved in the process, but who try to influence other participating actors by providing indirect incentives. An example is a higher public authority that stimulates certain policies by providing subsidies.
- An *intervention position* is taken by actors who have the means and power to change a course of action. Examples are an investor whose money is needed to realize a project or the owner of real estate who does not wish to sell his property and hence blocks new development.

About Relationships

Scharpf (1978) has performed a classic study about relations between actors. His definition of mutual dependencies is well known: see Table 15.1.

Table 15.1   Types of relations between actors

	A's dependency on B	
B's dependency on A	High	Low
High	Mutual dependent	Unilateral dependent
Low	Unilateral dependent	Mutual independent

Source: Scharpf 1978, 356.

In collaboration processes between actors the presence of trust is an important variable for arriving at consensus (Woltjer 2000). For instance, if an urban government does not have any trust in collaboration with a particular project developer, the government may choose not to collaborate at all with this actor and to open negotiations with other developers. Urban networks that are characterized by mutual dependencies can find stability in searching for consensus based on trust (Klijn 2002). Some refer to discourse coalitions where language and debate characterize the relationships between the actors (Hajer 1993 and 1997).

An Empirical Illustration

How workable is PNT for an analysis of the planning history of the city of Groningen in the northern Netherlands? This section will first briefly outline the decision making process of the Groningen's traffic circulation plan in the 1970s, and secondly analyse the process based on PNT.

Decision-making Process

The city of Groningen (population 180,908 in January 2006) introduced the traffic circulation plan (*Verkeerscirculatieplan*, VCP) in September 1977. In order to keep out through-traffic, this plan divided the inner city into four sectors by enforcing one-way traffic restrictions overall (Figure 15.2). Drivers had to go out to a ring road surrounding the inner-city to move from one sector to another. As a result, car traffic was cut by half, and the plan has created the possibility for urban design for pedestrians (not least as consumers), although it was criticized as 'devastating', 'catastrophic', or 'fatal' by business organizations (Groningen Entrepreneur Federation (GOF), Chamber of Commerce (KvK) and Business Circle Grote Markt) before its introduction.

One of the reasons for such huge criticism, which even now can be heard vividly, lies in the decision-making process. The first interim report was published

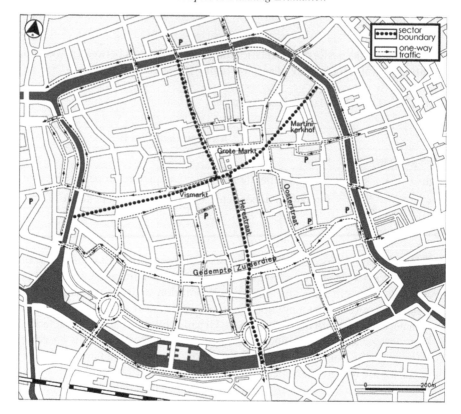

Figure 15.2 Traffic circulation plan of Groningen

in February 1975, the second report, the VCP Part II, in May, and the final plan was not published but revealed in July. Until this moment, there was literally no opportunity for the public to participate. This plan tried to expand the pedestrian area enormously. Tremendous opposition emerged from among business and commercial circles, and the municipality modified it in just '14 days'. Again there was no participation. The result was the 'sector-plan'. The first and last 'information evenings' were held for business people on 5 September, and for residents the next day. The plan was proposed to the municipal council on 15 September, and decided on 17 September.

This 'public' process of the VCP has already been described in detail by Tsubohara (2003). To illustrate the complexity of this process the attention in this chapter will be focussed on the 'intra party' decision-making process within the labour party (*Partij van de Arbeid*, PvdA) of Groningen. The PvdA has been the biggest party in Groningen since its establishment after the World War II, and as a result of the local election in May 1974, it boasted the most council seats in history, 18 out of 39 in total, when the VCP was decided. At this election, it obtained a dominating 40.4 per cent of the vote. The responsible *wethouder* (political executive) for urban planning was Max van den Berg, who was placed in the top position in the candidate list of the local PvdA.

The young new left activists of the PvdA, who emerged in the middle of the 1960s, advocated thorough 'democratization' of their own party as well as society as a whole. In order to realize this, according to the new left, rank-and-file party members had to be able to participate in party-policy making directly, while the public had to be able to participate in public decision-making process. The new left of Groningen had mainly three media or actors in mind to facilitate this intra-party participation. That is, 'district teams' (*wijkteams*), 'working groups' (*werkgroepen*) and 'general member meetings' (*ledenvergaderingen*). Through these media, every party member participates substantially in political discussion, and creates party policy. This was the socialist democratic ideal for them. The following will examine one by one how these worked in the planning of the VCP.

District team   Just after the local election in June 1970, which sent many young new left into the municipal council, each PvdA councillor was allocated to one or two districts as 'district councillor'. The local party bulletin, 'Onze Binding' of September 1970 publicized which councillors were responsible for which districts. The local party executive members were also allocated to districts.

The 'district team council' (DTC) was established in October 1972 in order to facilitate communication not only between party leaders and district teams, but also between district teams each other. This was to be held, in principle, monthly, and a few councillors, a few executive members and representatives from all the district teams were to attend and discuss or exchange information. Since December 1973, the 'description letter' had been sent to each district team a week before the DTC meeting. This letter included the agenda for the next DTC meeting, minutes of the last DTC meeting, notices and information from the party group and executive, and so on.

Despite the 'difficult start', district teams were organized in all but one district by 1972. There were 15 teams in 1973. The number of participants in total had increased from 'more than sixty people' in 1972 to about 180 in 1976.

Many district teams were rather autonomously engaged in not only various typical neighbourhood problems, such as the lack of playgrounds or traffic safety, but also city-wide projects and plans, like the Ring Road or the Policy Plan. Furthermore, in some cases, they cooperated with each other or even with other neighbourhood groups. However, for district teams to be active is one thing, for them to participate in party policy making is another. They could not necessarily systematically participate in party policy making, particularly in the planning of the VCP.

At the DTC, the VCP was for the first time talked about on May 21st, 1975. A description letter was as usual sent to each district team in advance. However, the VCP is not on the agenda in this letter, although the bus plan ('ring line') and 'Bicycle Plan' are separately listed as the subject of discussion. *Wethouder* Van den Berg attended this DTC meeting, while the district team Centrum, whose area was to be divided into four sectors in the VCP, did not attend. The VCP was not discussed, but only its schedule was given as information. Instead, the ring line and Bicycle Plan were widely discussed.

The description letter, which was issued for the DTC on 18 June, listed the VCP as one of agendas. However, according to the explanation of this topic, the VCP was not scheduled to be discussed at the DTC, nor did the explanation encourage discussion and preparation in advance. This was in contrast to the Bicycle Plan, for which the explanation encourages to 'try to work out these things for your own district'.

While the VCP was not discussed at the DTC, Part III was revealed by the police on 26 July by the local paper, *Gezinsbode*, on 4 August, modified for '14 days', and the BandW (*college van burgemeester en wethouder*, political executive office) decided upon this modified plan on 22 August. This plan was decided by the general member meeting of the PvdA on 26 August, and the VCP re-emerged in the description letter, which was intended for the DTC on 9 September. Now that the general member meeting accepted this plan, the next DTC meeting was not any more intended to be the place to discuss its content. In the explanation of the VCP, after pronouncing that the general meeting supported the VCP, the letter points out the fact that 'the general member meeting also decided that the party would go to the public with this plan'. So, 'we will have to tell the public why our party has supported this plan'. For the coming municipal council meeting, where the VCP would be proposed, 'we will have to consider at the district team council meeting to put a brochure into the post box in the greater part of the city.' The letter closes the explanation by saying, 'For this the help from district teams is necessary'.

District teams were placed utterly in the same circumstances as the public and other organizations in terms of direct participation in the planning process of the VCP. They were just mobilized to propagate the plan after the decision had been made.

Working group   The expected functions of the working groups were either to respond to the requests for advice from the party group or to examine various problems on their own initiative. The Binding of September 1971 proposed nine working groups in total, each of which had its own research topics, and called for participation by party members. For the working group 'Urban Development, Public Housing; Traffic, Public Transport', it listed the following topics as examples:

> Roads and green space. The car out of the inner city? City buses – where and how expensive? The next new neighbourhood. Housing and working separated? Old neighbourhoods destroyed? How can we realise affordable housing?

This working group was very quick to start. The Binding of November 1971 reported that 'the working groups Urban Development etc. and Culture have already long existed, and have now grown to big organs'. The party annual report 1971/72 also reported that 'The working groups Culture and Urban Planning held substantially attended meetings regularly.' The following is from the annual report 1974/75:

> Urban Planning/This working group consisted of 21 members. The subjects were, among others, bus line net, taxi plan, the Interim Report, bicycle plan.

For the Bicycle Plan, this working group organized the subgroup Bicycle. Asking for opinions from district teams, this subgroup published a report with 17 pages, Towards Better Bicycle Facilities in Groningen, in June 1975. It proposes the 'main bicycle routes', saying, 'bicycle routes of high quality have to be made between neighbourhoods, the inner city, employment centres and other objects attracting traffic'. This working group seems to have been engaged in research actively and produced well-founded proposals.

However, the working group Urban Planning was not integrated systematically into the planning process of the VCP. Concerning the relationship of the working group with the VCP, the group itself explains in the description letter of June 1975 as follows:

> Self-evidently the working group Urban Planning follows closely the development (of the VCP, by the author). The subgroup Bicycle tries to evaluate the proposed measures and routes for bicycle traffic, among others with the help of comments from district teams, and, if necessary, propose changes. Concerning public transport, the working group already agreed on what should be changed in the VCP.

The working group indeed examined the VCP in terms of public transport and bicycles, and submitted opinions, particularly based on the Bicycle Plan for the latter. However, it was not officially consulted by party leaders about the VCP itself.

General member meeting   The new left of Groningen naturally tried to 'democratize' the highest party organ, the general member meeting (GMM). Until the middle of the 1960s, the GMM was held only once a year and 50 to 60 members attended. Political matters were hardly discussed, and it had almost become a ceremonial gathering. The goal of the new left was to change the GMM into a place where many more party members attended and substantially discussed politics. For this, the executive let party members know the schedule of the GMM earlier, tried to devote as much time as possible to political discussion at the GMM, keeping organizational matters to a minimum, and so on.

Concerning the frequency, the GMM had been consistently held about ten times a year since 1972. On the other hands, the attendance at the GMM did not increase so much as party leaders expected. The annual report 1972/73 states that the average number of participants was 50 to 70, which were 20 to 30 less than the previous year. The report year 1974/75 saw on average about 150 participants, but the next year again experienced a decrease. In addition, political matters, including those at the municipal level, had been indeed raised on the agenda since the end of the 1960s. However, the GMM had not still gone so far as to become a forum where rank-and-file party members discussed politics substantially and created party policy. For example, the annual report 1972/73 reports that the intention of the executive 'to engage the general member meeting more intensively in discussion on important political problems' was 'practically not realized at all'.

The VCP was on the agenda of the GMM on 26 August 1975. In advance of this GMM, the Binding of August included the explanation of the VCP, covering six pages by the executive secretary and a councillor. They at first admit the lack of public participation, saying, 'We regret that it was not possible to undertake participation for this plan'. After citing the objectives of the VCP, they turn to the explanation of the plan's content. It drops a hint of the modified plan, 'sector-plan', but it is impossible to understand the content of the plan in detail from this description, particularly as no map of the plan was included. However, in the conclusion 'The plan and the party', they proclaim clearly, without waiting for the discussion at the GMM, that they support the VCP, saying, 'The undersigned are now already willing to say that they experience this traffic circulation plan as positive.' or 'We want to say out loud that we can clearly recognize the party standpoints in this plan.' At the GMM on 26 August, the executive submitted a 'draft-motion' for the VCP. According to this motion, the GMM supports the VCP, although it wants to put some 'remarks'. However, the BandW decided the sector-plan on 22 August, and the local newspaper Nieuwsblad van het Noorden published it, with a plan map, on 25 August, that is a day before the GMM. As a matter of course, the executive and party group were criticized by attendant party members for 'having paid too little attention to good information'. After all, the GMM approved this motion, it supported the VCP, and it also decided, as earlier mentioned, that party members should take action for the VCP. The GMM did not, or could not, discuss the VCP itself, and, as a result, did not ask for any modification of it.

Influence on party policy making　　The three actors, that is district teams, working groups, and general member meetings, all of which were intended to facilitate the intra-party democracy, did not function as media in which rank-and-file members participated directly in party-policy making. As a result, the party members could not participate directly in planning the VCP at all.

However, new left party leaders acted within frameworks which party members had created, and rules were introduced for leaders to take responsibility in terms of these frameworks. In addition, there was room for the public as a whole to influence the party policy through these frameworks. Therefore, the opinions of the public as well as party members were reflected indirectly in the VCP.

Election programme　　Until the 1970s, the Groningen PvdA, like other local PvdA and parties, did not make its own local election programmes. There were only nationwide party programmes. It was accepted or natural for the party to make municipal, not party, policies after the election, joining in the so-called '*afspiegelingscollege*' in which parties from left to right had seats, and compromising with each other. Or, policy making was rather the privilege of bureaucrats. The new left tried to break through this situation by emphasizing the difference from other parties, under the slogan, 'polarization'. As the foundation for this strategy, they introduced the election programme peculiar to the city.

For the election in 1974, the party executive commissioned a working group to draft the election programme. The district teams were asked for opinions and examined the draft and submitted opinions. The election programme was decided at the GMM on 15 January 1974. This Municipal Programme 74–78 included twelve chapters, like 'urban planning and housing', 'regional cooperation', 'economic development', and was sold at one guilder. The two sections 'traffic' and 'public transport' within the chapter 'urban planning and housing' explain the party traffic policy. As it is shown below in the quotation from the section 'traffic', the party clearly chooses particular traffic modes:

> It must be continued to keep out through traffic in the inner city and residential areas. Public transport and bicycles will acquire a clearly privileged position. ... Facilities for the car will be limited to those of the highest necessity. Existing short cuts will be closed. Concerning those plans that are not yet implemented, it will have to be examined whether the above principles were well considered.

As a result of campaigning with this election programme, the party won a historical victory by going from 13 to 18 seats.

After the election, the new so-called '*programcollege*', which consisted of only left wing parties, set out to draft the first comprehensive municipal plan, the Integrated Policy Plan 1975–1979. Each district team examined the draft of this plan and submitted opinions. The final draft, which took into account these opinions, was approved at the GMM on November 28th, 1974, with about 200 participants, and the municipal council decided this in December. The following is the quotation from '3.7.4. Traffic':

> The continuously increasing mobility requires intervening in traffic choice through facilitating the use of bicycles and mopeds and of public transport. The relationship between environment and traffic encourages the exclusion of through traffic out of the inner city and the creation of traffic calming areas in residential neighbourhoods. ... The facilities for the car have to be restricted to those of the highest necessity. ... The facilities for bicycles, pedestrians and public transport ... have priority.

Since the beginning of the 1970s, programmes or plans have been introduced, which worked as frameworks within which party leaders made each policy, and opinions of party members were reflected in those frameworks. The public as a whole were not involved in making election programmes, but, of course, in a representative democracy opportunities were guaranteed to express their approval or disapproval for them at the elections. The election programme Municipal Programme 1974–1978 obviously chose public transport, bicycles and pedestrians, and obviously rejected the increase of car use. It seems to be well-founded to say that 40 per cent, or more than 50 per cent if votes for other left parties included, of voters supported this policy.

Informal framework   Only election programmes or integrated plans might be not enough to direct what PvdA party leaders do. As frameworks for party leaders, an 'informal framework', which was created through daily dedicated activities by party members, must have been not less substantial than the officially published documents. District teams were engaged in neighbourhood problems in their own districts, and sent to party leaders various demands. Among those demands, there were many which were impractical or not well considered, like placing speed bumps on the trunk road. On the other hand, we can recognize radical proposals in the Bicycle Plan by the working group Urban Planning, such as adjusting the phasing of traffic signals to bicycles. Indeed, it was impossible to integrate these demands or proposals as they were into the party policy. However, these voices of party members could accumulatively contribute to creating a framework for party leaders, for example as a message that party members accept or even want drastic measures to restrain car use.

In addition, these voices did not reflect necessarily only the opinions of PvdA members. Many district teams cooperated with other neighbourhood groups, and there were even some district team members who were more active in these neighbourhood groups. The district teams functioned as an 'important link between the electorate and the elected', and, as a result, the opinions of the public as a whole influenced, at least to a certain extent, an informal framework.

The inquiry into district teams, whose result was reported in the description letter of February 1976, revealed the fact that 'councillors and executive members have visited the meetings of district teams very faithfully. No district team has complaints in this regard'. As an institution for communication, there was the DTC. Councillors and executive members, at least a few of them, always attended the DTC meetings, and members of the working group Urban Planning themselves attended this and explained its Bicycle Plan.

The opportunities were indeed limited to the party members to participate directly in party policy making. However, they participated or took the initiative in creating both formal and informal frameworks, and could take into account the pubic opinions in this process.

Party discipline   In addition, procedures or rules had been introduced in the 1970s to make sure that party leaders respected these frameworks. All the official standpoints of the Groningen PvdA had to be decided by the GMM. The executive and party group were obliged to submit annual reports to the GMM. Furthermore, party members got the chance to recall executive members and councillors. As pointed out, the GMM had not become a place for substantial discussion. However, at GMMs dealing with important matters, like annual reports, 100 or sometimes more than 200 party members attended. Judging from this fact the GMM must have played an important role in forcing party leaders to respect frameworks created by party members.

Moreover, at the election in 1974, it was made a prerequisite to defend frameworks as a united party group if he or she wanted to be placed in the candidate list. The following is the 'qualitative advice' for the candidates, which was approved by the GMM on 19 October 1973:

2. The (candidate) members of the party group defend the election programme which was approved by the division general member meeting of the PvdA, and are willing to test their policy continuously against it and against the decisions of the general member meeting.

3. The members of the party group take it upon themselves to make contact frequently and exchange information actively with members and parts of the organisation of the party, such as district teams, working groups, district team council and general member meeting. ...

5. The members of the party group must be willing to: ... dedicate themselves in a good team spirit for full four years. If this is no longer possible, this must be discussed in the party group, in which the continuation of the councillorship must be tested against the opinion of the party group.

Thanks to these procedures or rules, frameworks were not nominal, but functioned substantially. In addition, the new left got back the policy making from the bureaucrats, appealing for 'politicization', and were themselves engaged in policy making energetically. Even after the 1974 victory, the PvdA could not dominate alone the majority of the council. However, since 1972, they had chosen not the *afspiegelingscollege* but *programcollege*, and tried to realize their own policy, avoiding compromise as much as possible. For the *wethouders*, who were sent in the BandW by the party group, their responsibility to the party group was clearly stipulated in the decision by the GMM on 14 September 1972 as follows:

> The wethouders of the PvdA are obliged to refer to these statements of the party group in their political attitudes, and make an effort to realise these in the college van BandW and in the council. They are responsible to the party group for this. If a wethouder deviates from the views of the party group in matters that are essential for the party group, then he finds himself in conflict with the party group, and the party group can call him to account for this.

Party members were engaged in making not each policy but frameworks, and party leaders pursued each policy within these frameworks. In addition, these frameworks reflected public opinion.

Neither the public nor party members had any opportunities to participate directly in planning the VCP. However, again, this VCP was formulated within the formal and informal frameworks created by party members together with the public, and therefore reflected the opinions of the public as well as party members indirectly.

Application of PNT

The decision-making process of the VCP, which is partially outlined above, is translated into Tables 15.2 and 15.3 based on the structure mentioned in section 2.2. Table 15.2 represents an overview of the actors involved.

Table 15.2 Main actors in the VCP process

Actors	Goals/objectives	Means
Social framework		
Business and commercial organizations:	Promote interests of members, improve economic climate	Mobilize members, propose alternatives
– Groningen Entrepreneur Federation (GOF)		
– Chamber of Commerce (KvK)		
– Business Circle Grote Markt		
Citizens' organization Working Group Inner City	Inner city for pedestrians and bicycles	Investigation, direct action
Neighbourhood organizations	Improve neighbourhoods	Investigation, direct action
Local newspapers:	Maximize circulation	Provide local news and background information
– *Nieuwsblad van het Noorden*		
– *Groninger Gezinsbode*		
Organizational framework		
Ministry of Traffic and Transportation	Improve traffic circulation, influencing modal split	Reports and regulations, investment subsidies.
Police	Public safety and maintaining order	Police power, membership in municipal committees
Central Institute for Medium and Small Business (CIMK)	Investigation	Commissioned research
Royal Association of Entrepreneurs (KVO)	Promotion of entrepreneurial interests	Communication to members and press
Trade Unions (NVV)	Protection of employment	Political pressure
Legal framework		
Planning law	Balancing spatial interests	Plans and procedures
Traffic law	Traffic safety, smooth traffic	Plans and procedures
National Regulations	Facilitate national goals	Subsidies
Surroundings framework		

Table 15.2 cont'd

Actors	Goals/objectives	Means
Spatial situation	The central objective of planning has been to facilitate the 'meeting function' of inner city area since the 1970s.	Planning and implementation
Cultural situation	Groningen as regional capital always pays much attention to economic goals.	Planning and implementation
Demographic situation	As a university city the majority of population is less then 35 years. This results in an innovative, cultural and political climate.	Young politicians with innovative ideas and intentions
Political framework		
Local political parties, notably:		
– Labour Party (PvdA)	Promoting social democratic policies	Politicization, polarization, election programs, integrated policy plans, 'Onze Binding', intra-party procedures and rules
new left party leaders	Democratization	
party group	Advocate desires of party members	
party executive	Party management	
rank and file party members	Promoting social democracy	Vote at elections, help campaigns
district teams	Facilitate intra-party participation	District team council, description letter
working groups		
general member meeting		
– Christian Democratic Appeal (CDA)	Promoting middle class interests	
– Conservative Party (VVD)	Promoting entrepreneurial interests	
– Liberal Democrats (D66)	Promoting participation	
– Communist Party (CPN)	Promoting lower class interests	
Municipal council	Representative democracy	Decision making
Political executives (BandW)	Daily administration and implementation	Support from majority municipal council

Table 15.3   Network characteristics of VCP actors

Actors	Network position	Degree of influence
Business and commercial organizations:	Interaction	Low
Citizens' organization Working Group Inner City	Interaction	Low
Neighbourhood organizations	Interaction	Low
Local newspapers:	Incentive	High
– *Nieuwsblad van het Noorden*		
– *Groninger Gezinsbode*		
Ministry of Traffic and Transportation	Incentive and intervention	High
Police	Incentive	Modest
Central Institute for Medium and Small Business (CIMK)	Interaction	Low
Royal Association of Entrepreneurs (KVO)	Interaction	Low
Trade Unions (NVV)	Interaction	Low
Local political parties, notably:	Interaction	
– Labour Party (PvdA)	Intervention	High
new left party leaders	Intervention	High
party group		
party executive		
rank and file party members		
district teams	Interaction	High
working groups	Interaction	Modest
general member meeting	Interaction	Modest
– Christian Democratic Appeal (CDA)	Interaction	Low
– Conservative Party (VVD)	Intervention	Low
– Liberal Democrats (D66)	Intervention	Modest
– Communist Party (CPN)	Intervention	Modest
Municipal council	Intervention	High
Political executives (BandW)	Intervention	High

A Critical Assessment of PNT

Although policy network theory acknowledges the variation across and within political institutions, at the same time it simplifies public decision-making. The empirical illustration in the previous section revealed to us some practical and also fundamental problems. The first is the definition of the actors. Are they institutions/organizations or individuals? Should we consider connections between bureaus, agencies and groups; or examine individual, politicians, lobbyists, bureaucrats, experts and consultants?

Evidently, the finding that all public organizations are connected together is a truism. The fact that they are related is not relevant, but their impact on the final decision/outcome(s) is.

The application of PNT in urban planning only provides a snapshot of very fluid sets of relationships. What is missing is the feel of the policy process; the complexity of personal and professional connections and the multi-layered character of relationships between individuals (see also John 2004). PNT skims the surface with just an account of the decision-making context, an identification of the main actors and a discussion of the change in policy. This is too meagre for a meaningful ex-post planning evaluation. This needs an approach which shows how personal links can affect policy outcomes and the transfer of policy ideas in society.

PNT ignores that urban planning is essentially a goal-seeking process. Hence, problems and objectives may change during the process. This is usually omitted in PNT frameworks.

Ex-post planning evaluation is seen here as the systematic assessment of the effectiveness and sustainability of completed planning processes and as such contributes to planning history. In our opinion no new development activity should be planned or undertaken without first reviewing the lessons of past experience.

The aim of ex-post planning evaluation is to examine the underlying causes of activity outcomes in order to determine activity efficiency and impact, including their potential to be sustained in the longer-term. It assesses:

• the achievement of sustainable outcomes against the objectives identified during design and modified during implementation ('conformance');
• the impact of the activity on the actors, sectors and areas designated in the activity design ('performance').

Its main aim is to generate and distribute the lessons that can be learned from the activity experience to planners and policy makers.

Conformance and performance tests have been developed in ICT for determining the reliability of computer technologies (conformance) and their speed (performance). Faludi and Mastop (1997) have used these terms to denote two principally different methods of ex-post evaluation: conformance evaluation focusing on 'means-ends' logic; performance evaluation is to examine if the plan helped to improve the quality of subsequent actions.

PNT as applied in earlier also disaggregates into actors, goals and means and as such resembles a community impact evaluation framework (cf. Lichfield 1996). However, PNT is based on the assumption that problem formulation and objectives of actors may change in the course of action. Hence, there is a problem with evaluating success and failure of networks (see also Klijn and Koppejan 2000). Conformance evaluation can only be actor-based, but it doesn't include the dynamics of the network. Performance evaluations are in PNT only meaningful as a kind of satisfying analysis (Teisman 1992; Klijn and Koppejan 2000). It is conceivable that the actors involved are approached for an assessment of the process and its outcome together in relation to objectives formulated by actors and realized outcomes.

Some Concluding Remarks

It is illustrated in this chapter that *policy network theory* (PNT) provides a framework for evaluating urban policy-making processes. However, it appears to be difficult to make the role of actors and their interactions explicit in an unambiguous way. Social decision-making processes in practice are less neatly arranged than it is suggested in PNT.

It is fair to say that PNT, however interesting it is for its theoretical elegance, is not very appropriate as an evaluation methodology for planning history since it is unable to examine the underlying causes of activity outcomes for determining activity efficiency and impact. It doesn't offer any insights into their potential to be sustained in the longer-term. Besides, this chapter has shown that PNT as discussed is not suitable for revealing hitherto invisible planning practices and agendas.

References

Albrechts, L. and Mandelbaum, S. (eds) (2005), *The Network Society: A New Context for Planning*, Routledge, London.

Börzel, T. (1998), 'Organising Babylon – on the Different Conceptions of Policy Networks', *Public Administration* 76(2), pp. 253–73.

Dowding, K. (1995), 'Model or Metaphor? A Critical Review of the Policy Network Approach', *Political Studies* XLIII, pp. 136–58.

Hajer, M. (1993), 'Discourse Coalitions and the Institutionalisation of Practice: The Case of Acid Rain in Britain', in F. Fischer and J. Forester (eds), *The Argumentative Turn in Policy Analysis and Planning*, UCL Press, London.

Hajer, M. (1997) *The Politics of Environmental Discourse: Ecological Modernization and the Policy Process*, Oxford, Oxford University Press.

Ike, P., Linden, G. and Voogd, H. (2004), 'Environmental and Infrastructure Planning', in G. Linden and H. Voogd (eds), *Environmental and Infrastructure Planning*, Groningen, Geo Press.

John, P. (2004), 'Policy Networks', in K. Nash and A. Scot (eds), *The Blackwell Companion to Political Sociology*, Oxford, Blackwell Publishing, pp. 139–48.

Kakhee, A. (2001), 'Drama Democratic Discourse Policy Statement: An Evaluation of plan Texts', in H. Voogd (ed.), *Recent developments in Evaluation*, Groningen, Geo Press, pp. 235–52.

Kickert, W.J.M., Klijn, E-H. and Koppenjan, J.F.N. (eds) (1997), *Managing Complex Networks; Strategies for the Public Sector*, London/Beverly Hills, Sage Publications.

Klijn, E-H. and Koppenjan, J.F.N. (2000), 'Public Management and Policy Networks: Foundations of a Network Approach to Governance', *Public Management*, 2(2), pp. 135–58.

Klijn, E-H. (2002), 'Vertrouwen en samenwerking in netwerken: een theoretische beschouwing over de rol van vertrouwen bij interorganisatorische samenwerking', *Beleidswetenschap* 16(3), pp. 259–79.

Kräwinkel, M. (1997), 'Nieuw voegwerk voor de wijk', PhD dissertation University of Groningen, the Netherlands.

Lichfield, N. (1996), *Community Impact Evaluation*, London, UCL Press.

Mastop, H. and Faludi, A. (1997), 'Evaluation of Strategic Plans: The Performance Principle', *Environment and Planning B: Planning and Design* 24, pp. 815–32.

Roo, G. de (2003), *Environmental Planning in The Netherlands: Too Good to be True*, Aldershot, Ashgate.

Sandercock, L. (ed.) (1998), *Making the Invisible Visible: A Multicultural Planning History*, Berkeley, CA/London, University of California Press.

Scharpf, F.W. (1978), 'Interorganizational Policy Studies: Issues, Concepts, and Perspectives', in K. Hanf and F.W. Scharpf (eds), *Interorganizational Policy Making, Limits to Coordination and Central Control*, London/Beverly Hills, Sage Publications.

Stokman, F. and Zeggelink, E. (1996), 'Is Politics Power or Policy Orientated? A Comparative Analysis of Dynamic Access Models in Policy Networks', *Journal of Mathematical Sociology* 21(1–2), pp. 77–111.

Teisman, G.R. (1992), *Complexe besluitvorming*, Den Haag, VUGA.

Teisman, G.R. (2000), 'Models For Research into Decision-making Processes: On Phases, Streams and Decision-making Rounds', *Public Administration* 78(4), pp. 937–56.

Thatcher, M. (1998), 'The Development of Policy Network Analyses: From Modest Origins in Overarching Frameworks', *Journal of Theoretical Politics* 10, 389–416.

Tsubohara, S. (2003), 'Politicisation, Polarisation and Public Participation: Planning History of Groningen, the Netherlands, in 1970s, part 1 and 2', Ursi research reports 302/303, <http://www.rug.nl/ursi/publications/researchReports>.

Tsubohara, S. (2005), 'Intra-Party Democracy in Groningen Early in the 1970s', Ursi research report 310, <http://www.rug.nl/ursi/publications/researchReports>.

Tsubohara, S. and Voogd, H. (2004), 'Planning Fundamental Urban Traffic Changes: Experiences with the Groningen Traffic Circulation Plan', in C.A. Brebbia and L.C. Wadhwa (eds), *Urban Transport X – Urban Transport and the Environment in the 21st Century*, Southampton/Boston, WIT Press, pp. 287–96.

Woltjer, J. (2000), *Consensus Planning*, Aldershot, Ashgate.

Chapter 16

Evaluation for Accountability: Democratic Concern in the Review of Local Government's Environmental Policies in Sweden

Abdul Khakee and Anders Hanberger

Introduction

In planning evaluation the issue of accountability has not received much direct attention. Accountability has nearly always been related to public auditing of the budget and review of goal achievement and administrative efficiency in specific policy areas. The major democratic element in public auditing and public review is accountability – public officials are made accountable as to how they have managed a budget or a specific public activity.

Planning, on the other hand, has been concerned with the production and implementation of plans and evaluation has been either on the likely consequences (ex ante) of planned measures or the impact of implemented plans (ex post). The major democratic element in planning is public participation – planners are required to facilitate people's input of advice, critique and ideas during the preparation and implementation of plans. However, planners are not required to be accountable to people. In planning theory as well as planning practice accountability is assumed to have a political dimension i.e. it is the primary task of elected representatives who hold planners to account for their activities. Our hypothesis in this chapter is that *the simple notion that voters elect representatives who hold planners and administrators to account has become a less convincing theory.*

We put forward two major arguments to support this hypothesis. The first argument relates to the new democratic concern that maintains that representative democracy in its present form in the long run is untenable owing to the crisis of confidence and the loss of support for the political system. Environmental issues are put forward as one of the policy areas that require alternative ways to involve people in decision-making. In this context various democracy models have been put forward – dialogical, collaborative, deliberative democracy. All these models argue for more direct input of citizens' views and opinions in the public decision-making.

The other argument is based on the recent development towards political, administrative and spatial fragmentation in the way public policy is formed. As a result of the neoliberal trends, governments have lost their authority in setting the political agenda leading to political fragmentation (Bogason 1996). As a result of fiscal crisis several changes have taken place with regards to the production and delivery of public services. These include decentralization within the public sector, the introduction of New Public Management to make public services more competitive, the encouragement of 'voluntarism' through voluntary contributions from associations and individuals to activities that previously were carried out by the public sector (Amnå 1995), and the collaboration between public and private actors through networks and partnerships. This shift has been described as one from 'government' to 'governance' (cf. Björk et al. 2003; March and Olsen 1995; Pierre and Peters 2000). As a result of global competition for international capital, regional policy based on the objective that the regions could be economically coherent and politically administered in a uniform way has been replaced by a policy which means that the state directly or indirectly encourages regional differences and competition.

Planning fragmentation has been the response to this institutional fragmentation. This has in practice implied the existence of several planning modes co-existing side by side in practice. It is not unusual to find the use of rational, advocacy, deliberative, radical, negotiative and strategic planning being used to respond to different requirements with regards to the production and delivery of what were previously classified as public goods. For example, deliberative planning emphasizing public dialogue is used to prepare local Agenda 21. Negotiative strategic planning is used for large-scale space marketing investments. The co-existence of various planning approaches leads to crisis of confidence for planning e.g. people who participate in local Agenda 21 process and other environmental activities feel that their involvement has marginal or no consequences on decision-making especially in crucial development programmes that are planned in more or less exclusionary manner (Khakee and Barbanente 2003).

Recent concern in sustainable development puts a heavy share of responsibility on people with regard to managing household refuse, reducing energy consumption and changing wasteful consumption practices. Public input in the preparation and implementation of, for example, local Agenda 21 is of completely different dimension when compared to corresponding input in the preparation of traditional land-use and other plans. This development gives the issue of accountability a different status in environmental policy planning.

How does accountability look like in the current environmental policy planning in Sweden? What reflections can be made from the Swedish case study about the importance of accountability? More generally what lessons can be drawn about the role of accountability in planning evaluation? How does accountability strengthen democratic considerations in planning? The two main issues addressed in this chapter concern how evaluation for democratic accountability can be analysed and what lessons can be learned from prevailing forms of public reviews in the field of environmental policy. In order to answer these questions we analyse the entire review system of local environmental policy making. Our analysis is not limited to reviews carried out by public auditors but also private actors including the media and various people's organizations.

The term 'public review' is in this chapter used for various kinds of reviews and inquiries of public sector policies and programmes. Review implies some type of systematic and thorough analysis and/or assessment of public policies and programmes. Evaluation is mostly used interchangeably with review. Only when specific evaluations are discussed explicitly, evaluation refers to externally commissioned evaluations. Although both internal and external actors and institutions can carry out evaluations/reviews, this chapter primarily pays attention to external evaluations/reviews.

The chapter is divided into five sections including this introduction. In the following section we discuss various aspects of review and accountability with the help of democracy theory. The third section presents an account of local government's environmental policy, followed by a section on the review of environmental policy undertaken by various actors. This section includes a quantitative survey of review reports based on the examination of a selected number of review reports and a qualitative survey based on interviews with public auditors, local government officials and journalists in four municipalities – two municipalities in Northern Sweden (called in this chapter as 'Big North' and 'Little North') and two municipalities in southern Sweden (called 'Big South' and 'Little South'). The concluding section discusses the relationship between review, accountability and planning.

Different Notions of Democracy and Public Reviews

To attain more knowledge of the implications of public review for democracy in general, and for democratic accountability in particular, there is a need to delineate the democratic orientation that a public review can have, as well as to look carefully at accountability and analyse accountability in the light of different notions of democracy.

Democratic regimes are founded on laws and regulations specifying how citizens can hold the elected representatives accountable as well as how the elected can hold administrators (planners) to account. In addition, the media and citizens themselves and their organizations undertake public reviews for accountability. Although public review is embedded in some kind of a representative democratic order, democratic governance can adopt many faces. Hence, there is a need to distinguish different democratic orientations which democratic governance can be oriented towards. Three conditions associated with evaluating/reviewing governance are paid attention to in this chapter. First, in public policy (e.g. planning and evaluation) the actual democratic orientation can vary and the different orientations should be made explicit. Secondly, evaluation/public review is a significant component of democratic governance and can be conceived and realized as an integrated or a semi-detached component of governance. In both cases public review constitutes democratic governance. Thirdly, public review can be thought of as a governance component that changes along with governance or as an agent of change. A governance model's formal institutions for accountability may change due to incremental changes made in the governance model for example. When public reviewers pay attention to undemocratic policy processes benefiting those in power, or any dysfunction in the governance structure the review can be viewed as a potential agent of change. Hence, the implications for democracy of formal institutions for accountability (citizens holding the elected accountable, and politicians holding administrators to account) and self-organized public reviews (the media and citizens) need to be illuminated and discussed all together in times when democratic governance is changing.

From theories of democracy one can distinguish a number of democratic notions that are helpful to assess the democratic impact of public review.[1] Below some of the key differences of the three notions of democracy are highlighted with a focus on accountability.

There is a huge body of literature on democracy and democratic governance. According to the theory of *elitist democracy*, political elites compete for power in open societies (Schumpeter 1942). This theory, sometimes referred to as the liberal or Lockean view (Habermas 1996), assumes that citizens can control their government by choosing among competing elites. Ordinary citizens are encouraged to participate every three or four years in elections. Citizens are not given a direct role in the policy process, and democratization implies improving the elite's representation of the people. Decision-making is a task for those in power, the elected. In other words, the core idea here is that of representative democracy, but it is a very constrained elitist version. Arguments for this notion of democracy are that an elitist democracy is more effective and enlightened, and avoids uniformed impact on public policy (McCoy and Playford 1967). Good conditions for accountability, with clear principal-agent relations, are essential values in this notion of democracy.[2]

The *participatory* theory of *democracy* assumes that people's participation is the most important quality of a democracy. According to this view, the power of the people is exercised when they participate. Apathy and non-participation are seen as the major threats to democracy. Moreover, participation is assumed to foster democratic citizens. Participation is presumed to help in the creation of identity, to encourage a desire to participate further in common affairs, to develop responsibility, and so on. According to this view, it is only through participation that the idea of democracy can be realized (Pateman 1970). In contrast to an elitist democracy, participation between elections is assumed to strengthen democracy. Various notions of participatory democracy exist. The participatory model discussed here is a form of democracy *by* the people, associated with a self-governing society where citizens empower themselves or are delegated freedom of choice to decide what is feasible for them. The conditions for accountability are quite different in this notion of democracy. When the people govern themselves the

---

1 For a discussion on the implications of three broad democratic evaluation orientations see Hanberger (2006).

2 In addition, elite theory is helpful in clarifying what kind of elite is to be held accountable (Farazmand 1999a and b).

principle agent relation becomes less important or appears in a different light. A self-governed people act as principles and agents in public policy. Citizens actively involved in public policy are in relation to inactive citizens acting as informal 'representatives'. Thus, citizens taking responsibility are morally accountable for actions taken or not taken.

The third notion of democracy, the *discourse* theory of *democracy,* is also concerned with participation. However, this theory goes one step further in its participatory requirements. According to this view, the idea of democracy can only be realized through discussions among free and equal citizens. This type of democracy, sometimes called 'deliberative democracy' (Dryzek 2000; Elster 1998; Gutmann and Thompson 2004), is not an aggregation of opinion of the will of the majority but is a democracy founded on a common commitment to a mode of reasoning on matters of public policy. The discourse is open to those affected by collective decisions and/or their representatives (Dryzek 1990 and 2000; Elster 1998; Habermas 1996; House and Howe 1999). It is a mode of decision-making by 'means of arguments offered *by* and *to* participants who are committed to the values of rationality and impartiality' (Elster 1998, 8).[3] According to Gutmann and Thompson (2004, 7) it is

> a form of government in which free and equal citizens (and their representatives), justify decisions in a process in which they give one another reasons that are mutually acceptable and generally accessible, with the aim of reaching conclusions that are binding in the present on all citizens but open to challenge in the future.

Democratization, from this viewpoint, implies (re)creating meeting-places where communication without domination can take place. Accordingly, democracy is not primarily realized in elections, but rather in situations where policy-makers, citizens and other stakeholders have access to basic information, and are given enough time and trust to participate in practical reasoning to resolve social problems. What then are the conditions for accountability in this notion of democracy? The principle-agent relations are not the same as in the elitist or participatory model of democracy. The political-administrative elite is accountable to a people where some are actively involved and sharing responsibility in public policy and some are not. Thus, in situations where citizens participate in public policy the principle-agent relation is similar to a participatory democracy and self-governed society, but with the difference that citizens promote open debate on public affairs for example. The conditions for accountability are in this notion of democracy not so clear because many principles share responsibility, as in

---

3   This notion of democracy actually consists of two sub-categories, the liberal constitutionalist and the discursive (Dryzek 2000). Not all notions of deliberative democracy presume discursive processes, nor are they critical to established power structures and institutions. Therefore a distinction between the liberal and critical theorists' notions of deliberative democracy is helpful. John Dryzek (2000) suggests that the former should be referred to as the 'liberal constitutionalist' strand and the more critical and discursive strand should be called 'discursive democracy'.

the case of a partnership for example. Both active and inactive citizens view the elected representatives and administrators as accountable, although in different ways. Planners and administrators are not formally accountable but nevertheless perceived as morally responsible.

Furthermore, public reviews can have different implications for democracy depending on who undertakes the review. The commission given to, or taken by, a public reviewer will to some extent indicate the democratic impact of an inquiry. When a state inspector controls the environmental work in a municipality, for example, the democratic impact should be searched for in terms of its contribution to maintain or strengthen a national representative democracy, and most likely an elitist version. In contrast, if a local newspaper or a citizen organization evaluates the local government's responsibility in a partnership policy openly, supported by hearings, seminars and public debates, the democratic implications of the review are more associated with a discursive notion of democracy.

All three notions of democracy are founded on, and are complementary, to representative democracy as they are elaborated here, but the ideal types are heading in different directions. Democracy theory is not concerned with the role of administration and assumes administration's subordinated role vis-à-vis the elected representatives. However, one can assume that administration and planning developed in dialogical, deliberative, negotiative or strategic directions are not only formally accountable to the elected representatives, but also morally held to account by partners with whom they collaborate and by citizens. Thus, the condition for accountability varies in different modes of governance and because democratic governance is moving in different directions the question of accountability needs special attention.

Local Government's Environmental Policy

Local Government's environmental work is a multifaceted and highly fragmented policy area. As a result of the local government's obligation to prepare a local Agenda 21 the traditional environmental protection programme has become part of a broad interdepartmental process that involves even the local businesses and people. Unfortunately many local governments pursued Agenda 21 as a project that never managed to permeate the entire local administration. The most tangible result of Agenda 21 was the obligation for local government to participate in the implementation of national goals for environmental quality. These include global cooperation requiring goals for example, the reduction of green house gases to local goals for example environmentally congenial housing (Miljömålsrådet 2004).

The Swedish Environmental Law (*miljöbalken*) is the comprehensive legislation for the entire environmental policy area. The National Environmental Protection Agency, other state agencies, county administrative boards and local government, exercise supervision in accordance to what the national government decides. A vary large share of the work with securing the implementation of the legislation is exercised by the implementers themselves. Directions regarding internal control are specified for all public as well as private actors who pursue activities within the entire implementation field of the environmental legislation. Operative supervision is directly pursued towards enterprises and consists of both control and guidance. The operative supervisor can be a local government, a county administrative board, other regional agency or a national agency (see the table below).

Table 16.1   The major environmental operative agencies in Sweden and their tasks

Operative agency	Tasks
Local government auditors	Review the operations of the local environmental boards, which includes local government's environmental programme and its implementation
Local Environmental Boards	Monitoring of other local government agencies and public companies as well as of all private businesses
County Administrative Boards	Exercise control of major hazardous activities, handling of chemicals, water management and nature reserves
National Environmental Protection Agency	Coordinates and evaluates all the operative monitoring in Sweden. Provides guidance to all other agencies
Other national agencies e.g. the National Food Administration	Coordinate, guide and evaluate monitoring in specific policy areas e.g. food administration

The demand for environmentally conscious provision of housing, transportation and other technical and social services has resulted in some local authorities amalgamating their environmental and housing boards. In other local authorities the environmental board is either merged with one or more of local government boards or have extensive cooperation with these boards. Regardless of the organization adapted, the environmental board has the major responsibility for local environmental policy. Local government research shows that other local government boards and offices are not active participants in this work. This applies even to local businesses and popular movements (e.g. Ricklander 2000).

Review of the environmental board differs from similar review of other boards since its main task is operative supervision of both local companies (restaurants, grocery shops, industries, etc.) and local public agencies. This supervision takes place partly through delegation from state agencies and county administration boards and partly through individual responsibility to control the implementation of local environmental goals. In short the review of local environmental policy is an issue of review of how the environmental board carries out its supervisory role.

Review of Environmental Policy

The review of environmental policies in the four case studies is presented in three sub-sections – the first discussing the major aspects of the review followed by a quantitative analysis of the auditors' reports and finally an analysis of local government officials' experience of the reviews.

Main Aspects of the Review

The model that we have applied for reviewing environmental policy is much broader than is normally applied. Our model includes four major actors:

- local government auditors;
- state agencies and county administrative boards;
- media;
- citizens (popular movements, associations, etc.).

The local government auditors carry out the most extensive and systematic review of local government's environmental policy. The annual review focuses on several issues:

- What environmental questions does the environmental board focus on?
- Which decisions are taken?
- How does the board implement the supervisory functions that are delegated to it?
- How does the board carry out its steering and monitoring or internal control function?
- How does the board use tax revenue allotted to it?

Professional auditors usually assist local government's own auditors. Most local governments make use of the same national company specialized in municipal revision – KomRev. The company has developed a review model that evaluates the board's management system, follow-up, reporting and goal-achievement. The evaluation is carried out with special attention to the development of the boards' activities, economy and consumer satisfaction.

Looking more closely at the local government review in the four case studies, we find shifts in emphasis in the review process depending on local circumstances:

- in Little North the reviews specially dwell on follow-up and goal achievement;
- in Big North the reviews draw attention to the difficulties in defining local operative goals and policies from national environmental goals;
- in Little South the reviews take up the difficulties in goal- and framework-management in the decentralized organization style adopted in local government;
- in Big South reviews highlight proper economic management of tax revenue and integration of environmental goals with other local government objectives as major issues.

Besides the annual review of the environmental boards, local government in all the four case studies carry out varying forms of internal quality control. The major emphasis in the internal control is the extent to which environmental boards live up to the national legislation, and implement the nationally determined environmental goals.

Nowhere in local government's policy areas is the state supervision as intensive as in the environmental field. This depends partly on the national commitment to develop Sweden in an ecologically sustainable society. This not only finds its expression in the national environmental goals but also in the long tradition with regards to food production, animal husbandry, water management, etc.

The county administrative boards act as the regional extension of the state. They monitor local governments' environmental work closely with the help of questionnaire surveys and occasional visits. The boards examine local governments' resources, maintenance of qualified personnel and management of places of national concern. They put forward a programme of measures in order to achieve national environmental goals. The boards also provide guidance for local government's control of private and public activities.

The State agencies exhibit varying degrees of activity. The National Food Administration is regarded as the most active state agency which through visits and other means scrutinizes how local authorities carry out their control of restaurants, food stores, etc. Other agencies are less active and only carry out occasional inspections of local government's environmental protection work in their domains.

Among the media, local newspapers play a major role in the scrutiny of local government. However, in the last few years, resource scarcity has diminished this role. Newspapers, TV and radio receive prior information when local and state reviews are published. Moreover it is customary in all municipalities to invite journalists to a press conference after every meeting of the environmental board. Our four case studies show that media scrutiny of environmental policy is more intense in the big municipalities than the smaller ones. Moreover the newspapers more often report on incidents involving local governments' supervision of specific activities e.g. restaurants, food shops and local industries.

Local associations, popular movements and individual citizens are the most infrequent of all the reviewers. Trade unions and tenants associations have been popular bodies to which proposed measures were referred for consideration. The same does not apply to environmental movements like the Greenpeace and local societies for the conservation of nature. The latter become active only in sporadic cases involving environmental accidents.

Local Auditors' Reviews 1995–2002

Although Swedish local government has a large amount of freedom of choice, they also serve as an implementation level in the state (Amnå and Montin 2000; Hanberger 2003). The scope for municipality initiated environmental policy is considerable, that is, after state policies have been implemented. In practice, however, the willingness to expand this policy field at the local level varies a great deal because municipalities make different prioritizations. Some local governments have made extensive progress in the field of environmental policy whereas others lag behind (Eckerberg and Forsberg 1998).

All performance audit reports produced by the local auditors in the same four municipalities discussed in the previous section have been analysed. For the purpose of this chapter, only the reports on environmental policies, compiled between 1995 and 2002, are paid attention to. The number of reports indicates the relative importance shown by auditors for different policies.

Table 16.2 indicates that the number of environmental reports is very small. For the whole period only 12 environmental reports have been produced. As indicated not all auditors have undertaken a specific review and compiled an audit report on environmental policy. The table shows that most reports do not refer to any of the four policy sectors (social, school, environmental or industrial policy). Instead, the majority of reports have scrutinized the municipalities' internal control.

Table 16.2   Number of performance audit reports in four municipalities divided on policy sectors, 1995–2002

Reports	1995	1996	1997	1998	1999	2000	2001	2002
Social	0	2	2	3	5	3	8	9
School	1	2	2	4	3	7	6	5
Environmental	1	0	1	3	1	2	2	2
Industrial	1	1	1	0	6	1	4	5
Other	16	27	29	16	29	35	47	56

The number of environmental reports is, in comparison with the number of reports produced for the other three policy sectors and in relation to the relative size of local environmental policy, neither better nor worse. Besides, we know from the interviews that local auditors have reviewed all policy sectors each year, but site visits and answers to the auditors' questions are not always compiled in the form of a report.

The growth of internal control reviews is a trend in local auditing in Sweden. This trend is mainly captured by 'other' reports in Table 16.2. Our first observation is that the content of the reports tells us that auditors mainly have been checking whether the procedures for steering, the conditions for goal and result based management, monitoring and internal control have been institutionalized. Surprisingly, not much attention is paid to the outcomes of the implementation processes. During the period 1995 to 2002 local auditing has moved from reviewing how local government programmes and policies perform, including outcomes, towards ensuring that systems and routines to steer, implement and control local government activities are in place.

In all four municipalities local auditors have commissioned professional auditors, KomRev, to do the actual auditing.[4] KomRev belongs to the international enterprise PricewaterhouseCoopers. A second observation is that a specific model for reviewing municipality steering, monitoring and control is often

---

4   Every four years local auditors commission a company to carry out the financial and performance auditing until the next election. According to the framing-contracts professional auditors are then commissioned to undertake a number of reviews each year.

applied. The use of review models indicates that auditing is standardized. Whether standardization is an advantage or disadvantage depends on the perspective from which one views the situation. Obviously, a review model provides coherence and stability on the one hand, but on the other hand standardized reviews tend to be developed in a mechanistic way. Standardized devices imply that auditors are looking for the same matters, in the same way. But to pay attention to the same deficiencies in the same way is not always a wise thing to do. The accountable learn what to expect, for example. A critical question is what is being standardized and who decides about the review model. From a democratic point of view one could say that the auditors in these four municipalities have given away the democratic control instrument to an international firm. Hence, local auditors' reviews of (environmental) policy are partly developed from a market perspective. The trend towards internal control and oversight reviews can also be a result of the limited amount of the money spent by local governments on auditing. Only 0.0005 per cent of the municipality budget is spent in auditing.

A third observation is that the critique in the reports becomes more positive over time. In the Swedish auditing culture auditors strive to combine two roles. On the one hand they control local governments in various ways and on the other they have a commission to assist them in finding solutions to deal with the shortcomings that they have identified. Some municipal auditors argue that what is most important is that the reviews lead to better practice, and better practice is achieved by supporting processes heading in the right direction. The review process can be described as a collaborative process for arriving at better practice. The auditors' strategy is often to initiate processes of change or support sound processes. Review reports consist of stories of problems and shortcomings, action taken in the right direction, and some critique. Although local governments to some extent have improved the administrative routines as a result of auditors' critique and requests, also auditors' pedagogical effort to support steps taken in the right direction can also explain the trend to more positive critique.

A fourth observation is that the review reports are not written in a language accessible to ordinary citizens. The audience is elected representatives and administrators. Few citizens know that there are performance audit reports compiled, and those that know do not know how to find them. It would be very difficult to put a report in context for an ordinary citizen and to use these reports to evaluate local government. The reports indicate that the auditors implement the review commission as 'ombudsman' for citizens by speaking directly to the political –administrative elite.

The State of Environmental Policy Review

More generally local government officials feel that local government auditing often comes with critical comments about how the board's work has deteriorated and suggestions on how to improve the board's management activities. It enjoys a high level of confidence. However, the quality of review varies depending on personal qualifications of the auditors and greater consistency in auditing is considered desirable.

Evidence of informal contacts between the auditors and the members of the environmental boards does not in any way affect the impartiality of local government auditing. Since the main aim of auditing is to improve environmental management rather than criticize, an open dialogue between the auditors is considered to frame premises for innovation and does not affect the essence of public auditing. There is also a general impression among the officials that the auditors' reports do not have political bias nor do they arouse political sentiments despite the fact that the local government council appoints the auditors.

The officials view the county administrative boards review as less controlling and more as advisory. These reviews are focussed on important environmental issues and are useful for a broader local debate about the implementation of national environmental goals. The state-agencies' reviews of local government's environmental policies and programmes are generally considered to complement local government's own auditing and provide guidance for intricate issues.

The general view among local government officials is that the guiding principle for the publication of environmental news is only its 'news value'. If a local government successfully implements the policy to reduce chemical discharge from local industries, newspapers hardly pay any attention to such news. If however the news is about salmonella poisoning from a local kebab restaurant, this receives extensive publicity. Local newspapers are widely read and people's views on local governments' environmental policies are naturally affected by what they read. On the other hand the publication of local government auditor reports hardly gets any publicity and arouses very little interest.

There is a dearth of reviews carried out by popular organizations. Two opposing views emerge: some officials feel that people have direct contacts with local officials if they find faults in environmental work. Other officials believe that representative democracy works well and people can contact their own parties and air grievances if they want to do so.

The state of review is summarized in the following table that also includes our own comments especially with reference to the democratic concern of accountability.

Concluding Reflections

This chapter has emphasized the increasing importance of accountability in planning evaluation because of increasing impact of institutional and planning fragmentation as well as the growing demand for democratic reforms. The four case studies of the review of local government's environmental policies in Sweden provide evidence for the enhanced importance of accountability in planning evaluation.

Useful as the local and state auditing is for improving the daily management process of the environmental boards in local government, it does not contribute towards a broader dialogue on environmental goals and goal achievement. Most of the auditors' reports are not accessible both with regard to the format of the reports as well as the information provided to the public. There is an extensive need to improve the channels of information to the population at large in order

Table 16.3 Current status of environmental policy review

Reviewer	Characteristics	Our comments
Local government auditors	Auditor reports are mainly used to improve the management of the environmental boards and act as an incentive for making alterations in environmental policies.	Auditor reports are too terse to excite popular interest. Need to develop local government auditing to involve people in the achievement of environmental goals.
County Administrative Boards	Provides guidance by focusing on essential issues.	County administrations work in water management, hazardous industries and national nature reserves should be reorganized to involve local population.
State agencies	Allow outsider evaluation of how local authorities carry out the control of activities that have environmental implications.	Bilateral communication between state agencies and local government should be made multilateral in order to involve local businesses, popular movements and people more generally
Local newspapers	Draw attention to environmental failures.	More resources are needed to develop investigative reporting on a more continual basis in order to maintain interest in local/global environmental development.
Popular movements, local associations and individual citizens	Sporadic involvement in the preservation of neighbourhood environment.	Local government and county administration should encourage greater participation in environmental policy review process.

to ensure a wider participation in the review processes. The reviews are more or less for internal consumption in the local government. Moreover, a majority of reviews have a functional emphasis – highlighting specific policy management aspects. Perhaps a more territorial emphasis e.g. by relating policy management to neighbourhood environment may arouse the interest of local population.

The far-reaching amalgamation of municipalities leading to a radical reduction in their number, a drastic reduction in the number of political representatives and professionalization of local councillors have been contributory factors towards the lack of civic culture in Swedish municipalities. Accountability can act as an incentive for revitalizing popular participation provided *citizen auditors* or *citizen auditor panels* are organized and allowed to participate at all stages of local government auditing. The achievement of goals towards sustainable development requires novel approaches towards people's involvement in the review of environmental policies. Accounting in planning evaluation should be introduced in such a way as to avoid excessive complexity in the mode of planning evaluation and focus less on formal links and institutions and more on civic dialogue to create high visibility and directness of face-to-face accountability.

Owing to the current fragmentation in planning practice, it is necessary to develop a mix of political and community accountability that focuses on the performance of plans in order to address the issue of how citizens can make a real and meaningful input into the planning process. Creation of citizen auditors and/or citizen auditor panels may be a way of promoting the mix in accountability. Planning can not be judged by the same standards as in public budgeting. Perhaps it may be useful to develop performance indicators emphasizing planners' morale and integrity in face of the increasing fragmentation in planning practice.

References

Ahlbäck, S. (1999), 'Revisionens roll i den parlamentariska demokratin' ['The Role of Auditing in a Parliamentary Democracy'], in *Demokratiutredningens forskarvolym 1* [*Research Report 1 on the Public Commission on Democracy*], Stockholm, Fritzes pp. 179–223.

Amnå, E. and Montin, S. (2000), *Towards a New Concept of Local Self-government? Recent Local Government Legislation in Comparative Perspective*, Bergen, Fagbokforlaget.

Amnå, E (1995), 'Det mångtydiga mellanrummet. Några utländska forskningsimpulser' ['The Ambiguous Inter-space: Some Foreign Research Impulses'], in E. Amnå (ed.), *Medmänsklighet att hyra? Åtta forskare om ideell verksamhet* [*Human Feelings to Hire: Eight Researchers on Voluntary Work*], Örebro, Liber, pp. 119–71.

Björk, P., Bostedt, G. and Johansson, H. (2003), *Governance*, Lund, Studentlitteratur.

Bogason, P. (1996), 'The Fragmented Locality', in P. Bogason (ed.), *New Modes of Local Political Organizing: Local Government Fragmentation in Scandinavia*, New York, Nova Science Publishers, pp. 169–90.

Dryzek, J.S. (1990), *Discursive Democracy – Politics, Policy, and Political Science*, Cambridge, Cambridge University Press.

Dryzek, J.S. (2000), *Deliberative Democracy and Beyond. Liberals, Critics, Contestations*, Oxford, Oxford University Press.

Eckerberg, K. and Forsberg, B. (1998), 'Implementing Agenda 21 in Local Government: The Swedish Experience', *Local Environment* 3(3), pp. 333–47.

Elster, J. (1998), *Deliberative Democracy*, Cambridge University Press, Cambridge.

Farazmand, A. (1999a), 'Globalization and public Administration', *Public Administration Review* 59(6), pp. 509–22.

Farazmand, A. (1999b), 'The Elite Question. Toward a Normative Elite Theory of Organization', *Administration and Society* 31(3), pp. 321–60.

Gutmann, A. and Thompson, D. (2004), *Why Deliberative Democracy?*, Princeton, NJ, Princeton University Press.

Habermas, J. (1996), 'Three Normative Models of Democracy', in S. Benhabib (ed.), *Democracy and Difference. Contesting the Boundaries of the Political*, Princeton, NJ, Princeton University Press.

Hanberger, A. (2003), 'Public Policy and Legitimacy: A Historical Policy Analysis of the Interplay of Public Policy and Legitimacy', *Policy Sciences* 36, pp. 257–78.

Hanberger, A. (2005), 'Evaluation of and for Democracy', *American Journal of Evaluation* 26(4), pp. 532–43.

Hanberger, A., Khakee, A., Nygren, L. and Segerholm C. (2005), *Den kommun granskande aktörernas betydelse* [*The Importance of Municipal Reviewers*], Research Report No. 1, Umeå, UCER.

House, E.R. and Howe, K.R. (1999), *Values in Evaluation and Social Research*, Thousand Oaks, CA, Sage.

Khakee, A. (1999), 'Demokratin i samhällsplanering' ['Democracy in Urban Planning'], in *Medborgarnas erfarenheter* [*Citizens' Experiences*], Government Public Report, Stockholm, Fakta Info Direkt.

Khakee, A. and Barbanente, A. (2003), 'Negotiative Land-use and Deliberative Environmental Planning in Italy and Sweden', *International Planning Studies* 8(3), pp. 181–200.

Miljörådet (2004) *Miljömälen – når vi dem?* [*Environmental Goals – Can We Achieve Them?*], Stockholm, Naturvårdsverket.

MacCoy, C. and Playford, J. (1967), *Apolitical Politics*, London, Cromwell.

March, J.G. and Olsen, J.P. (1995), *Democratic Governance*, New York, The Free Press.

Pierre, J. and Peters, G. (2000), *Governance, Politics and the State*, New York, Macmillan.

Ricklander, L. (2000), *Grönt ljus för kommunerna. Miljöledningssystem för kommunal verksamhet* [*Green Light for the Municipalities. Environmental Management System for Local Administration*], Malmö, Komrev.

Schumpeter, J. (1942), *Capitalism, Socialism and Democracy*, New York, Harper.

Index